Chad Kjorlien

VISUALIZATION AND VERBALIZATION OF DATA

Chapman & Hall/CRC
Computer Science and Data Analysis Series

The interface between the computer and statistical sciences is increasing, as each discipline seeks to harness the power and resources of the other. This series aims to foster the integration between the computer sciences and statistical, numerical, and probabilistic methods by publishing a broad range of reference works, textbooks, and handbooks.

SERIES EDITORS
David Blei, Princeton University
David Madigan, Rutgers University
Marina Meila, University of Washington
Fionn Murtagh, Royal Holloway, University of London

Proposals for the series should be sent directly to one of the series editors above, or submitted to:

Chapman & Hall/CRC
Taylor and Francis Group
3 Park Square, Milton Park
Abingdon, OX14 4RN, UK

Published Titles

Semisupervised Learning for Computational Linguistics
Steven Abney

Visualization and Verbalization of Data
Jörg Blasius and Michael Greenacre

Design and Modeling for Computer Experiments
Kai-Tai Fang, Runze Li, and Agus Sudjianto

Microarray Image Analysis: An Algorithmic Approach
Karl Fraser, Zidong Wang, and Xiaohui Liu

R Programming for Bioinformatics
Robert Gentleman

Exploratory Multivariate Analysis byExample Using R
François Husson, Sébastien Lê, andJérôme Pagès

Bayesian Artificial Intelligence, Second Edition
Kevin B. Korb and Ann E. Nicholson

Published Titles cont.

Computational Statistics Handbook with MATLAB®, Second Edition
Wendy L. Martinez and Angel R. Martinez

Exploratory Data Analysis with MATLAB®, Second Edition
Wendy L. Martinez, Angel R. Martinez, and Jeffrey L. Solka

Clustering for Data Mining: A Data Recovery Approach, Second Edition
Boris Mirkin

Introduction to Machine Learning and Bioinformatics
Sushmita Mitra, Sujay Datta, Theodore Perkins, and George Michailidis

Introduction to Data Technologies
Paul Murrell

R Graphics
Paul Murrell

Correspondence Analysis and Data Coding with Java and R
Fionn Murtagh

Pattern Recognition Algorithms for Data Mining
Sankar K. Pal and Pabitra Mitra

Statistical Computing with R
Maria L. Rizzo

Statistical Learning and Data Science
Mireille Gettler Summa, Léon Bottou, Bernard Goldfarb, Fionn Murtagh, Catherine Pardoux, and Myriam Touati

Foundations of Statistical Algorithms: With References to R Packages
Claus Weihs, Olaf Mersmann, and Uwe Ligges

Computer Science and Data Analysis Series

VISUALIZATION AND VERBALIZATION OF DATA

EDITED BY
JÖRG BLASIUS
UNIVERSITY OF BONN, GERMANY
MICHAEL GREENACRE
UNIVERSITAT POMPEU FABRA, BARCELONA, SPAIN

CRC Press is an imprint of the
Taylor & Francis Group, an **informa** business

A CHAPMAN & HALL BOOK

MATLAB® and Simulink® are trademarks of The MathWorks, Inc. and are used with permission. The Math-Works does not warrant the accuracy of the text or exercises in this book. This book's use or discussion of MATLAB® and Simulink® software or related products does not constitute endorsement or sponsorship by The MathWorks of a particular pedagogical approach or particular use of the MATLAB® and Simulink® software.

CRC Press
Taylor & Francis Group
6000 Broken Sound Parkway NW, Suite 300
Boca Raton, FL 33487-2742

© 2014 by Taylor & Francis Group, LLC
CRC Press is an imprint of Taylor & Francis Group, an Informa business

No claim to original U.S. Government works

Printed on acid-free paper
Version Date: 20140224

International Standard Book Number-13: 978-1-4665-8980-3 (Hardback)

This book contains information obtained from authentic and highly regarded sources. Reasonable efforts have been made to publish reliable data and information, but the author and publisher cannot assume responsibility for the validity of all materials or the consequences of their use. The authors and publishers have attempted to trace the copyright holders of all material reproduced in this publication and apologize to copyright holders if permission to publish in this form has not been obtained. If any copyright material has not been acknowledged please write and let us know so we may rectify in any future reprint.

Except as permitted under U.S. Copyright Law, no part of this book may be reprinted, reproduced, transmitted, or utilized in any form by any electronic, mechanical, or other means, now known or hereafter invented, including photocopying, microfilming, and recording, or in any information storage or retrieval system, without written permission from the publishers.

For permission to photocopy or use material electronically from this work, please access www.copyright.com (http://www.copyright.com/) or contact the Copyright Clearance Center, Inc. (CCC), 222 Rosewood Drive, Danvers, MA 01923, 978-750-8400. CCC is a not-for-profit organization that provides licenses and registration for a variety of users. For organizations that have been granted a photocopy license by the CCC, a separate system of payment has been arranged.

Trademark Notice: Product or corporate names may be trademarks or registered trademarks, and are used only for identification and explanation without intent to infringe.

Library of Congress Cataloging-in-Publication Data

Visualization and verbalization of data / [edited by] Jorg Blasius, Michael Greenacre.
   pages cm. -- (Chapman & Hall/CRC computer science & data analysis)
   Includes bibliographical references and index.
   ISBN 978-1-4665-8980-3 (hardback)
   1. Information visualization. 2. Correspondence analysis (Statistics) 3. Multiple comparisons (Statistics) I. Blasius, Jorg, 1957- editor of compilation. II. Greenacre, Michael J., editor of compilation.

QA76.9.I52V57 2014
001.4'226--dc23   2014001306

Visit the Taylor & Francis Web site at
http://www.taylorandfrancis.com

and the CRC Press Web site at
http://www.crcpress.com

In memory of Paul Lewi (1938–2012).

Contents

Foreword xi
Preface xiii
Editors xix
Contributors xxi
Prologue: Let the Data Speak! xxvii
Richard Volpato

Section I: History of Correspondence Analysis and Related Methods

1 Some Prehistory of CARME: Visual Language
   and Visual Thinking 3
   Michael Friendly and Matthew Sigal

2 Some History of Algebraic Canonical Forms and Data Analysis 17
   John C. Gower

3 Historical Elements of Correspondence Analysis and Multiple
   Correspondence Analysis 31
   Ludovic Lebart and Gilbert Saporta

4 History of Nonlinear Principal Component Analysis 45
   Jan de Leeuw

5 History of Canonical Correspondence Analysis 61
   Cajo J. F. ter Braak

6 History of Multiway Component Analysis and Three-Way
   Correspondence Analysis 77
   Pieter M. Kroonenberg

7 Past, Present, and Future of Multidimensional Scaling 95
   Patrick J. F. Groenen and Ingwer Borg

8 History of Cluster Analysis 117
   Fionn Murtagh

Section II: Contribution of Benzécri and the French School

9 Simple Correspondence Analysis 137
   Pierre Cazes

10 Distributional Equivalence and Linguistics 149
   Mónica Bécue-Bertaut

11 Multiple Correspondence Analysis 165
   François Husson and Julie Josse

12 Structured Data Analysis 185
   Brigitte Le Roux

13 Empirical Construction of Bourdieu's Social Space 205
   Jörg Blasius and Andreas Schmitz

14 Multiple Factor Analysis: General Presentation and
   Comparison with STATIS 223
   Jérôme Pagès

15 Data Doubling and Fuzzy Coding 239
   Michael Greenacre

16 Symbolic Data Analysis: A Factorial Approach Based on Fuzzy
   Coded Data 255
   Rosanna Verde and Edwin Diday

17 Group Average Linkage Compared to Ward's Method
   in Hierarchical Clustering 271
   Maurice Roux

18 Analysing a Pair of Tables: Coinertia Analysis and Duality
   Diagrams 289
   Stéphane Dray

References 301

Index 339

Foreword

Scientific progress is built not only on individual breakthroughs, but likewise on the accumulation of prior achievements and efficient networking of capable analytic minds and institutions that can provide the resources and intellectual environment for the respective research orientation. This insight from extended experience in the management of scientific institutes is nicely illustrated by the events to come.

It was May 21–24, 1991, that the first international conference devoted exclusively to correspondence analysis took place at the former Central Archive for Empirical Social Research at the University of Cologne (ZA), nowadays Data Archive for the Social Sciences, part of GESIS–Leibniz Institute for the Social Sciences, where I was the executive manager until my retirement in 2009.

As I learned later, the idea for this conference can be traced back to March 1990, when Walter Kristof (formerly of the University of Hamburg, Germany) was invited to give a one-week lecture on correspondence analysis during the Cologne spring seminar. Walter Kristof trained his student Jörg Blasius in statistics, and he was a friend of Michael Greenacre. By 1984, Walter had already invited Michael to give a series of lectures at Hamburg University—Jörg belonged to the small group of interested students who attended these lectures.

I met Walter and Jörg for the first time at a conference of the Section for Methods of Empirical Social Research of the German Sociological Association at Lüneburg in 1986. They gave a presentation on a survey employing telephone interviews, which was a hot new topic in those days. While listening to their presentation I was impressed by both their scrutiny of their data collection approach and their statistical analysis skills. At that time we had to fill a position for a scientific assistant in the team of Erwin K. Scheuch at the University of Cologne. When I reported my impressions of the conference, we quickly agreed to invite Jörg to Cologne and offer him the job. He started in 1986.

Walter asked Jörg, now working at the Central Archive of the University of Cologne, whether there was any possibility of inviting Michael to collaborate on a conference about correspondence analysis. Jörg showed me the letter, as well as Michael's book *Theory and Applications of Correspondence Analysis*, and asked if I would agree to invite Michael to run the first international conference on correspondence analysis.

At that time ZA had a highly visible record of inviting scholars who were internationally known as protagonists of innovative advanced data analysis methods to its international workshops and spring seminars, which had been conducted on an annual basis since 1972. These seminars and workshops were designed to demonstrate the uses of new research techniques, applying them to data sets from the ZA holdings with the intention to create visibility and make the new approaches widely available in Germany and beyond. Correspondence analysis had not much visibility in Germany, given that the focus was on everything new in data analysis coming from America. I learned from Erwin K. Scheuch, however, that one of his Japanese colleagues, Chikio Hayashi, had developed a remarkably useful multidimensional method for analysing questionnaire data in 1956. This was then known as Hayashi's quantification method III and was later found to be equivalent to Jean-Paul Benzécri's independently developed method for analysing data in a contingency table format, which became widely used as correspondence analysis predominantly in the French-speaking world.

Looking at the book and the first tentative plan, there was no doubt that Jörg, teaming up with Michael, would be fit to shoulder such a high-level conference in line with the objectives of ZA. What looked like an ideal match right from the beginning proved to be the start of a successful series of conferences to come, which also added to the already existing international scholarly networks of ZA.

It should be noted that more than 60 researchers from Europe and North America attended this conference in 1991, and, notably, a group of students from the former East Germany. Since the great success of the conference and the ensuing edited book *Correspondence Analysis in the Social Sciences: Recent Developments and Applications*, published in 1994, Michael and Jörg decided to organize another conference four years later in 1995, again hosted by the Central Archive for Empirical Social Research, which again resulted in an edited book, *Visualization of Categorical Data*, published in 1998. Since then these so-called CARME (correspondence analysis and related methods) conferences have been held successfully every four years, in Cologne again but also in Barcelona (leading to the book *Multiple Correspondence Analysis and Related Methods*, published in 2006) and Rotterdam, with the most recent conference in 2011 hosted in Rennes, France. The present book, *Visualization and Verbalization of Data*, is a fruit of this latest meeting, which celebrated the 50th anniversary of correspondence analysis, and fittingly the 20th anniversary of the first CARME conference in Cologne.

CARME has made a significant contribution to expanding the boundaries of data literacy and analytic skills—the art of data analysis—into visualization of data. Today corrrespondence analysis is well established in the fields of data analysis and textual analysis (see, e.g., special routines available as part of the Hamlet text analysis system, in Brier and Hopp, 2011). Given the continuing digitization of all facets of modern societies, we are fed rapidly growing masses of data. At that pace visualization is one decisive step to go beyond the factual knowledge of distributions to the recognition of conceptual spaces by advanced analytic methods. The big challenge will continue to be the verbalization of what we see as results of the analyses. This book lays the groundwork for significant advances on the way ahead.

Ekkehard Mochmann
Cologne, Germany

Preface

The sixth CARME conference on correspondence analysis and related methods took place at the Agrocampus in Rennes, France, continuing the tradition of the quadrennial CARME conferences started in 1991. This sixth edition of CARME was again a major event for all those interested in correspondence analysis and multivariate exploratory analysis in general. The conference was special in that Rennes is the birthplace of so-called *analyse des données* and, in particular, *analyse des correspondances*, where Jean-Paul Benzécri and his doctoral student, Brigitte Escofier, first developed this approach to data analysis. In the early 1960s Benzécri was working on his ideas of classification and dimension reduction, mostly for large sets of linguistic data, focusing on a method now called correspondence analysis, an approach that resulted in data visualizations that lend themselves naturally to the verbalization of the structures revealing themselves in the data.

This CARME conference in Rennes celebrated at the same time not only the 50th anniversary of correspondence analysis, but also the 40th anniversary of Ruben Gabriel's 1971 paper in *Biometrika* on the biplot, and the 25th anniversary of canonical correspondence analysis, published in 1986 by Cajo ter Braak in *Ecology*. This was also the reason that, besides the latest advances on the subject as in previous conferences, special sessions were programmed where the very first protagonists who played major roles in the history of correspondence analysis and related techniques gave their testimonies: Pierre Cazes, Jan de Leeuw, Michael Friendly, John Gower, Patrick Groenen, Pieter Kroonenberg, Ludovic Lebart, Brigitte le Roux, Fionn Murtagh, Jérôme Pagès, Maurice Roux, Edwin Diday, Gilbert Saporta, and Cajo ter Braak, among others.

This book has two objectives and two sections. The first section explains the historical origins of correspondence analysis and related methods. The second section concentrates more specifically on the contributions made by the Benzécri school and related movements, such as social space and geometric data analysis. Here we were not looking for an existing standard way of explaining these topics, but a perspective from the French viewpoint. The Benzécri school published almost all of its work in French, which made diffusion outside the French-speaking world difficult. The idea of this book is to bring as many of these ideas as possible to an international audience.

The book starts with a prologue by **Richard Volpato**, who gives his personal philosophy on data visualization and verbalization, having met Benzécri in the 1980s and having been impressed and heavily influenced by his revolutionary approach to data analysis.

Section I, 'History of Correspondence Analysis and Related Methods', starts with a chapter coauthored by **Michael Friendly** and **Matthew Sigal** on the prehistory of CARME, on visual language and visual thinking, a topic on which Michael Friendly has been working for a long time. The chapter contains examples from more than 400 years, discussing, among others, a world map by Guillaume de Nautonier de Castelfranc in 1604 that shows isogons of geomagnetism, as well as the work of André-Michel Guery in the 19th century, whose spatial visualization of multivariate data anticipates the later approaches in biplot methodology.

In the second chapter **John C. Gower** discusses the history of algebraic canonical forms and data analysis, a topic that has accompanied him for most of his working life. In this chapter his main focus is on the history of algebraic concepts underpinning those methods of data analysis that depend strongly on visualization.

The third chapter is a joint paper by **Ludovic Lebart** and **Gilbert Saporta**, both of them belonging to the original school of Jean-Paul Benzécri. The authors discuss historical elements of both correspondence analysis and multiple correspondence analysis, both methods being core elements of *analyse des données*.

The fourth chapter is on the history of nonlinear (or categorical) principal component analysis (NLPCA), written by **Jan de Leeuw**, who originally headed the Data Theory Group in Leiden, from which emanated the so-called Gifi system, of which NLPCA is an important part. In this chapter he gives an overview from the early beginning of principal components analysis (PCA) up to logit and probit PCA, including the new developments of 'pavings' and 'aspects'.

The fifth chapter covers the history of canonical correspondence analysis (CCA), written by its originator, **Cajo J. F. ter Braak**. CCA was first introduced in ecology as a new multivariate method to relate species communities to known variations in the environment. The author gives a detailed overview of different aspects of the method and concludes with some recent extensions and ramifications.

The sixth chapter is on the history of multiway data analysis, written by **Pieter M. Kroonenberg**, one of the most well-known researchers in this area. His historical overview covers the development of multiway components analysis with a special focus on three-way correspondence analysis.

In Chapter 7 **Patrick J. F. Groenen** and **Ingwer Borg**, authors of the book *Modern Multidimensional Scaling*, discuss the past, present, and future of multidimensional scaling (MDS). They pay tribute to several important developers of MDS while giving an overview of milestones in the history of this wide field.

Chapter 8 is the last of this historical part of the book, written by **Fionn Murtagh**, who is one of the few English-speaking persons to have been a doctoral student of Jean-Paul Benzécri. His chapter testifies to the rich history of cluster analysis, mostly in the computer science literature, and extends from the 1950s onward.

Section II, 'The Contribution of Benzécri and the French School', starts with Chapter 9 by **Pierre Cazes** on simple correspondence analysis (CA). Cazes was a key figure in Benzécri's laboratory at the Institut de Statistique in Paris. In this chapter he provides a formal description of the dimensional analysis of a cloud of weighted points in Euclidean space, of which CA is a special case.

Chapter 10, written by **Mónica Bécue-Bertaut**, focuses on the distributional equivalence principle in linguistics. In her first example she shows the effects of distributional equivalence, and in the second example she illustrates how a large textual data set can be transformed into numerical data that can be analysed using CA.

Chapter 11, on multiple correspondence analysis (MCA), is written by **François Husson** and **Julie Josse**. Using different examples from social science research, this chapter discusses the basics in interpreting MCA solutions from a French perspective.

Chapter 12, written by **Brigitte Le Roux**, is concerned with the study of Euclidean clouds of respondent points, typically from a social survey, taking into account sociodemographic characteristics as supplementary variables. This comparison of respondent characteristics in the multidimensional space afforded by MCA is called structured data analysis. This chapter is very close to the work of Pierre Bourdieu and his ideas of social spaces.

Chapter 13, written by **Jörg Blasius** and **Andreas Schmitz**, discusses the empirical construction of Bourdieu's social space, using both simple and multiple correspondence analysis. The authors also describe the close personal relation between Jean-Paul Benzécri and Pierre Bourdieu.

Chapter 14 covers multiple factor analysis (MFA), written by **Jérôme Pagès**. The aim of MFA is to analyse tables of individuals according to several groups of quantitative or qualitative variables. The approach is demonstrated on an example of different orange juices, where there are two sets of quantitative variables (chemical and sensory data) and a set of qualitative variables (for example, the juices' origin).

In Chapter 15, **Michael Greenacre**, a former doctoral student of Benzécri, discusses different possibilities of recoding data prior to visualization by CA, focusing on data doubling and fuzzy coding. These ways of recoding data, which are important aspects of Benzécri's *analyse des données* approach, allow different data types to be analysed by CA, including continuous data, rank-order data, and paired comparisons.

Chapter 16 focuses on symbolic data analysis (SDA), which is a factorial approach based on fuzzy coded data, and is written by **Rosanna Verde** and **Edwin Diday**. SDA includes standard exploratory techniques of multivariate data analysis for studying a more general class of so-called symbolic variables such as histograms and intervals.

Chapter 17, written by **Maurice Roux**, is a comparison of two competing methods of hierarchical cluster analysis: group average linkage clustering and Ward's method, the latter having a close connection to CA. Using a simple example, he shows how these methods function, their relation to CA, and their advantages and disadvantages.

The final chapter, written by **Stéphane Dray**, shows how to analyse a pair of tables using co-inertia analysis and duality diagrams. He presents the duality diagram theory and its application to the analysis of a contingency table by CA, and he shows that this framework can be generalized to the analysis of a pair of tables, in the scheme of co-inertia analysis.

This book is dedicated to the memory of Paul Lewi, an enthusiastic member of the CARME community who died in August 2012. Paul studied chemical engineering and mathematics and obtained a PhD in pharmaceutical sciences. He was the head of the Information Science Department and vice president of the Centre for Molecular Design, Janssen Pharmaceutica NV, in Beerse, Belgium. His prolific research record includes over 200 publications in the fields of information science, chemometrics, and drug design, and he is coinventor of 23 patents in the field of HIV-antiretroviral compounds. After his retirement, he remained very active and served as a guest professor at the Faculty of Medicine, Catholic University of Leuven (KUL); the Faculty of Medicine and Pharmacy, Free University of Brussels (VUB); and the University of Antwerp (UA). His major interests were in statistics, multivariate data analysis, prevention of HIV, and research management.

With respect to CARME, Paul will be remembered for his contribution to what he called spectral mapping, a multivariate visualization method used specifically in the analysis of biological activity spectra. This method, developed originally in the late 1970s, has been used extensively and with much success at the Janssen laboratories in testing new molecules in the design of new drugs. Paul was working on double-centred matrices of log-transformed spectral data, and being aware of Benzécri's work in correspondence analysis, he realized that in his application he also needed to weight the rows and columns of the table by the margins of the data matrix. This method of spectral mapping has been renamed more generally as weighted log–ratio analysis—see Greenacre and Lewi (2009)—and is a clear competitor to correspondence analysis when the data are strictly positive. It is sometimes difficult for CARME researchers to imagine that such visualization methods can lead to results beyond what might be expected anyway—as Benzécri often said: 'The results have to surprise; otherwise, the analysis is not worth it'. In his tribute to Paul for the LinkedIn group 'In Memoriam Paul Lewi', his colleague Koen Andries writes:

> In 1990, Paul used Spectramap [the software performing spectral mapping] to analyse a huge amount of data (17 antivirals tested against 99 rhinovirus serotypes) and we discovered that there appeared to be two groups of rhinoviruses, each having different susceptibilities to certain antivirals (Google 'Andries Lewi rhinovirus'). Based on the availability of just 7 rhinovirus genome sequences, we hypothesized that the antivirals were sniffing the amino acid composition of the antiviral binding

site and that the Spectramap analysis was a reflection of a phylogenetic tree. Almost 20 years later, the genomes of all 99 strains were cracked and used to build a phylogenetic tree. It matched the Spectramap analysis for 94 out of the 99 serotypes (Google 'Palmenberg Science').

The many dimensions of Paul's life, scientific, philosophical, and artistic, can be found on his personal website, http://www.lewi.be. Paul recounted his years of experience with spectral mapping in the pharmaceutical industry as keynote speaker at the CARME conference in Barcelona in 2003, and also attended CARME 2007 in Rotterdam. In addition to his brilliance as a statistician, which he used to great profit in conjunction with his chemical and mathematical background, Paul was a warm and friendly person, always willing to share a debate about the benefits of multivariate data analysis to humanity. He will be missed by us all.

Finally, we thank those persons without whom the book would not have been possible. Here we mention first Jérôme Pagès, as well as his colleagues François Husson and Julie Josse, for organizing the successful conference in Rennes where the idea for this book was born. We are grateful to Fionn Murtagh, who has been responsible for scanning and placing online the whole content of *Les Cahiers de l'Analyse des Données*, the journal that disseminated the work of Benzécri, and the many students and colleagues who passed through his laboratory in Paris. We thank all our authors who offered their precious time to this project and who suffered patiently through the several revisions of their chapters. Silvia Arnold and Nico Schäfer at the University of Bonn prepared the unified common reference list from all the various contributions, which forms a rich resource for anyone working in this field. And thanks to Rob Calver and his team at Chapman & Hall/CRC Press for once again supporting the CARME enterprise in publishing this book.

Jörg Blasius and Michael Greenacre
Bonn and Barcelona

MATLAB® is a registered trademark of The MathWorks, Inc. For product information, please contact:

The MathWorks, Inc.
3 Apple Hill Drive
Natick, MA 01760-2098 USA
Tel: 508 647 7000
Fax: 508-647-7001
E-mail: info@mathworks.com
Web: www.mathworks.com

Editors

Jörg Blasius is professor of sociology at the Institute for Political Science and Sociology, University of Bonn, Germany. His research interests include exploratory data analysis, especially correspondence analysis and related methods, data collection methods, sociology of lifestyles, and urban sociology. Together with Michael Greenacre, he has edited three books on correspondence analysis. Together with Simona Balbi (Naples), Anne Ryen (Kristiansand), and Cor van Dijkum (Utrecht), he is editor of the Sage series 'Survey Research Methods in the Social Sciences'.

Email: jblasius@uni-bonn.de

Web: www.politik-soziologie.uni-bonn.de/institut/lehrkoerper/blasius

Michael Greenacre is professor of statistics at the Universitat Pompeu Fabra, Barcelona, Catalonia. His current research interest is principally in multivariate analysis of ecological data, and he is participating in several research projects in Mediterranean and Arctic marine ecology, as well as giving workshops on multivariate analysis for ecologists. He has published five books on correspondence analysis and related methods, and the present book is the fourth coedited book with Jörg Blasius.

Email: michael.greenacre@upf.edu

Web: www.econ.upf.edu/~michael

Contributors

Mónica Bécue-Bertaut is associate professor of statistics at the Universitat Politècnica de Catalunya, Spain. Her research interests include multidimensional analysis methods and their applications to mixed textual and contextual data (open-ended and closed questions in surveys, free comments as a complement to sensory data). She is currently participating in a project about argumentative and narrative structure in law domain and giving workshops on textual statistics for statisticians and sensory analysts. She has published two books on textual analysis.
    Email: monica.becue@upc.edu
    Web: recerca.upc.edu/liam/menu1/monique-becue

Ingwer Borg is professor emeritus of psychology at the University of Giessen, Germany. He now works for Org Vitality, New York, as an HR consultant. His research interests are in scaling, facet theory, survey research (especially employee surveys), and various substantive topics of psychology. He has authored or edited 19 books and numerous articles on scaling, facet theory, survey research, and various substantive topics of psychology, from psychophysics to work attitudes and social values.
    Email: ingwer.borg@gmail.com

Pierre Cazes is professor emeritus of statistics at CEREMADE (Centre de Recherches en Mathématiques de la Decision) of the University Paris–Dauphine in Paris, France. His research interests include exploratory data analysis, especially correspondence analysis and related methods. He has been editor in chief of the *Revue de Statistique Appliquée* and is coauthor with J. Moreau and P. Y. Doudin of the book *L'analyse des correspondances et les techniques connexes. Approches nouvelles pour l'analyse statistique des données*.
    Email: cazes @ceremade.dauphine.fr

Jan de Leeuw is distinguished professor and founding chair of the Department of Statistics at the University of California, Los Angeles. His research is mainly in computational statistics, multivariate analysis, and statistical software development. He is the editor in chief of the *Journal of Multivariate Analysis* as well as founding editor and editor in chief of the *Journal of Statistical Software*. Over the years he has published widely in correspondence analysis, multidimensional scaling, and matrix calculus. He is a fellow of the American Statistical Association and of the Institute of Mathematical Statistics, and a corresponding member of the Royal Netherlands Academy of Arts and Sciences.
    Email: deleeuw@stat.ucla.edu
    Web: gifi.stat.ucla.edu

Edwin Diday is professor emeritus of computer science and mathematics at the University of Paris–Dauphine (CEREMADE Laboratory). He is also the scientific manager of the SYROKKO company that produces the SYR software. He is author or editor of 14 books and more than 50 refereed papers. His most recent contributions are in spatial pyramidal clustering, mixture decomposition of distributions data by copulas, stochastic Galois lattices, complex data fusion, and symbolic data analysis. He has been recognized by the Montyon Award given by the French Academy of Sciences.

Email: diday@ceremade.dauphine.fr

Stéphane Dray is a research scientist at the CNRS based in the Biometry and Evolutionary Biology Lab, Université Lyon 1, France. His research interests are in statistical ecology, mainly multivariate methods and spatial statistics. He is the author of several software, including the ade4 package for R. He is associate editor of *Methods in Ecology and Evolution*.

Email: stephane.dray@univ-lyon1.fr
Web: pbil.univ-lyon1.fr/members/dray

Michael Friendly is professor of psychology and chair of the graduate program in quantitative methods at York University, Toronto, Canada, where he is also a coordinator of the Statistical Consulting Service. His recent work includes the further development and implementation of graphical methods for categorical data and multivariate linear models, as well as work on the history of data visualization. Dr. Friendly is the author of the books *SAS for Statistical Graphics* and *Visualizing Categorical Data*, coauthor (with Forrest Young and Pedro Valero-Mora) of *Visual Statistics*, and an associate editor of the *Journal of Computational and Graphical Statistics* and *Statistical Science*.

Email: friendly@yorku.ca
Web: datavis.ca

John C. Gower is professor emeritus of statistics at the Open University, UK. His interests are in applied multivariate analysis, especially methods concerned with graphical representations. His book on the general principles of biplots (Chapman & Hall) with David Hand was followed by a more friendly exposition (Wiley) with Niel Le Roux (Stellenbosch) and Sugnet Lubbe (Cape Town). The first book (Oxford University Press) on Procrustes problems, written with Garmt Dijksterhuis, links with methods, such as generalized canonical analysis and individual differences scaling.

Email: j.c.gower@open.ac.uk

Patrick J. F. Groenen is professor of statistics at the Econometric Institute, Erasmus University, Rotterdam, the Netherlands. His research interests include multidimensional scaling, multivariate analysis, computational statistics, visualization, algorithms, and the applications of these techniques in the social sciences and epidemiology. He is author of more than 40 scientific articles. Together with Ingwer Borg, he has written four books on multidimensional scaling.

Email: groenen@ese.eur.nl
Web: people.few.eur.nl/groenen

François Husson is professor of statistics at Agrocampus Ouest, Rennes, France. His current research interests are principally in multivariate exploratory data analysis and especially on handling missing values in (and using) exploratory principal components methods. He has published three books and has created two R packages in multivariate exploratory data analysis.

Email: husson@agrocampus-ouest.fr
Web: www.agrocampus-ouest.fr/math/husson

Julie Josse is an associate professor of statistics at Agrocampus Ouest, Rennes, France. Her research interests focus mainly on handling missing values in (and using) exploratory principal components methods such as PCA, MCA, or multiple factor analysis for multitables data. She is also interested in suggesting regularized versions of these methods. She is involved in the development of R packages associated with these topics and has collaborated in the publication of the book *R for Statistics*.

Email: josse@agrocampus-ouest.fr
Web: www.agrocampus-ouest.fr/math/josse

Pieter M. Kroonenberg occupies the chair in multivariate analysis, in particular of three-way data, at the Faculty of Social and Behavioural Sciences, Leiden University, the Netherlands. His research interest is primarily multiway data analysis and its application to empirical data. He is the author of two books, one on three-mode principal components analysis and one on applied multiway data analysis, both featuring applications from a large number of different disciplines.

Email: p.m.kroonenberg@fsw.leidenuniv.nl
Web: three-mode.leidenuniv.nl (The Three-Mode Company); www.socialsciences.leiden.edu/educationandchildstudies/childandfamilystudies/organisation/staffcfs/kroonenberg.html

Ludovic Lebart is senior research fellow at Telecom-ParisTech, Paris, France. His research interests concern the statistical analysis of qualitative and textual data and the methodology of survey design and processing. Together with Alain Morineau, he has been at the origin of the software SPAD. He has published several books about multivariate data analysis, textual analysis, and survey methodology.

Email: ludovic@lebart.org
Web: www.lebart.org

Brigitte Le Roux is a researcher at the Laboratoire de Mathématiques Appliquées MAP5/CNRS, Université Paris Descartes, France, and at the political research centre of Sciences-Po Paris (CEVIPOF/CNRS). She is an assistant director for the *Journal Mathématiques and Sciences Humaines*, and she serves on the editorial board of the *Journal Actes de la Recherche en Sciences Sociales*. She completed her doctoral dissertation with Jean-Paul Benzécri in 1970 at the Faculté des Sciences de Paris. Dr. Le Roux has contributed to numerous theoretical research works and full-scale empirical studies involving geometric data analysis.

    Email: Brigitte.LeRoux@mi.parisdescartes.fr
    Web: www.mi.parisdescartes.fr/~lerb/

Ekkehard Mochmann is a board member of the German Association for Communications Research. He was director of GESIS, the German Social Science Infrastructure Services, and executive manager of the Central Archive for Empirical Social Research at the University of Cologne, where he founded the European Data Laboratory for Comparative Social Research. He has been expert advisor to the European Science Foundation—Standing Committee for the Social Sciences (ESF-SCSS) for the European Data Base. He was president of the International Federation of Data Organizations for the Social Sciences (IFDO) and president of the Council of European Social Science Data Archives (CESSDA).

    Email: E.Mochmann@web.de

Fionn Murtagh is professor and head of the School of Computer Science and Informatics at De Montfort University, Leicester, UK. His research interests include digital content analytics, computational science, analysis of narrative, and multiscale morphological modelling. All areas are underpinned by the metric and ultrametric mapping provided by correspondence analysis, clustering, and related methods. Dr. Murtagh has published 25 authored and edited books and 267 articles. He has been president of the Classification Society, and president of the British Classification Society.

    Email: fmurtagh@acm.org
    Web: www.multiresolutions.com/home

Jérôme Pagès is professor of statistics at Agrocampus, Rennes, France, where he is the head of the applied mathematics department. His research interest is mainly in exploratory data analysis, especially the analysis of multiple tables. Together with Brigitte Escofier, he is the originator of multiple factor analysis. He has published six books on statistics, mainly in exploratory data analysis.

    Email: pages@agrocampus-ouest.fr
    Web: math.agrocampus-ouest.fr

Maurice Roux is a retired professor of statistics at the Aix-Marseille Université, Marseille, France. His research interests are in cluster analysis in relation to correspondence analysis. His applications of these methods are mainly in botany, zoology, and ecology. Apart from several publications in the French statistical and biological literature, he has published a French textbook on cluster analysis.

Email: mrhroux@yahoo.fr

Gilbert Saporta is professor of applied statistics at the Conservatoire National des Arts et Métiers (CNAM), Paris, France. His research interests are in multidimensional data analysis and data mining, including functional data analysis and sparse methods. He is the leader of the research team 'Statistical Methods for Data Mining and Learning' at CEDRIC (CNAM lab for computer science, statistics, and signal processing). He has published two books about exploratory data analysis and statistics in general. Dr. Saporta has been ISI vice president and president of IASC (International Association for Statistical Computing).

Email: gilbert.saporta@cnam.fr
Web: http://cedric.cnam.fr/~saporta/indexeng.html

Andreas Schmitz is a research associate at the Institute for Political Science and Sociology, University of Bonn, Germany. His research interests are mainly in relational sociology, methodology, and applied statistics. He is currently working on habitus and psyche, the field of German universities, Bourdieu's field of power, and digital mate markets.

Email: andreas.schmitz@uni-bonn.de
Web: www.politik-soziologie.uni-bonn.de/institut/lehrkoerper/andreas-schmitz

Matthew Sigal is a doctoral student in the quantitative methods area of psychology at York University in Toronto, Ontario, Canada. He is primarily interested in methods of data visualization, especially for truly multivariate models, and using graphical displays as a pedagogical technique for improving statistical education.

Email: msigal@yorku.ca
Web: www.matthewsigal.com

Cajo J. F. ter Braak is professor of multivariate statistics in the life sciences at Wageningen University and senior statistician at Biometris, Wageningen, the Netherlands. His research interests are in statistical ecology, multivariate analysis of ecological and genetic data, resampling, and Bayesian computational methods. He is the senior author of the computer program Canoco for visualization of multivariate data, which started as a program for canonical correspondence analysis. He is among the most cited statisticians worldwide in Google Citations.

Email: cajo.terbraak@wur.nl
Web: scholar.google.nl/citations?user=dkqwZxkAAAAJ&hl=en

Rosanna Verde is professor of statistics at the Faculty of Political Sciences of the Second University of Naples, Italy. Her main research areas of interest are multivariate data analysis, symbolic data analysis, classification, data mining, data stream analysis, and functional data analysis. She has participated in several European projects and European and Latin American scientific cooperation programs. Dr. Verde is author of more than 80 scientific publications.

Email: rosanna.verde@unina2.it
Web: scholar.google.it/citations?hl=it&user=m5ffRB8AAAAJ&view_op =list_works

Richard Volpato is the manager of allocations and analytics at the Copyright Agency, Australia. In the 1980s he developed social research teaching labs in both the University of Tasmania and the University of Melbourne and introduced correspondence analysis. Outside academia, he has also undertaken data analysis projects in forestry, public service management, superannuation, and creative economies. He has also help established the Sydney R group and a national best practice in analytics group, AnalystFirst.com.

Email: Richard@Volpato.net
Web: richard.volpato.net

Prologue: *Let the Data Speak!*

Richard Volpato

CONTENTS

Introduction xxvii
Orientation: Data as Cultural Carrier xxviii
Data and Information xxix
Four Modes of Informing through Correspondence Analysis xxx
Four Tones of Writing from a Correspondence Analysis xxxii
    Wisdom xxxii
    Data xxxiii
    Experience xxxiv
    Knowledge xxxiv
Expositional Steps xxxv
    Large Plots, Big Pens, Bright Colours xxxv
    Writing Blind xxxvi
    Naming a Dimension by Sliding along It xxxvi
    Writing in E-Prime: Eliminating *Is* .... *xxxvi*
    Report Structures and Styles xxxvii
    The Drama of Ecology xxxvii
    Software Stack for Data Verbalization xxxviii
Applied Correspondence Analysis as Social Interventions xxxviii
    Wonder of Wood (Forestry Campaign) xxxix
    Localities: Boundaries and Backbones xxxix
    Saving for Retirement: Customer or Member? xl
    Creative Class Mobilization xl
Legacy of Jean-Paul Benzécri xli

Introduction

Some people love books; others enjoy modern media, like movies. Does anyone yet know how to love data? The writings of Jean-Paul Benzécri and many of his students have opened up a world where data can not only stimulate the mind but also stir the heart. Take a book like Bourdieu's *Distinction*, wherein the seemingly innocent pursuit of culture operates to reproduce much larger systems of social life. Yet it depicts these worlds with finesse worthy of Jane Austen. Of course the dialogue of protagonists does not fill the pages; instead you can imagine all kinds of dialogue. But better still, you can collect similar data to describe other forms of cultural capital and its deployment as a social boundary and a social escalator. Does the famous film *Casablanca* look the same after seeing Fionn Murtagh's correspondence analysis and clustering of the film's dialogue, where the plot actually follows Murtagh's map? While correspondence analysis has long turned words into data that then become maps, which in turn offer new meanings to words, the reverse flow—of data into words—has not been so extensively attempted. This prologue offers one such overview of various attempts.

I will focus on the verbalization of data. After a brief cultural context for data and some comments about the nature of information, I will explore aspects of Benzécri's correspondence analysis, one of the main topics of this book, in terms of distinct ways in which it can inform an audience or agent. These aspects also provide four *tones*, as it would call them, of *expression*, which in turn an analyst can sustain through a series of expositional tactics. I end with a summary of examples where correspondence analysis has guided some social interventions that I have undertaken. I conclude with a return to broad cultural dynamics, which I think will carry Benzécri's legacy well beyond his well-deserved fame regarding specific techniques of data analysis, many aspects of which I have learned through the lucid texts of Michael Greenacre.

To imagine that data can drive expressive verbalization might seem doomed to fail. After all, data are mute. A data set collects observations (nouns) and describes them with variables (adjectives). No verbs! Yet the whole adventure of data analysis as Benzécri developed it distinguishes itself from hypothesis testing statistics by letting data determine descriptions. Yet the techniques used have attracted attention. Less clear are the various styles and facets of interpretation.

Orientation: Data as Cultural Carrier

As a carrier of culture, data still lack the kind of gloss associated with mass media, printed text, or even basic writing. Yet in times gone by, earlier material carriers of culture started on the margins of social life. Alphabets did not receive high esteem from the ancient Greeks. The *pharmakon* in Plato captured both the potency and the poison that alphabetic writing might infuse a culture of oral tradition. Yet, it also gave birth to the Platonic 'forms', fixating cultural constants at a new level, protected from the ravages of dumb alphabets (Derrida, 1981). Similarly, printing presses threatened the authority of priests to mediate the relation between believer and God while also opening up the power and presence of personal conscience and the compelling allure of certainty, particularly with the printing of maps and reference works (Eisenstein, 1979). Finally, and more recently, mass media fuelled the intensities of ideologies while also unleashing the never-ending venture of 'finding myself' (Luhmann, 2000).

In each case, the cultural carrier starts on the cultural and social margins, builds an edifice of cultural and institutional support, which a newer, material carrier disrupts. Data today disrupt how we live and how we will come to understand living. Of course, like alphabets, print, and media, data initially appear as a corrosive force: 'Lies, damn lies, and statistics!' as Disraeli put it. More fundamentally, data get treated as a kind of cultural detritus left behind by the modern mania for quantification and surveillance! Descriptions of overload abound: data deluge, data explosion, data overload, or, more recently, 'big data'. The verbs that capture the engagement with data suggest tenacious toils with obdurate material: data marshalling, cleaning, transforming, coding, reducing, mining, excavating, etc. Even statisticians can be heard talking this way. Data have yet to appear as a rich semantic ingredient to discourse. Benzécri understands this predicament. As his writings weave between questions of data, coding, and visualization, myriad asides about interpretation pepper his texts, for example, his urging analysts to

> introduce continually concrete details of interest to the reader, and [thus] sustain his imagination without taxing his patience. (Benzécri, 1992, p. 417)

Data and Information

In a commentary on the nature of induction, data, and ideas, Benzécri follows classical tradition and splits the notion of idea across two independent directions: comprehension (or what an idea means) and extension (or what 'falls under' the idea as instances) (Benzécri, 1980). The philosopher Peirce in the 1860s developed a similar contrast with the corresponding terms of *informed depth* and *informed breadth* (De Tienne, 2006). A data set encapsulates this duality: columns of attributes (depth) describe rows of cases (breadth). The task of correspondence analysis runs in both directions: to reduce the columns to a few *parsimonious dimensions* that capture significant covariation of attributes across cases and to reduce cases to a few, often hierarchically ordered and *distinct clusters* (Greenacre, 2007). Qualitative research works with few cases but seeks out all the depth these cases can offer through the many attributes that together help to constitute significant contrasts between cases (or within a case over time). Quantitative research focuses on attributes in the hope of expanding the breadth of their reach through dimensions.

In the same meditation on the nature of ideas, Benzécri articulates a separate distinction independent of the first. In discussing William of Occam and his 'razor', Benzécri provocatively claims:

> Statistics is not only a means of knowledge, draining into an idea the flood of facts; it is a mode of being. (Benzécri, 1980, Section 6: 'Statistical Existence')

Here statistics functions as a *mode of knowing* or a *mode of being*. As a mode of knowing, statistics places the analyst *outside* of the dynamics being investigated: on the banks of the river watching the flow, discerning patterns (of currents, cross-winds, eddies, etc.). As a mode of being, by contrast, statistics situates the analyst *inside* the flow of events trying to find an orientation that enables a clear, if momentary, picture to emerge and actions to be formulated. Insofar as the witnessing can inform actions, then an interplay can unfold between knowing and being. The work of Bourdieu (1979, 1984) illustrates this kind of interplay between knowing and being through his concept of practice.

A different approach to this interplay can be developed through the notion of information as an event rather than as stuff. Information happens to the extent that it informs someone. Thus, the noun *information*, like the noun *explosion*, refers to analogous events: information dissolves symbolic structures of belief, just as an explosion dissolves physical structures (Pratt, 1998). An explosion requires explosives, and correspondingly, information requires, as it were, 'informatives' (Clark and Carlson, 1982). Benzécri, in a sense, aims to show how to inform beliefs with believable information using data rather than raw eyewitness accounts (journalism), the cleansed results from the laboratory (science), the immersion in a different culture (fieldwork), or the authority of tradition (heritage).

Four Modes of Informing through Correspondence Analysis

We can combine these two sets of directions that determine an idea: comprehension versus extension on the one hand, and knowing versus being on the other. Four distinctive information vehicles (Nauta, 1970) or informatives (Clarke, 1982) emerge, each of which has a role to play in verbalizing insights emerging from a correspondence analysis that can inform beliefs:

Prologue: Let the Data Speak!

	Comprehension	Extension
Mode of Being	Wisdom	Data
Mode of Knowing	Experience	Knowledge

With respect to *modes of knowing*, in any conversation that attempts to inform via experience, a speaker will draw on rich distinctions so that each new case can be aligned to all the other similar cases: imagine a midwife who has 'seen a lot' pointing out how 'this case' is like some other in its peculiarities (depth) and needs special care. How much an experience informs relates to what has been called pragmatic information (Ackoff, 1958). Similarly, experts may inform us convincingly by invoking established knowledge (as they do, for instance, in climate science or a medical diagnosis). The extent that a discovery alters the established views has been linked to semantic information (D'Alfonso, 2011). Extension aims for abstraction, leaving detail behind. In terms of correspondence analysis, these two information vehicles take the form of reducing data to clusters and dimensions, respectively.

With respect to *modes of being*, the flux and flow of events leave traces behind: data. Data get housed in data matrices, which of their nature makes them 'mute': you cannot easily read through a data matrix aloud and make sense of it! The amount of information in such a matrix might be called syntactic information, as it can only measure the extent of deviation from pure randomness (uncertainty). At a basic level, the formulae of correspondence analysis that decompose a data matrix into dimensions operate at this syntactic level. But unlike other techniques, the interpretative effort continues to convert syntactic into semantic information and that into pragmatic information. Many approaches to data only deal with data as part of a measurement model. For Benzécri, *potential meaning* rather than *predictive model* requirements defines the condition of data admissibility.

Benzécri does not use the concept of wisdom explicitly in his writings. Instead, he focuses on a concept of a mathematical filter (Benzécri, 1980), which I think does the same work. This refers to finding a minimal set of characteristics that encompass a local dispersion (e.g., a cluster), but also means determining dimensions that summarize the total dispersion within which all distinctions that differentiate clusters find a common containment. The drive to filter might also be called a *quest*, specifically, a *questioning quest*. Such a quest brings ignorance to the surface. Ignorance can also be depicted as an *empty* data matrix: rows waiting to be filled by observations, and questions about attributes waiting to be transcribed as they get answered. A more complex problem will more likely reveal a greater scale and cost of ignorance. While Benzécri's use of data can appear imaginative and explorative, his use of filter provides the compass in data exploration. Without deep and penetrating questions, data will remain mute!

Four Tones of Writing from a Correspondence Analysis

We can move through these same four information vehicles in terms of the tones of writing:

Wisdom	→	Enlighten
Data	→	Encapsulate
Experience	→	Exemplify
Knowledge	→	Explicate

By tone I mean the writing ought to carry a feeling that orients the reading to the situation you want the readers to face.

Wisdom

Wisdom poses questions *that matter* to some actor caught in some practice. Hence, seeking out relevant data presupposes the analyst aligning to a *quest* and questioning it and eliciting what questions would in turn matter to such a quest. Wisdom would then transform qualms about *uncertainty* into *doubts* about specific actions or contrasts. The specification of ignorance thus becomes more feasible: in terms of how much such ignorance costs some actors and how it might be reduced through some analytic intervention.

To enlighten means to enliven a debate or a research area by posing questions, which can then be linked to potential data. Thus, in a data-drenched world, wisdom cannot function as being 'in the know', as used to occur: It aims to create new conditions for credence (in the reader) precisely through the fearless delineation of doubts. Problems posed might also be placed on a scale of complexity: puzzles, problems, wicked problems, and mysteries. Thinking in terms of correspondence analysis widens the scope for data that can prove informative, particularly in the area of so-called wicked problems or what Ackoff calls a mess (Ackoff, 1974, p. 21).

Three lines of questioning can partition uncertainties: questions about composition, causes, and functions. The first opens up the space of variation for an outcome variable. For instance, some outcome variables immediately suggest components: rates of truancy, industrial strikes, ceremonies, meals, discussions, and domestic disputes all have aspects of *intensity* (how long it lasts?), *extensity* (how many involved?), and *frequency* (how recurrent?). Thus, rates in the volume of such events can be decomposed, and the variations in these cases across three component measures of volume can be opened out into illuminating plots (Greenacre and Lewi, 2009). These questions open up a single dependent variable into a multidimensional space of variation.

Similarly, many causal claims can be tested if formulated in terms of narrowly defined variables. For instance, links between cholesterol and blood pressure might be obvious, but from a correspondence analysis view, questions about the *global* impact of diets and healthy heart (in all its facets) open up a manifold of influences and contingencies, which begin to determine a *wide* data matrix waiting to be filled so it can inform.

Finally, functional relationships, for instance, in determining the resilience of an individual, organization, or institution to adapt to changing environments, open up the capacity to pose ecological questions about how social and organizational life thrives or withers within competitive dynamics and across gradients of resource and obstacles. This sounds like a metaphor and could introduce a study. However, correspondence analysis, particularly in the form of canonical correspondence analysis (ter Braak, 1995), provides a powerful way to organize a complex weave of influences.

Data

Data too often does not get much space as 'itself' in text. Data comes across as boring. A first point of engagement with data lies in describing the adventure of its acquisition. This keeps the reader close to some flux and flow of some domain and the intervention to harvest data from it. From a correspondence analysis point of view, for example, a study on backpacker tourists, one might, instead of using interviews or questionnaires, do inventories of the contents of their backpacks—in all their heterogeneous glory, or get copies of, or links to, their resulting photo image libraries. Staying close to data sources makes it easier to find encapsulations of some issues with some data point that, as it were, offers iconic value in addressing the questions.

Correspondence analysis has specific ways of organizing data. A fundamental form is the so-called *disjunctive* (or *indicator*) *matrix*. Here, *every* variable has *specific* value category, with continuous variables quantized. Observations become lists of 0s and 1s. Where uncertainty exists regarding an appropriate value label for a case, fuzzy coding can distribute proportions of the 1 over the likely values. Encapsulation with respect to this matrix has a specific meaning: a simple cross-tabulation of counts of two dichotomous variables in which a dependent variable of prime interest (smokes versus does not smoke) lies at the top of the table, and a presumed predictor variable (e.g., manager versus nonmanger) on the side can generate a number and an encapsulating sentence. If percentages are calculated *across* the rows and the difference obtained *within* the column of prime interest (smoking), then the most primitive verbalization possible of a data matrix emerges that can inform: 'senior managers lead junior managers in smoking by a 15 percentage point difference'. Given that disjunctive data matrices can be massive; it might seem pathetic to start playing with just two variables! However, dichotomous variables can make easy sense and can lead into the large multidimensional spaces as the narrative expands. Further, mathematically, such a percentage difference equals a *regression coefficient*, and the subtracted category, 'junior managers' (within the column of interest, 'smoking'), equals an *intercept*.

In another tradition, stemming from Yule and developed by Lazarsfeld as the elaboration model, and later by Davies (1984) in the United States and Boudon (1974) in France, the introduction of a third and more variables enabled a narrative to tease out the impact of other variables on the original percentage difference. Other variables may reinforce, attenuate, or condition the original association. The narrative correspondingly grows (in the idiom of a casual model directed graph) while the underlying 'intercept' becomes an ever more extended collection of values (one value from each of the variables in the directed graph) that define a 'syndrome' of the *least* likely group of attributes that continue to do the action (e.g., 'even among junior managers who are educated, health aware, etc., a small x% still smoke'). Encapsulation presents a *node* of topical interest linked to a presumed predictor around which a narrative can grow.

Since only a few, typically dichotomous, variables are used in an elaboration model, they can be highlighted within a wider dimensional space (usually developed off a Burt matrix of the fuller set of variables—see Chapter 4 by de Leeuw and Chapter 11 by Husson and Josse in this book). From such a correspondence analysis plot, it becomes possible to *select* the variables that will deliver significant percentage differences, since these are simply larger distances between values of variables that look like they belong in a causal narrative. Of course, the resulting percentage differences have to matter to a reader, by encapsulating some insight or doubt he or she had.

Experience

Experience trades on familiarity of objects. Exemplification provides concrete bearings as ignorance gets reduced. From within correspondence analysis, the development of clustering using the same distance metric often produces clusters that exemplify a 'life world': what it is like to be in some part of the social (or, indeed, ecological) world. A simple technique here lies in using 'quotations' from open-ended questions that capture the objects, issues, and moods associated with some individuals in a specific cluster.

Similarly, popular misconceptions typically suggest clusters of characteristics that come bundled unthinkingly, as it were. Much of the appeal of Bourdieu's work lies in opening out alternate ways in which obvious social differences might be reproduced (e.g., via cultural capital). A compelling way to illustrate this aspect of correspondence analysis is to have the quotes from respondents associated with clusters to be heard in headphones as the reader/user moves the cursor over specific clusters on a screen, or the quotes to scroll past, or photo collages to appear.

Knowledge

Knowledge removes the 'cases' from view. Hence, its tone in writing operates within a patient *explication* of aspects of some outcome variable. With correspondence analysis, such explication can take on a more gritty quality since the purity of causal links can be embedded within distinct subsystems (of components, causes, or functions). For instance, a link between parental income and educational opportunities for their children barely rates mentioning. But in a correspondence analysis the full range of indicators of what gets *accessed* with income (e.g., consumption profiles of parents) and indicators of education *engagement* of students can be interlaced in a common space. Such granular details need not be reduced to the core causal link between two variables; rather, the details can help explicate the many ways in which income and education interact. Explication finds its strongest support through the use of supplementary variables in a correspondence analysis: instead of a single dependent variable, all the aspects of outcome can define a space, and all the likely causes can be overlaid as supplementary variables in that space. Alignment reveals causal texture. Explication ought to feel to the reader like a process of patiently unfolding a crumpled structure.

These four tones orient analysis (Benzécri's idea) toward informing the reader. To enlighten, define doubts relative to actions; to encapsulate, start with a data point or a contrast (e.g., a percentage difference), the words of which address some doubt; to exemplify, get the reader to experience 'what it is like' to be in a cluster; and finally, to explicate, depict the world as a weave of causal influences.

Expositional Steps

In teaching correspondence analysis to both students and professionals needing to use data, two transformations have struck me: first, the expansion of imagination (even though the object of engagement is data) and sociability engaging data in this way creates. Trying to teach LISREL (software for structural equation modelling), for example, has produced opposite effects! I list here various tools that have produced these effects.

Large Plots, Big Pens, Bright Colours

Regardless of the data preparation issues that can be reviewed and improved later, having large plots put on walls with values of variables linked and clusters (relative to individuals) forces a visceral engagement far more compelling than that of screen-based or individualized exercises. Having to go close to a large plot on a wall, draw or label things, then walk back to 'see the picture' at a distance creates a yearning for encapsulation. One can only marvel at the way students and professionals who, given the chance to engage the totality of data through a correspondence analysis discover the joy of discovery.

Writing Blind

Too often correspondence analysis plots have too much *encoded* in them and the readers will not be able to *decode* the results for themselves. Accordingly, to counter this, students look at plots, but then have to turn their screen *off* but also write. This initially causes a collapse in writing ('I cannot see!'), but eventually it forces a layering of content and associated writing (Klauser, 1987). In fact, good graphs are *lean* and ought to leave a strong 'afterimage' that can inform writing. Reading data-enriched discourse can thus have a certain oscillation—for as the reader comes to a full stop (of a sentence), he or she can allow his or her eyes to look back at the plot and return to the next sentence, replenished.

Naming a Dimension by Sliding along It

While large plots on a wall (even with the second- and third-dimensional plot at right angles to the first wall) can provoke curiosity and a questioning quest, the tendency to label dimensions from the graph directly runs into problems. First, disentangling the 'pull' of points with large masses or high inertia can get complicated if these work on several dimensions at once. Second, there are different kinds of influence (cosine, contribution, supplementary, and jitter). *Jitter* refers to the movement of points in a correspondence analysis plot due to the reiterative subsampling of the data.

A simple procedure that can assist letting the data nudge the analyst to correct names involves focusing on a dimension *one at a time* and not two or three at a time. Writing the labels that fall on a dimension with rows and column labels to the left and right of the dimension can orient the analyst. A spreadsheet with dimension labels written down columns can be illuminating. Labels with high cosines (i.e., high associations with dimensions) are in bold, high contributions also in bold and italic, for instance. Supplementary variables can be particularly powerful in this regard. For instance, if age groups operate as supplemental and they fall across the full range of a single dimension, then the values of the *other* variables also aligned to the dimension will give indications of whether the dimension in question relates to aging (as a process) or cohorts (journeying through change) or the legacy that a period (in life) has left its mark. Each of these orientations to age can dramatically alter the narrative. Herein lies the guiding role of questioning quest—the labels of dimensions really matter relative to agencies that will be informed by them.

Writing in E-Prime: Eliminating *Is*

Benzécri never strays far from ecological dynamics in his interpretations. Such dynamism requires active verbs, actions, and intimations of likely outcomes. A useful discipline here comes from banning the use of *is* in sentences. When fully accomplished, the writing style has been called E-prime

(Bourland and Johnston, 1991). This forces active voice. Later, the rule need not apply so rigorously. But steps 1–4 move the writing toward thinking *with* data rather than *about* data. In some contexts, dimensions can be described as gradients of difficulty: the values of variables along the dimension suggest active verbs, for instance, a dimension of cultural capital might express a lure for learning or culture (from one end, going up, as it were), while from the top (i.e., high cultural capital) the values might suggest the leveraging of culture for other purposes. Depending on what questions drive the writing, different active verbs might operate depending on whether the discussion is moving from one end to the other of a dimension or vice versa.

Report Structures and Styles

The earlier list of the tones information vehicles can provide ways to structure reports, and this can also lead to structuring interactive reports and indeed finally to quasi-automation of report (or report snippets).

In some cases, each of the four tones can be sections of a report. In other cases, *each* paragraph might move *between* each of the four tones: a provocative data item revealing some difference encapsulates an issue, some explication follows, exemplified with more familiar cases, and finally more questions can be posed seeking further enlightenment that prompts the next paragraph. Alternatively, each tone's text can iterate within the others. For instance, a cluster might itself be opened up into its local dimensions. Now the exemplification receives further explication through subspace dimensions. In my experience, reports carrying correspondence analyses tend to try to reveal too much too soon. Given that a simple four-way tabulation of dichotomous variables can sustain 20 pages of elaboration, correspondence analyses can deliver orders of magnitude more information (syntactic, semantic, and pragmatic) and yet fail because of poor *pacing*. Academic protocols for research papers and templates for market or policy impact studies all risk constraining the full power of what can be said from data released into view and discussed in a correspondence analysis.

The Drama of Ecology

Narratives create drama. While correspondence analysis plots typically do not show time, the nature of the duality between observations and attributes engenders a sense of ecological process. This explicitly happens in ecological studies using correspondence analysis (their proliferation indicating the affinity between this method and ecological processes). The same kind of thinking can apply to social surveys. In all these cases, an implicit temporal aspect beckons to take centre stage in the narrative. The values of variables can be imagined as activities roaming through the terrain of, as it were, hapless individuals: they thrive, survive, or fade within various niches defined by permutations of values (of variables).

The more such individuals can be informed of these predicaments, the more they can develop a questioning quest of their own and imagine paths through a correspondence analysis space. In this context, narrative becomes reflexive. For instance, carpooling as an activity may spread into more individual lives, but along which vector of contagion? A plot of time budgets by household, location, and demographic characteristics might show several directions a marketing campaign might take, and within the granularity of the plot, each campaign can define benchmarks of accomplishments, in terms of the cluster of car pooling individuals expanding in specific directions. More generally, Badiou's (2007) concept of forcing wherein a vanguard group 'pushes the envelope' with new ways of framing a mode of life can in this context be given a more specified terrain within an ecology where moves in particular directions within a correspondence space may reveal easy wins, resistance, or reactions.

Software Stack for Data Verbalization

Although many kinds of software offer ways to structure writing or even to create graphs, correspondence analysis generates some specific challenges and offers some particular benefits to data-enriched writing. First, as with large plots on a wall, correspondence analysis plots on screens benefit immensely from collaborative annotation systems where researchers, perhaps widely dispersed, can label values, dimensions, and clusters as they see fit and then discuss. Second, correspondence analyses carry far too much for a reader to engage. Rather, they need to enter the plots with a verbal commentary. This might be going along dimensions, hopping between clusters, labelling quadrants from a two-dimensional display—each of which may require different layers of plot to be made visible in a series (or animation). Third, beyond labels, actual textual commentary might emerge from several writers. Correspondence analysis needs a software stack to support a collaborative interpretative venture, just as it has a rich variety of software programs to service its algorithmic requirements. Such a verbalization software stack might build on R (statistical programming language), linked to some of the newer collaborative outlining tools and powering an interactive graphical framework, like D3 (Data-Driven Document). D3 has its origins with Jacques Bertin's *Semiology of Graphics* (2010), but built with JavaScript and SVG. The new movement toward data journalism (Gray et al., 2012) may soon discover correspondence analysis, which might in turn focus efforts on this software stack.

Applied Correspondence Analysis as Social Interventions

Most of the correspondence analyses I have done have related to practical settings over years, rather than through a one-off consultancy or research project. While I helped to introduce correspondence analysis in Australia in the 1980s, with powerful pedagogical results in social science courses, it also proved subversive: students became rapidly more self-assertive about grounding what they said in data, rather than being deferential about quoting from literature. In applied settings correspondence analysis had interesting political and not simply research effects: searching out any quest naturally connected to stakeholders. Questioning all quests with an unwavering curiosity elicited a level of respect that gave research venture more autonomy. In contexts, verbalization of results mattered, for while plots can be arresting and individuals may want to dwell on them, the impact of such analyses came from the verbalization used to frame options for actors often with *opposing* interests. Some examples follow.

Wonder of Wood (Forestry Campaign)

In the context of sustained conflicts between 'greens' and so-called logging, ideological stances and stereotypes naturally proliferate. Through six years of surveying, it became clear that two dimensions discriminated the value of wood: (1) ambient value versus utility and (2) artificial use versus natural use. The resulting quadrants provided four faces of the wonder of wood—shifting focus from trees to something actually implicated in many activities. Thus, wood could be a resource (e.g., wood chips to make paper or synthetics), a setting (e.g., a forest to explore), an object (of carpentry), or a built backdrop (e.g., floors, walls, shelters). The data comprised series of surveys that questioned *behavioural* encounters with *attitudes* toward and *perceptual* distinction between various uses of wood, forest practices, forest scenes, and stakeholders in forests (government bodies, local saw millers, tourists, etc). In this particular case, the resulting quadrants, rather than clusters or dimensions in particular, could structure the narrative. From this, a number of interventions and mediations could be formulated to focus on the wood in its distinct modes (so rural schools could get involved with monitoring, craftsmen could get at woods neglected, scientists could formulate forest practices, the general public view of pristine nature could be addressed, and so on). Some of the interventions might have happened regardless, but their coordination through a common frame meant some of the heat of debate could morph into hopes for (distinct) futures. That in turn enabled otherwise competing interests to talk specifics rather than retreat to rigid positions.

Localities: Boundaries and Backbones

In other research—again over years—the need to amalgamate local government inevitably sparked controversy. But one of the perceptual distortions that survey results showed lay in the articulation of hopes that presupposed the collective use of facilities that *could* be backbones to a region (of amalgamated local governments), but which currently functioned as boundaries between local government areas (e.g., roads, rivers, hills, industries). Three distinct dimensions emerged: an orientation toward lifestyles rather than productive activity, a focus on basic services (garbage collection) versus values (parks, business hubs, cultural events), and fear of change versus hopefulness. While such plots and commentary might appear a worthy research outcome (of semantic information), the real challenge lay with framing these dimensions with respect to local idiom and geography. The point of such research lay with finding ways to achieve regional consolidation without losing services or trust in local officials. Some local governments could merge by converting *boundaries* (that could be collectively mobilized) into *backbones* of comparative advantage.

Saving for Retirement: Customer or Member?

Superannuation funds inevitably drive toward a customer view of their depositors. Yet they in turn (and in part) see depositors as members. A variety of surveys and qualitative studies showed the usual concerns with the future (for about-to-retire individuals). From a customer point of view, the marketing pitch had been to 'save more with us' with some pitch about 'income returns and service'. But by contrast, once *clusters of interest* could be found, it became possible to formulate strategies for activating these clusters as membership groups, spread within three dimensions of retirement orientations: retreat versus activism, commercial versus voluntary activity, and familial versus nonfamilial. While questionnaires provided the first source of data later, various specialized websites and communication initiatives could activate members with their interactions leaving data trails that could enable recalibration of earlier correspondence analyses and further refinement of strategies. Thus, in contrast to many financial institutions, which reduce individuals into solitary customers, in this setting, groups could be addressed and far more socialization of insights and debate about retirement could be modulated through them. Again, without the verbalizations involved, many of these strategies could not have been articulated against the normal backdrop of marketing practices and hype.

Creative Class Mobilization

A growing body of thought suggests that capitalism has moved into a new mode, so-called cognitive capitalism (Vercellone et al., 2010). A component of this relates to the role of individuals who create content and how they relate to the wider economy. At one extreme, some consider such creative a vanguard, called the creative class (Florida, 2005). At another extreme, the work of creation operates as a nonremunerated shadow labour. Marginalized content creators may not even see themselves as part of some nebulous creative class. These include teachers, health workers, tourist guides, fitness trainers, etc., who, in actually creating content (for a variety of reasons), reveal a distinctive cultural consumption profile (e.g., they have more focused and filtered consumption of specific content, more articulated filing/tagging, and actively reproduce or disseminate content). Due to a particularly access-enabling copyright regime in Australia, instructors and government employees can copy (modest) portions of *whatever* they require by way of texts or images (from the *whole world* of content) to do their jobs. A modest fee gets collected (e.g., about 50 cents per week per student). This *secondary* money flow not only keeps a local publishing sector in a vibrant state, but also remunerates many thousands of others (either directly or via publishers) who develop niche engagements with educational or service delivery contexts. Opportunities abound for mobilizing collaborative creativity through correspondence analyses of

1. Word co-occurrences in content, to determine subject areas (which can then be mapped to library and tag cloud services) to then enable easier discoverability of content and its authors
2. Creator by content tabulations relative to subject, level, and style of learning engagement, to enable aggregation of creative effort and optimal location of creative hubs
3. Textbooks (as environments) relative to copies of *other* content that goes with them, so content curation relative to publishing can be partially automated
4. The overall ecology of the distribution of survival of content (as copies) by works, their imprints, publishers across subjects, locations

This last area, by developing a metric of distance between cultural objects, would enable genuine measure of diversity (Leinster and Cobbold, 2012) to be calculated so that questions might be answered: diversity mix of regional, national, and international content; profiles of educational bodies increasing or decreasing diversity of content.

Legacy of Jean-Paul Benzécri

The horizon of possible writing practices and styles that data might engender has yet to fully dawn on the intelligentsia. Indeed, they may be the last to notice. Software may get there first!

Benzécri, clearly a very erudite scholar, nevertheless has had a constant interest in practical matters. Indeed, correspondence analysis came alive after the 1968 events, as students returned to reopened universities now keen to go to internships in the wider worlds. The previous year Benzécri had had difficulty mustering interest.

If I had to summarize his thinking (I have only met him once, in 1985), I would hazard that he is one of the few who can blend Hellenic and Judaic thinking within the medium of data. The tension or dialogue between knowledge and experience (or practically, the framing of clusters by dimensions) has a Hellenic quality to it: Plato's eternal forms have become emergent dimensions, while Aristotle's *energeia* (actualizations) have become thriving/declining clusters. The data themselves and the questioning quest that enable data to frame their own dispersion (i.e., wisdom) have a Judaic quality to them: the data, like Job, are demanding to be heard, demanding to illuminate the human condition. Job was a successful person and a paragon of virtue who found himself stripped of all possessions, health, and family, and who thus demanded that God explain himself for letting this happen to one so righteous. His friends counselled that he should just remain mute and meek, lest he provoke further divine wrath. Data too have been stripped of normal cultural eminence. Do we hear data as giving off Job's plea? And do we engage data as God engages Job: 'It is my turn to ask questions and yours to inform me' (Book of Job, Jerusalem Bible, 1966). The questioning quest that answers Job's pleas can be found in St. Paul's articulation of faith: a quest to reframe a situation through questions which, in being asked as part of a quest, and not some idle curiosity, become universal in appeal and ultimately never stop seeking out more data. We may still love older cultural forms like books or movies, but it may turn out that only data as a cultural medium can now enable us to approach matters of destiny.

Section I

History of Correspondence Analysis and Related Methods

1

Some Prehistory of CARME: Visual Language and Visual Thinking

Michael Friendly and Matthew Sigal

CONTENTS

1.1 Visual Language 6
    1.1.1 Rise of Visual Language 6
    1.1.2 Maps 8
    1.1.3 Graphs and Diagrams 10
1.2 Visual Thinking 12
    1.2.1 Graphic Vision of Charles Joseph Minard 12
    1.2.2 Francis Galton's Visual Discoveries 14
1.3 Conclusion 15

> If statistical graphics, although born just yesterday, extends its reach every day, it is because it replaces long tables of numbers and it allows one not only to embrace at glance the series of phenomena, but also to signal the correspondences or anomalies, to find the causes, to identify the laws.
>
> **Émile Cheysson, c. 1877**

Correspondence analysis and related methods (CARME), as described in the preface, includes simple and multiple correspondence analysis (CA and MCA), biplots, singular value decomposition (SVD) and principal components analysis (PCA), canonical correspondence analysis (CCA), multidimensional scaling (MDS), and so forth. The commonalities shared by these methods can be grouped in relation to the features of hypothesized lateralized brain functions. The left-brain elements are more logical, formal, and mathematical: matrix expression, eigenvalue formulations, and dimension reduction, while the right-brain features are more visual: (point) clouds, spatial data maps, geometric vectors, and a geometric approach to data analysis. This lateralization of brain function is often exaggerated in popular culture, but it resembles a conjecture I have long held regarding data analysis (see Friendly and Kwan, 2011):

> **Conjecture** (bicameral minds). There are two kinds of people in this world—graph people and table people.

The term *bicameral mind* comes from Julian Jaynes's (1978) book on the origin of consciousness, in which he argued that ancient peoples before roughly 1000 BC lacked self-reflection or meta-consciousness. For bicameral humans, direct sensory neural activity in the dominant left hemisphere operated largely by means of automatic, nonconscious habit schemas, and was separated from input of the right hemisphere, interpreted as a vision or the voice of a chieftain or deity.

We don't fully believe the strong, two-point, discrete distributional form of the above conjecture, but rather, a weaker claim for bimodality or clearly separated latent classes in the general population. That being said, we also believe that the CARME community is largely composed of 'graph people', who, despite their interest in formal mathematical expression, can still hear the voice of a deity proclaiming the importance of data visualization for understanding.

With these distinctions in mind, this chapter aims to sketch some of the historical antecedents of the topics that form the basis of this book. As self-confessed graph people, we confine ourselves to the right-brain, deity side, and consider developments and events in the history of data visualization that have contributed to two major revolutions: the rise of visual language for data graphics and successes in visual thinking.

To further align this chapter with the themes of this book, we focus largely on French contributions and developments to this history. We rely heavily on the resources publicly available via the Milestones Project (Friendly and Denis, 2001; Friendly, 2005; Friendly, Sigal, and Harnanansingh, 2013). A graphical overview appears in Figure 1.1, showing birthplaces of 204 authors who are important contributors to this history. Of these, 36 were born in France, second only to the UK. The Google map on the http://datavis.ca/ milestone site is global, zoomable, and interactive, with each geographic marker linked to a query giving details about that individual.

Some Prehistory of CARME: Visual Language and Visual Thinking

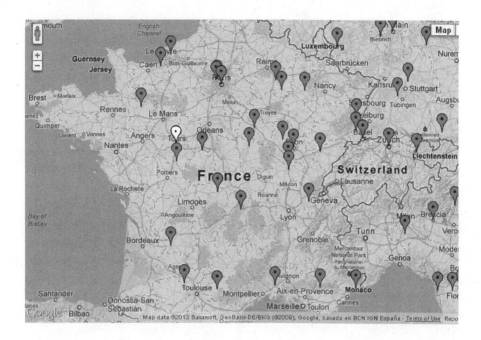

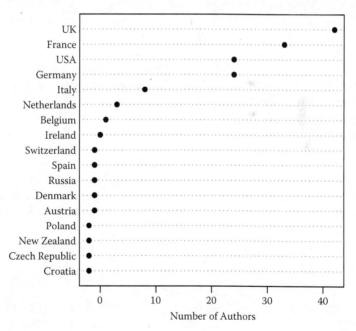

FIGURE 1.1
Birthplaces of Milestones authors. Top: Portion of an interactive Google map, centred on France. The highlighted point is that for André-Michel Guerry, born in Tours, December 24, 1802. Bottom: Frequencies, by country of birth.

1.1 Visual Language

Data and information visualization, particularly for the descriptive and exploratory aims of CARME methods, are fundamentally about showing quantitative and qualitative information so that a viewer can see patterns, trends, or anomalies in ways that other forms—text and tables—do not allow (e.g., Tufte, 1997; Few, 2009; Katz, 2012; Yau, 2013). It is also important to realize that data displays are also communication tools—a visual message from a producer to a consumer—designed to serve some communication goal: *exploration and analysis* (to help see patterns and relations), *presentation* (to attract attention, illustrate a conclusion), and for *rhetoric* (to persuade). As such, effective data displays rely upon commonly understood elements and shared rules of visual language.

1.1.1 Rise of Visual Language

Such rules were developed through use and experimentation. In data-based maps, as well as statistical graphs and diagrams, this modern visual language arose over a period of time, largely the 18th and 19th centuries. This era also witnessed the rise of quantification in general, with many aspects of social, political, and economic life measured and recorded (Porter, 1995); visual summaries were necessary to take stock of and gain insight about the growing body of data at hand.

With this increase in data, the graphical vocabulary for thematic maps surged. New features were introduced to show quantitative information, such as *contour lines* that revealed the level curves of a surface (de Nautonier, 1602–1604; Halley, 1701; von Humboldt, 1817), *dot symbols* that could be used to represent intensities such as population density (Frère de Montizon, 1830), *shading*, as in choropleth and dasymetric maps, to show the distribution of data variables such as education or crime (Dupin, 1826; Guerry, 1832), and *flow maps*, to show movement or change on a geographic background, such as those developed by Minard (1863).

Likewise, while the modern lexicon of statistical graphs stems largely from the work of William Playfair (1801) with the line graph, bar chart, and pie chart, other methods soon followed, such as the scatterplot (Herschel, 1833), area charts (Minard, 1845) and other precursors to modern mosaic displays (Friendly, 1994), polar area diagrams (Guerry, 1829; Lalanne, 1845) or 'rose diagrams' (Nightingale, 1858), and so forth.

In the second half of the 19th century, a period we call the golden age of statistical graphics (Friendly, 2008), the International Statistical Congress began (in the third session, Vienna, 1857) to devote considerable attention to standardization of this graphical language. This work aimed to unify disparate national practices, avoid 'babelization', and codify rules governing conventions for data display (see Palsky, 1999).

However, absent of any overarching theory of data graphics (what works, for what communication goals?), these debates faltered over the inability to resolve the differences between the artistic freedom of the graph designer to use the tools that worked, and the more rigid, bureaucratic view that statistical data must be communicated unequivocally, even if this meant using the lowest common denominator. For example, many of Minard's elegant inventions and combinations of distinct graphical elements (e.g., pie charts and flow lines on maps, subtended line graphs) would have been considered outside the pale of a standardized graphical language.

It is no accident that the next major step in the development of graphical language occurred in France (extending the tradition of Émile Cheysson, and Émile Levasseur) with Jacques Bertin's (1967, 1983) monumental *Sémiologie graphique*. Bertin codified (1) the 'retinal variables' (shape, size, texture, colour, orientation, position, etc.) and related these in combination with (2) the levels of variables to be represented ($Q:$ = quantitative, $O:$ = ordered, $\neq:$ = selective (categorical), $\equiv:$ = associative (similar)), (3) types of 'impositions' on a planar display (arrangement, rectilinear, circular, orthogonal axes), and (4) common graphic forms (graphs, maps, networks, visual symbols).

Moreover, Bertin provided extensive visual examples to illustrate the graphical effect of these combinations and considered their syntax and semantics. Most importantly, he considered these all from the perceptual and cognitive points of view of readability (elementary, intermediate, overall), efficiency (mental cost to answer a question), meaningfulness, and memorability.

The most recent stage in this development of graphical language is best typified by Lee Wilkinson's (2005) *Grammar of Graphics*. It considers the entire corpus of data graphics from the perspectives of syntax (coordinates, graphical elements, scales, statistical summaries, aesthetics, etc.) and semantics (representations of space, time, uncertainty). More importantly, it incorporates these features within a computational and expressive language for graphics, now implemented in the Graphics Programming Language (GPL) for SPSS (IBM Corporation, 2012) and the `ggplot2` (Wickham, 2009) package for R (R Development Core Team, 2012).

This is no small feat. Now consumers of statistical graphics can learn to *speak* (or write) in this graphical language; moreover, contributors to these methods, as in the present volume, can present their methods in computational form, making them more easily accessible to applied researchers. A leading example is the *Understanding Biplots* book (Gower et al., 2011) that provides R packages to do all of the elaborate graphical displays in 2D and 3D that comprise a general biplot methodology related to PCA, CA, MCA, CCA, and more.

The historical roots of these developments of visual language are firmly intertwined with those of data-based maps and statistical graphics. In the remainder of this section we highlight a few important contributions, largely from a French perspective.

1.1.2 Maps

In this subsection, there are many important French contributions we could emphasize. For example, amongst the earliest uses of isolines on a map was the world map by Guillaume de Nautonier de Castelfranc (1602–1604) showing isogons of geomagnetism. This considerably predated Halley (1701), who is widely credited as the inventor of this graphic form.

Among many others, Phillipe Buache (1752) deserves mention for an early contour map of the topography of France that would later lead to the first systematic recording of elevations throughout the country by Charles Lallemand, mentioned later in this chapter. Moreover, although Playfair is widely credited as the inventor of the bar chart, the first known (to me) exemplar of this graphic form occurred in a graphic by Buache (1770), charting the ebb and flow of the waters in the Seine around Paris over time.

However, there is only one contribution of sufficient importance to describe and illustrate in any detail here, and that must be the work of André-Michel Guerry (1801–1864) on 'moral statistics', which became the launching pad for criminology and sociology and much of modern social science. Guerry's work is especially relevant for this volume because it considers multivariate data in a spatial context. Beyond Guerry's own work, his data have proved remarkably useful for modern applications and demonstrations. For example, in Friendly (2007a) biplots, canonical discriminant plots, HE plots (Friendly, 2007b), and other CARME-related methods were used to provide a modern reassessment of his contributions and suggest other challenges for data analysis.

The choropleth map, showing the distribution of instruction in the French regional zones, called *départements* (departments), was invented by Charles Dupin (1826). Shortly after, Guerry, a young lawyer working for the Ministry of Justice, began the systematic study of relations between such variables as rates of crime, suicide, literacy, illegitimate births, and so forth, using centralized, national data collected by state agencies. Guerry's lifelong goal was to establish that constancies in such data provided the basis for social laws, analogous to those in the physical world, and open discussion of social policy to empirical research.

In 1829, together with Adriano Balbi, he published the first *comparative* moral maps (Balbi and Guerry, 1829) showing the distribution of crimes against persons and against property in relation to the level of instruction in the départements of France, allowing direct comparison of these in a 'small multiples' view (see Friendly, 2007a, Fig. 2). Surprisingly, they seemed to show an inverse relation between crimes against persons and property, yet neither seemed strongly related to levels of instruction.

Guerry followed this line in two major works (Guerry, 1833, 1864), both of which were awarded the Montyon prize in statistics from the Académie Française des Sciences. The 1833 volume, titled *Essai sur la Statistique Morale de la France*, established the methodology for standardized comparisons of

Some Prehistory of CARME: Visual Language and Visual Thinking

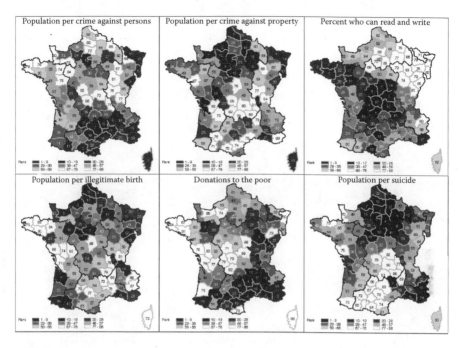

FIGURE 1.2
Reproduction of Guerry's (1833) maps of moral statistics of France. Shading, as in Guerry's originals, is such that darker shading signifies *worse* on each moral variable, ranked across *départements* (shown by numbers in the map).

rates of moral variables over time and space, and the rationale for drawing conclusions concerning social laws. In addition to tables, bar graphs, and an innovative proto-parallel coordinates plot (showing relative ranking of crimes at different ages [Friendly, 2007a, Fig. 9]), he included six shaded maps of his main moral variables. A modern reproduction of these is shown in Figure 1.2.

Guerry wished to reason about the relationships among these variables and, ultimately (in his final work, Guerry, 1864), about causative or explanatory social factors such as wealth, population density, gender, age, religious affiliation, etc. This is all the more remarkable because even the concept of correlation had not yet been invented.

We can give Guerry a bit of help here with the biplot of his data shown in Figure 1.3. This two-dimensional version accounts for only 56.2% of total variation, yet contains some interesting features. The first dimension aligns positively with property crime and illegitimate births (*infants naturelles*) and suicides, and negatively with literacy. The second dimension weights strongly on personal crime and donations to the poor. Using this and other dimension reduction techniques (e.g., canonical discriminant analysis), Guerry could have seen more clearly how the regions of France and individual départements relate to his moral variables and underlying dimensions.

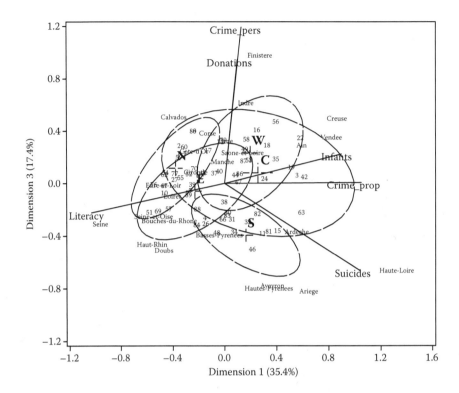

FIGURE 1.3
A symmetric 2D biplot of Guerry's six moral variables shown in maps in Figure 1.2. The points for the départements of France are summarized by region (N, S, E, W, C) with 68% data ellipses, and points outside their ellipse are labelled by department name.

1.1.3 Graphs and Diagrams

Aside from the standard, and now familiar, methods to display quantitative data, French civil and military engineers made another important contribution to graphic language: nomograms and computational diagrams. These arose from the need to perform complex calculations (calibrate the range of field artillery, determine the amount of earth to be moved in building a railway or fortification) with little more than a straightedge and a pencil (Hankins, 1999).

Toward the end of the 19th century these developments, begun by Léon Lalanne (1844), gave rise to a full-fledged theory of projective geometry codified by Maurice d'Ocagne (1899). These ideas provide the basis for nonlinear scales used in nonlinear PCA (Chapter 4 by de Leeuw in this book), linear and nonlinear biplot calibrations (Chapter 2 by Gower in this book), contribution biplots (Greenacre, 2013), and the modern parallel coordinates plot, whose theoretical basis was also established by d'Ocagne (1885). This includes the principles of duality, by which points in Cartesian coordinates map into lines in alignment diagrams with parallel or oblique axes and vice versa, polar transformations of curves and surfaces, and so forth.

Among the most comprehensive of these is Lalanne's 'universal calculator', which allowed graphic calculation of over 60 functions of arithmetic (log, square root), trigonometry (sine, cosine), geometry (area, surface, volume), and so forth (see http://datavis.ca/gallery/Lalanne.jpg for a high-resolution image). Lalanne combined the use of parallel, nonlinear scales (as on a slide rule) with a log-log grid on which any three-variable multiplicative relation could be represented.

Charles Lallemand, a French engineer, produced what might be considered the most impressive illustration of this work with the multigraphic nomogram (Lallemand, 1885) shown in Figure 1.4. This tour de force graphic was designed to calculate the magnetic deviation of the compass at sea, which depends on seven variables through complex trigonometric formulas given at the top of the figure. It incorporates three-dimensional surfaces, an anamorphic map with nonlinear grids, projection through the central cone, and an assortment of linear and nonlinear scales. Using this device, the captain could follow simple steps to determine magnetic deviation without direct calculation, and hence advise the crew when they might arrive at some destination.

Lallemand was also responsible for another grand project: the *Nivellement Général de la France*, which mapped the altitudes of *all* of continental France.

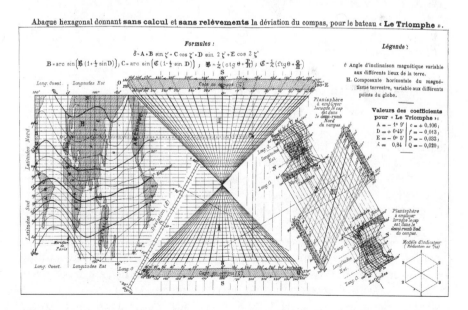

FIGURE 1.4

Nomograms: Computational diagrams and axis calibration. This tour de force nomogram by Charles Lellemand combines diverse graphic forms (anamorphic maps, parallel coordinates, 3D surfaces) to calculate magnetic deviation at sea. (From École des Mines, Paris. Reproduced with permission.)

Today, you can still find small brass medallions embedded in walls in many small towns and villages throughout the country, indicating the elevation at that spot.

1.2 Visual Thinking

The development of graphic language through the end of the 19th century and the widespread adoption of graphic methods by state agencies did much more than make data graphics commonly available, in both popular expositions and official publications. For example, the *Album de Statistique Graphique*, published under the direction of Émile Cheysson by the Ministère des Traveaux Publiques from 1879 to 1897, represents a high point in the use of diverse graphic forms to chart the development of the modern French state.

It also presented a concrete means to plan for economic and social progress (where to build railroads and canals, how to bolster international trade) to reason and perhaps draw conclusions about important social issues (e.g., the discussion above of Guerry) and make some scientific discoveries that arguably could not have been arrived at otherwise.

We focus here on two aspects of this rise in visual thinking that characterize the golden age of statistical graphics: visual explanation, as represented by the work of Charles Joseph Minard, and visual discovery, typified by the work of Francis Galton.

1.2.1 Graphic Vision of Charles Joseph Minard

Minard, of course, is best known for his compelling and now iconic depiction of the terrible losses sustained by Napoleon's Grande Armée in the disastrous 1812 Russian campaign (Minard, 1869). However, the totality of Minard's graphic work, comprising 63 *cartes figuratives* (thematic maps) and *tableaux graphiques* (statistical diagrams), is arguably more impressive as an illustration of visual thinking and visual explanation.

Minard began his career as a civil engineer for the École Nationale des Ponts et Chaussées (ENPC) in Paris. In 1840, he was charged to report on the collapse of a suspension bridge across the Rhone at Bourg-Saint-Andèol. The (probably apocryphal) story is that his report consisted essentially of a self-explaining before-after diagram (Friendly, 2008, Fig. 4) showing that the bridge collapsed because the riverbed beneath one support column had eroded.

Minard's later work at the ENPC was that of a visual engineer for planning. His many graphics were concerned with matters of trade, commerce, and transportation. We illustrate this with another before-after diagram (Figure 1.5), designed to explain what happened to the trade in cotton and wool as a consequence of the U.S. Civil War. The conclusion from this pair

Some Prehistory of CARME: Visual Language and Visual Thinking

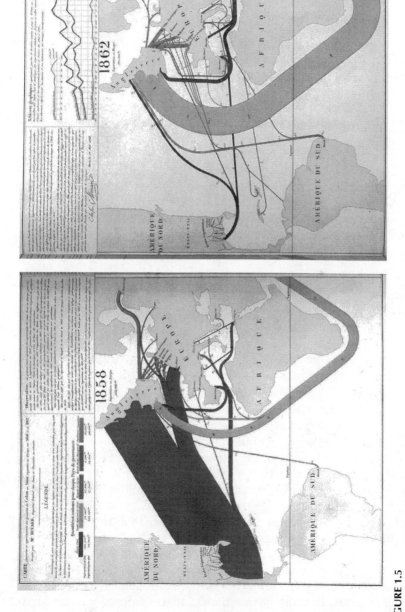

FIGURE 1.5
Visual explanation: What happened to the trade in cotton and wool from Europe in the U.S. Civil War? Left: In 1858, most imports to Europe came from the southern U.S. states. Right: By 1862, U.S. imports had been reduced to a trickle, only partially compensated by increased imports from India, Brazil, and Egypt. (From Minard, C. J., Carte figurative et approximative des quantités de coton en laine importées en Europe en 1858 et en 1861, Lith. (868 535), ENPC: Fol 10975, 1862. Image from École Nationale des Ponts et Chaussées, Paris. Reproduced with permission.)

of *cartes figuratives* is immediate and interoccular: Before the war, the vast majority of imports came from the southern U.S. states. By 1862, the Union naval blockade of the Confederacy reduced this to a tiny fraction; demand for these raw materials in Europe was only partially met by greater imports from Brazil and Egypt, but principally from India.

1.2.2 Francis Galton's Visual Discoveries

De Leeuw (Chapter 5 in this book) points out that the early origin of PCA stems from the idea of principal axes of the 'correlation ellipsoid', discussed by Galton (1889), and later developed mathematically by Pearson (1901). It actually goes back a bit further to Galton (1886), where he presented the first fully formed diagram of a bivariate normal frequency surface together with regression lines of $E(y|x)$ and $E(x|y)$, and also with the principal axes of the bivariate ellipse. This diagram and the correlation ellipsoid can arguably be considered the birth of modern multivariate statistical methods (Friendly, Monette, and Fox, 2013).

What is remarkable about this development is that Galton's statistical insight stemmed from a largely geometrical and visual approach using the smoothed and interpolated isopleth lines for 3D surfaces developed earlier by Halley, Lalanne, and others. When he smoothed the semigraphic table of heights of parents and their children and found that isolines of approximately equal frequency formed a series of concentric ellipses, Galton's imagination could complete the picture, and also offer the first true explanation of 'regression toward mediocrity'. Pearson (1920, p. 37) would later call this 'one of the most noteworthy scientific discoveries arising from pure analysis of observations'.

However, Galton achieved an even more notable scientific, visual discovery 25 years earlier, in 1863—the anticyclonic relation between barometric pressure and wind direction that now forms the basis of modern weather maps and prediction. This story is described and illustrated in detail in Friendly (2008, §3.2) and will not be relayed here. In the book *Meteorographica* (Galton, 1863), he describes the many iterations of numerical and graphical summaries of the complex multivariate and spatial data he had elicited from over 300 weather stations throughout Europe at precise times (9 a.m., 3 p.m., 9 p.m.) for an entire month (December 1861).

The result was a collection of micromaps (Figure 1.6) in a 3×3 grid of schematic contour maps showing barometric pressure, wind direction, rain, and temperature by time of day, using colour, shape, texture, and arrows. From this he observed something totally unexpected: whereas in areas of low barometric pressure, winds spiraled inward rotating counterclockwise (as do cyclones), high-pressure areas had winds rotating clockwise in outward spirals, which he termed anticyclones. This surely must be among the best exemplars of scientific discovery achieved almost entirely through high-dimensional graphs.

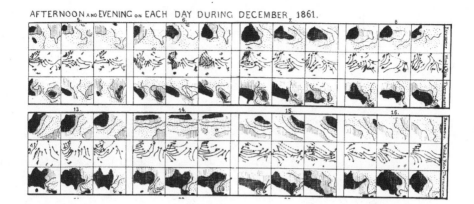

FIGURE 1.6
Visual discovery: Top portion of Galton's (1863) multivariate schematic micromaps. Each 3 × 3 grid shows barometric pressure, wind, rain, and temperature (rows) by time of day (columns). (From Galton, F., *Meteorographica, or Methods of Mapping the Weather*, Macmillan, London, 1863, Appendix, p. 3. Image from a private collection.)

1.3 Conclusion

This chapter demonstrates how the underlying attitudes of CARME—data exploration and analysis (largely model-free), reduction of complex, high-dimensional data to comprehensible low-dimensional views, and an emphasis on visualization—are rooted in a long, primarily European history that gave rise to the elements of visual language and visual thinking. Along with the rise of quantification and novel methods for visualization came new ways to think about data and mathematical relationships, and to express them graphically.

Many of these innovations came from France, and were popularized and taught through works like *La Méthode Graphique* (Marey, 1885). The spirit of CARME, embodied in this volume, gives due attention to these historical developments we consider commonplace today.

2

Some History of Algebraic Canonical Forms and Data Analysis

John C. Gower

CONTENTS

2.1 Canonical Forms 18
    2.1.1 Diagonalization of a Symmetric Matrix 19
    2.1.2 Decomposition of a Square Matrix 19
    2.1.3 Simultaneous Diagonalization of Two Symmetric Matrices (Two-Sided Eigendecomposition) 20
    2.1.4 Results for Orthogonal Matrices, Skew-Symmetric Matrices, and Hermitian Matrices 20
    2.1.5 Singular Value Decomposition of a Rectangular Matrix 22
2.2 Twentieth Century—Before 1950 23
2.3 Twentieth Century—After 1950 26
2.4 Epilogue 29

This chapter is about the history of the algebra underpinning those methods of data analysis that depend on a strong appeal to visualization. It is impossible to do justice to this end in the limited space we have here, so we shall just mention the milestones. For fuller information see the excellent book of Macduffee (1946) and the paper of Stewart (1993). In fact, the algebra came long before its use in data analysis, its antecedents stretching back into history. However, it was only toward the end of the 18th century that interest began to expand in problems involving sets of linear equations, quadratic forms, and bilinear forms. These were in response to problems arising in applied mathematics, especially those concerned with geodetics and the conical forms of planetary motion. Famously, Gauss derived the least-squares equations for estimating the parameters (coordinates) of linear systems and an algorithm for solving the equations. Legendre (see, e.g., Gantmacher,

1959) gave his method of expressing a quadratic form as a sum of squares of linear terms, thus initiating the further steps to requiring the coefficients of the linear terms to be orthogonal. Throughout the 19th century, much algebraic research was devoted to studying the properties of determinants. This aspect, although of interest, has only marginal relevance to the visualization applications of data analysis, and so will not be considered here.

That this early work was mainly set against the three-dimensional measurements of physical space was not conducive to multidimensional generalization. It took time to realize that there was commonality in the methods used. Sylvester (1850) was the first to use the term *matrix*, though the idea in various forms had been around, e.g., tableau, array. It was also Sylvester (1852) who defined principal axes of inertia, a terminology derived from dynamics that is still common in the correspondence analysis literature. It was not until the middle of the 19th century that it came to be realized, that after defining a matrix inverse, there was a basic algebra of matrices that was very similar to the familiar algebra of scalars (Cayley, 1858). As was noted by Cayley, the only difference was that the product of two matrices was not associative ($\mathbf{AB} \neq \mathbf{BA}$). It was with the adoption of matrix notation that the matrix components of quadratic and bilinear forms, and many other problems, could be studied in a unified way. But it was not until the 1950s that matrix notation was used with the freedom to which we are now accustomed.

In the following years, the nascent results discovered in the first half of the 19th century took on their matrix forms, which, together with newer discoveries, began the development of matrix algebra as a recognized part of mathematics. Kronecker, Frobenius, and Weierstrass, working in Berlin, were foremost in the development of a rigorous treatment. Weierstrass (1868) published the first book on matrix algebra.

The above gives a very sketchy account of how matrix algebra developed from attempts to solve a variety of problems in applied mathematics to the point where a unified matrix algebra was recognized. It is not my attention to focus on the relative national contributions, but we have already mentioned Gauss and Weierstrass, while the British were very much on the sidelines, although Cayley and Sylvester (both lawyers) made important contributions in unification, as mentioned above. Overwhelmingly, the main early contributions were given by scientists working in the French tradition, so it is fitting to acknowledge that this tradition continues to survive, as is evidenced in the other chapters in this book.

2.1 Canonical Forms

Many algebraic decompositions were discovered in the 19th century, often in the context of linear transformations and bilinear forms. These decompositions were often expressed in forms that antedate matrix notation. Nowadays, results are presented in simpler matrix forms. Mathematical theory and applicable mathematics are not necessarily well matched. So we may ask, what part of this matrix algebra had any relevance to data analysis? Of course, one is never sure what recondite algebra may become useful, but it seems to me that the vast majority of results useful in data analysis concern canonical forms, especially those concerned with the eigenstructure of symmetric matrices and the singular value decomposition of rectangular matrices. We also need algorithms to compute these forms and algorithms for matrix inversion.

2.1.1 Diagonalization of a Symmetric Matrix

This is the most basic tool used in data analysis (Cauchy, 1829–1830). Any real symmetric matrix $\mathbf{S}$ may be written as $\mathbf{S} = \mathbf{V\Lambda V^T}$, where $\mathbf{V}$ is an orthogonal matrix and $\mathbf{\Lambda}$ is a real diagonal matrix of eigenvalues and the columns of $\mathbf{V}$ are corresponding eigenvectors. Equivalently, $\mathbf{V^T S V} = \mathbf{\Lambda}$ is often referred to as diagonalization or the spectral decomposition of $\mathbf{S}$. Both forms may be written in an eigenvalue/eigenvector equation:

$$\mathbf{SV} = \mathbf{V\Lambda} \qquad (2.1)$$

Repeated eigenvalues determine an r-dimensional subspace of permissible eigenvectors, and any independent set, usually determined orthogonally, may be used in $\mathbf{V}$. When $\mathbf{S}$ is positive semidefinite the eigenvalues are nonnegative. It follows that any positive semidefinite quadratic form $\mathbf{x^T S x}$ may be written as $(\mathbf{x^T V})\mathbf{\Lambda}(\mathbf{V^T x})$ whence transformation to $\mathbf{y} = \mathbf{V^T x}$ gives the quadratic form as a sum-of-squares $\mathbf{y^T \Lambda y}$ of linear canonical variables.

2.1.2 Decomposition of a Square Matrix

A similar transformation $\mathbf{T^{-1} A T}$ for a square nonsymmetric matrix $\mathbf{A}$ is given by the Jordan form (Jordan, 1870), but in general $\mathbf{A}$ cannot be expressed in terms of diagonal matrices. The Jordan form says that we may write $\mathbf{T^{-1} A T} = \mathbf{J}$, where $\mathbf{T}$ is nonsingular and $\mathbf{J}$ is a Jordan block-diagonal matrix, with

$$\mathbf{J} = \begin{bmatrix} \mathbf{J}_1 & & & \\ & \mathbf{J}_2 & & \\ & & \ddots & \\ & & & \mathbf{J}_k \end{bmatrix} \text{ where } \mathbf{J}_r = \begin{bmatrix} \lambda_r & 1 & & \\ & \lambda_r & 1 & \\ & & \ddots & \ddots \\ & & & \lambda_r \end{bmatrix} \qquad (2.2)$$

Note that the same (repeated) eigenvalue may occur in more than one Jordan block. The rth Jordan block (2.2) determines r repeated eigenvalues but only a single eigenvector. This major structural difference from that of symmetric matrices is easily verified by determining the eigenvectors of J_r, which turn out uniquely to be the vector $(0, 0, \ldots, 1)$. When all the eigenvalues are distinct we continue to have a conventional eigenvalue/eigenvector relationship as in (2.1):

$$AT = TJ$$

where J is diagonal. Note that T is not orthogonal and both eigenvalues and eigenvectors may be complex numbers, even when A is real.

2.1.3 Simultaneous Diagonalization of Two Symmetric Matrices (Two-Sided Eigendecomposition)

The simultaneous diagonalization of two positive semidefinite symmetric matrices A and B is very important in data analysis. The basic result is that it is possible to find independent eigenvectors L such that

$$BL = AL\Lambda \qquad (2.3)$$

where Λ is a diagonal set of eigenvalues normalized so that $L^T AL = I$ and $L^T BL = \Lambda$. The normalization is equivalent to the simultaneous diagonalization of A and B (Weierstrass, 1868, earliest known reference). When $A = I$, then L is orthogonal and the two-sided eigenvalue problem reduces to the symmetric eigenvalue problem. A ratio form, referred to as Raleigh's quotient, was used by Lord Rayleigh (c. 1870) in his study of vibrating strings, a terminology still used in current applied mathematical literature. Initially in (2.3) it was required that the matrix A be definite, and it was not until 1961 that this condition was relaxed to allow both matrices to be positive semidefinite. Even more recently, this condition has been further relaxed to allow B to be indefinite (Albers et al., 2011). Of course, if A is indefinite we cannot have $L^T AL = I$ and have to replace some of the diagonal elements of I by zeroes.

2.1.4 Results for Orthogonal Matrices, Skew-Symmetric Matrices, and Hermitian Matrices

The results in this section are useful either as the basis of some less used data analytical methods, or as steps in interpretation, or as essential for understanding the foundations of some of the ideas used in data analysis. We have already met orthogonal matrices as the eigenvectors of symmetric matrices. Orthogonal matrices are the foundation of many results in Euclidean geometry. Euclidean distances are invariant to orthogonal transformations. Orthogonal projection onto subspaces, perhaps spanned by orthogonal eigenvectors, is a tool that underpins many visualizations. Distance measured in a metric and projections in a metric are associated with the two-sided eigenvalue problem. Then there is the decomposition:

$$Q = V \begin{bmatrix} H_{11} & & & \\ & \ddots & & \\ & & H_{rr} & \\ & & & \ddots \\ & & & & \pm I \end{bmatrix} V^T \quad (2.4)$$

which gives a decomposition of an orthogonal matrix Q in terms of elementary Jacobi rotations H_{rr} and projections $-I$ and identity matrices $+I$ corresponding to the columns of V (Vitali, 1929). The matrix V is orthogonal, and its columns give the planes of rotation and projection of the corresponding vectors. This has potential for data analytical applications.

Skew-symmetric matrices N, which are defined by the property $n_{ij} = -n_{ij}$, have interesting properties. Unlike symmetric matrices, they have imaginary eigenvalues that occur in pairs ($i\lambda$, $-i\lambda$). In fact, it is easy to calculate these from the positive semidefinite symmetric matrix $-N^2$, whose eigenvalues occur in equal pairs (λ, λ).

Skew-symmetric matrices have data analytical interest (see Gower, 1977) and also have interesting links with skew-symmetric matrices through the Cayley formulae that relate orthogonal and skew-symmetric matrices:

$$Q = (I+N)^{-1}(I-N), \; N = (I+Q)^{-1}(I-Q)$$

with the proviso that the inverses exist.

We end this miscellaneous section with mention of Hermitian matrices. These are the counterpart in the complex domain of symmetric matrices in the real domain. A Hermitian matrix H may be written

$$H = M + iN$$

where M is symmetric and N is skew-symmetric, and has conjugate $H^T = M - iN$. Hermitian matrices have real eigenvalues.

Unitary matrices are the counterpart in the complex domain of orthogonal matrices in the real domain. If $U = X + iY$ is unitary it satisfies

$$(X + iY)^{-1} = (X - iY)^T \text{ and } (X - iY)^{-1} = (X + iY)^T$$

or, in terms of conjugate matrices, denoted by a tilde:

$$U^{-1} = \tilde{U}^T \text{ and } \tilde{U}^{-1} = U^T$$

Many of the properties of symmetric and orthogonal matrices carry over to Hermitian and unitary matrices. In particular, we have the decompositions:

$$M + i N = U \Lambda U^{-1} = U \Lambda \tilde{U}^{-1} \text{ and } M - iN = \tilde{U} \Lambda \tilde{U}^{-1} = \tilde{U} \Lambda U^T$$

The properties of Hermitian and orthogonal matrices have been used as a basis for data analysis.

2.1.5 Singular Value Decomposition of a Rectangular Matrix

It is a relief to have a decomposition that refers directly to rectangular arrays. The singular value decomposition (SVD) of a $p \times q$ matrix X is

$$X = U \Sigma V^T \tag{2.5}$$

where U is an orthogonal $p \times p$ matrix, V is an orthogonal $q \times q$ matrix, and Σ is $p \times q$, zero except that σ_{ii} is nonzero for $i = 1, 2, \ldots, q$. Thus, when $p = q$, Σ is diagonal, and when $p \neq q$, Σ may be said to be pseudodiagonal. The diagonal values of Σ are called singular values, and the columns of U and V are called singular vectors. Singular values are nonnegative but may be zero, and indeed, when $p > q$, considerations of rank show that at least $p - q$ singular values must be zero. The singular vectors corresponding to zero singular values are arbitrary, but they are usually set to form consistent orthogonal matrices U and V.

The SVD was formulated independently in the 19th century by Beltrami (1873), Jordan (1874), and Sylvester (1889). The extension to complex matrices, where orthogonal matrices are replaced by unitary matrices, was given by Autonne (1915).

There are several ways of presenting the SVD, but all are equivalent. The differences are associated with how to dispose of redundant zero singular values. Assuming, without loss of generality, that $p > q$, we can write (2.5) as

$$X = {}_p U_q \Sigma_q V_q^T \tag{2.6}$$

where the singular vectors in U corresponding to zero singular vectors are ignored; if $\text{rank}(X) < q$, then we replace q with a lower value equal to the rank.

Another convenient presentation is to define $U = (u_i, \ldots, u_q)$ and $V = (v_i, \ldots, v_q)$ and write

$$X = \sum_{i=1}^{q} \left(\sigma_i u_i v_i^T \right) \tag{2.7}$$

which expresses the SVD as a sum of q rank 1 matrices. Throughout, we assume that the singular values are ordered from largest to smallest, i.e., $\sigma_1 \geq \sigma_2 \geq \ldots \geq \sigma_q > 0$.

Finally we arrive at the important problem of approximating a matrix. The basic result says that if **X** is a rectangular array, then the best r-dimensional approximation $\hat{\mathbf{X}}_r$ to **X** obtained by minimizing

$$min\left\|\mathbf{X}-\hat{\mathbf{X}}_r\right\|^2 \text{ is given by } \hat{\mathbf{X}}_r = {}_p\mathbf{U}_r\Sigma_r\mathbf{V}_q^T \quad (2.8)$$

where the singular values are ordered as above.

$$\text{Equivalently } \hat{\mathbf{X}}_r = \sum_{i=1}^{r}\left(\sigma_i\mathbf{u}_i\mathbf{v}_i^T\right) \quad (2.9)$$

Thus, we are concerned with least-squares approximation. The result (2.8) is usually attributed to Eckart and Young (1936), which appeared in the first volume of *Psychometrika*. However, Stewart (1993) points out that the essential result was found by Schmidt (1907), of Gram-Schmidt fame, in studies of the kernel of integral equations.

I recall a conference in the 1980s where a distinguished colleague disparagingly referred to the SVD as 'an obscure result in algebra', but Greenacre (2012) refers to the 'joy of' SVD. Both are right. As an isolated algebraic result the SVD has limited appeal, but the applications of the SVD in approximating matrices have far-reaching consequences for data analysis. Stewart (1993) neatly echoes these sentiments in two comments: (1) Schmidt 'establish[es] what can properly be termed the fundamental theorem of the singular value decomposition', and (2) 'in doing so he transformed the singular value decomposition from a mathematical curiosity to an important theoretical and computational tool' (Stewart, 1993, pp. 558 and 561).

2.2 Twentieth Century—Before 1950

The 20th century conveniently splits into two equal parts: the first 50 years before computers were available and the last 50 years when computers became increasingly available, increasingly reliable, increasingly fast, and able to handle increasingly large data sets. Statistics has its own long history (see Hald, 1990, 1998), but the modern study was begun with Karl Pearson at University College, London. In 1884 he was appointed to the chair of applied mathematics and mechanics, where he met Wheldon (zoologist) and Francis Galton (wealthy polymath and eugenicist). Their collaboration shaped many statistical ideas that first appeared in their journal, *Biometrika*, founded in 1901. In a chair endowed by Galton, in 1911 Pearson became the first UK professor of statistics (actually eugenics).

Pearson is especially important for the history of data analysis. Pearson (1894) give the first discussion of a distribution-mixture problem, in which he sought to separate a large sample of (presumed) male and female bones, thus initiating one kind of clustering problem of importance in the second part of the century. Pearson (1901) is titled 'On lines and planes of closest fit' and describes principal components analysis (PCA) in terms of orthogonal projection of sample points onto a best-fitting plane that passes through the centroid of the samples. This naturally leads to the plane spanned by the r dominant eigenvectors of the $\mathbf{X}^T\mathbf{X}$ covariance matrix and the projections $\mathbf{X}\mathbf{V}_r\mathbf{V}_r^T = \hat{\mathbf{X}}$. Substituting the SVD for $\mathbf{X}$ we find that $\hat{\mathbf{X}} = \mathbf{U}\Sigma_r\mathbf{V}_r^T$, which is the approximation referred to above, at the heart of the SVD approximation. Whether Pearson knew this property from his applied mathematical background is possible, but I suspect that he developed it independently. Indeed, we shall see below that R. A. Fisher did something similar 20 years later.

Some 30 years later Hotelling (1933) wrote an influential paper titled 'Analysis of a complex of statistical variables into principal components'. This paper presented principal components as a factor analysis method and was concerned with approximating $\mathbf{S} = \mathbf{X}^T\mathbf{X}$ rather than $\mathbf{X}$ itself. Of course, approximating $\mathbf{S}$, or more commonly the corresponding correlation matrix, is acceptable, but the difference between Pearson's PCA and Hotelling's approach should be clearly recognized. The difference is that (2.9) is now replaced by an SVD of $\mathbf{S}$, equivalent in this case to an eigendecomposition:

$$min\left\|\mathbf{S}-\hat{\mathbf{S}}_r\right\|^2 \text{ given by } \hat{\mathbf{S}}_r = {}_p\mathbf{V}_r\Sigma_r^2\mathbf{V}_q^T \qquad (2.10)$$

and the interest is in approximating $\mathbf{S}$ (possibly in correlational form) by the inner products between pairs of rows of ${}_p\mathbf{V}_r\Sigma_r$. The same SVD is used in both approaches, but the components are interpreted differently. Another difference is that because $\mathbf{S}$ is a symmetric matrix, all values formed from the off-diagonal elements are summed twice, while off-diagonal elements are summed once, thus giving a weighted sum of squares. The weighting is in the right direction, but it would be better to ignore altogether the contributions from the diagonal terms. This is essentially what Greenacre (1988a) did in the multiple correspondence analysis (MCA) context when formulating joint correspondence analysis (JCA) (see below).

Fisher had come to Rothamsted in 1919, and he set about reforming the design and interpretation of agricultural experiments, and much besides, making several important contributions to what may be termed data analysis (see J. F. Box, 1978, for a biography). Fisher and Mackenzie (1923) were the first (1) to fit a biadditive model to data forming a two-way table, (2) to do an extended analysis of variance, and (3) (or at least among the first) to use dummy variables in a least-squares context. It is (1) to which we draw special attention. When considering how to analyse data from experiments, Fisher and Mackenzie considered using either an additive model or an alternative biadditive model. The additive model was well understood, and especially with balanced experiments designs, computation was easy, whereas biadditive models were new. Fisher and Mackenzie found the normal equations and solved them as an eigenvalue problem, essentially using the SVD approximation. At the time, the biadditive model was not followed up, mainly because eigenvalue decompositions were not practicable with the primitive mechanical and electromechanical computing equipment of the time. Nevertheless, as we shall see, in the second half of the 20th century biadditive models came into their own.

Hotelling (1936) described canonical correlation defined as the linear combinations of two sets of variables that have the greatest correlation. There are several ways of developing canonical correlation, some more elegant than others. We shall use the most direct, if not the most elegant. If X and Y denote sample matrices for the two sets, we have to maximize

$$\rho^2 = \frac{(s^T X^T Y t)^2}{(s^T X^T X s)(t^T Y^T Y t)}.$$

Writing $A = X^T X$, $B = Y^T Y$, and $C = X^T Y$ and differentiating shows that ρ satisfies

$$\left.\begin{array}{l} \rho A s = C t \\ \rho B t = C^T s \end{array}\right\}$$

which in multidimensional form may be arranged into

$$\begin{bmatrix} A & C \\ C^T & B \end{bmatrix} \begin{bmatrix} S \\ T \end{bmatrix} = \begin{bmatrix} A & \\ & B \end{bmatrix} \begin{bmatrix} S \\ T \end{bmatrix} R \qquad (2.11)$$

where R is a diagonal matrix of the canonical correlations, ρ increased by one. We use this two-sided eigenvalue form for comparison with generalized canonical correlation (see below).

One year earlier Hirschfeld (1935) (whose name was later anglicized to Hartley) sought the quantifications of two categorical variables that maximized their correlation. This was a notable paper in three respects: (1) it was a forerunner of Hotelling's canonical correlation, (2) it was a forerunner of methods for analysing categorical variables by first transforming them into numerical form (i.e., quantification), and (3) it was a forerunner of correspondence analysis. It is noteworthy that both Hirschfeld and Hotelling, both of German origin, were among those scientists who enriched Anglo-American statistics in the 1930s—Jerzy Neyman was another. Both used the two-sided eigenvalue problem, and one wonders whether they brought it from Germany, or at least popularized its use. Certainly, it was known in applied mathematics since the mid-19th century, as the work of Lord Rayleigh testifies, and it was certainly used by Fisher (1938) in his optimal scores quantification method for handling categorical response variables in two-way tables and by Guttman (1941) in his optimal scores quantification for a data matrix of multivariate categorical variables.

We cannot leave the first half of the 20th century without mentioning the mathematically impressive work on the derivations of several multivariate statistical distributions. Fisher's (1924) derivation of the distribution of the sample correlation coefficient when the population correlation is zero or nonzero led to Wishart's (1928) multivariate generalization. The Wishart distribution was at the basis for finding the distributions of the eigenvalues and the maximal eigenvalue of sample sums-of-squares matrices. Multivariate generalizations of the D, F, and t distributions are part of this effort. These developments are little concerned with data analysis, and many would say that they have little to do with more general statistical practice. At the end of this chapter we return to the roles of data analysis, statistics, and mathematical statistics.

2.3 Twentieth Century—After 1950

Thus, at the beginning of the second half of the 20th century the foundations of many methods of multivariate analysis had been laid. However, applications were few, and it was a tour de force to do any data analysis. This changed in the early 1950s, when the first electronic computers became available to researchers. Early computers had small program and data capacities, but nevertheless were a huge advance on what was currently available. Computing rapidly developed, the milestones being (1) availability of Fortran and other universal programming languages, (2) increase in storage capacity, (3) availability of graphical interfaces, and (4) networking. The development of computing entailed increased management responsibilities, based on batch mode operation. This was a step backwards, but fortunately it was a passing phase. In the mid-1980s personal computers began to appear and the era of big computer service departments was confined to activities such as managing networks, providing and maintaining supporting software, managing access and security, etc. The beginnings of interactive computing had been explored.

The increases in speed and capacity supported corresponding increases in algorithmic complexity and database magnitude. Bigger is not necessarily better, and I have reservations about the efficacy of 'data mining' and 'big data' and think that too little effort is currently given to the design of data collection both with observational data and with experimental data. Increases in speed have opened the door to data sampling methods such as the bootstrap, jackknife, and permutation tests. Interactive computing and the ready access to reliable graphic capabilities can only be seen as welcome. Indeed, graphical presentation, or 'maps', as Greenacre aptly terms them in the context of dimension reduction, pervades almost everything mentioned in the following and will be taken for granted. In the remainder of this chapter it is only possible to give a superficial account of everything that has been achieved, so I shall confine myself to giving the flavour of some major areas of growth.

We begin with hierarchical classification (see Chapter 8 from Murtagh and Chapter 17 from Roux in this book). Hierarchical classification has long been used in plant and animal taxonomy, for both forming classes and assigning new objects to existing classes. Numerical approaches to hierarchic classification were among the first novel applications of computers in the life sciences. Blashfield (1980) regarded Sokal, Sneath, and Tryon as having made seminal contributions. It is interesting to note that two of these were taxonomists and the other a psychometrician. Their objective was definitely to form classes. Biologists were torn between whether classes should have evolutionary status or be separated by descriptive differences. Statisticians criticized the new methods because they (1) associated classification with discrimination and hence assignment, and (2) they had no stochastic basis and tended to favour multivariate development of Pearson's (1894) mixture-distribution model. Mathematicians found them a playground for deriving new theorems based on graph theory and axiomatic properties. As I see it, the truth is that there are many different reasons for an interest in classification that give rise to different formulations, different algorithms, and different mathematics. Users of software do not necessarily respond satisfactorily to different circumstances: for example, the distinction between a data matrix whose rows refer to a single population, each with one representative, and a matrix with samples all of which are from the same population—and everything in between. Neither may software distinguish between different types of variables (numerical, categorical, ordered categorical, ranks, etc.). The difference between units is often expressed in terms of (dis)similarity coefficients, a few of which go back to the early part of the 20th century, but the definition of new measures has proliferated. One should not forget that classification does not have to be based on exhaustive hierarchies. There is plenty of room for confusion.

Correspondence analysis (CA) and multiple correspondence analysis (MCA) have already been mentioned as having their origins in the first half of the 20th century. Gower (1990) summarized this early work. It was Benzécri et al. (1973) who developed and strongly promoted the method, first in France and then more widely by several of his collaborators and students (for example, Lebart et al., 1984; Greenacre, 1984). Benzécri (1977b) has written his personal historical account. Correspondence analysis has variant forms, depending on how axes are scaled, or how weights are interpreted. The differences between CA and MCA are analogous to those between approximating X or X^TX in PCA and have the same problem of whether to approximate the diagonal of X^TX or not, thus leading to JCA (see Greenacre, 1988a). I have written about these issues previously (Gower, 2006a; Gower et al., 2011) and shall not do so again. Suffice to say that whatever differences there are make rather little difference to final results and their graphical interpretation.

CA is concerned with approximating the nonnegative information contained in contingency tables. Computationally, it is sufficient for the row and column sums to be nonnegative. A trend of some interest is the way that the nonnegative property can be exploited. One use is with ranked data that can be treated as if they were numerical. It turns out that it is more interpretable if each rank is supplemented by its complementary rank, a process termed *dédoublement* in France, or doubling. This extends CA to handle data in rank form. The motive seems to be a desire to cast problems into familiar forms. There is no reason why this course should not be taken, and there are many precedents where people have been more comfortable to turn their current problem into a familiar form. Nishisato (1994) believes that the structure of the data should form the basis for a proper analysis and terms the range of problems concerned as 'dual scaling'. Nishisato distinguishes several forms of data that are often treated as if appropriate to correspondence analysis: contingency/frequency data, multiple choice data, paired comparisons data, rank-order data, rating data, sorting data, multiway data.

Finally, we come to Galton's manservant Gifi (c. 1889). Thus, we leave the 20th century as we began it, but much has happened in the meantime. Gifi's name was adopted, in his honour, by a group working in Leiden. The Gifi group began in the 1970s initially under the auspices of Jan de Leeuw, but from the 1990s was headed by Willem Heiser and Jacqueline Meulman. However, many people made substantial contributions. Gifi (1990) is the major publication that sums up the contribution of the Gifi school. The Gifi work is characterized by (1) unifying and extending many methods that already had some currency in the psychometric literature (e.g., generalized canonical correlation, nonlinear PCA, homogeneity analysis, alias multiple correspondence analysis, multidimensional scaling), (2) adopting a unified programming policy adopting least-squares and inner-product criteria (loss functions), (3) extensive use of monotonic transformations of data, either in step function form or using monotonic splines, and (4) making geometric methods the principal vehicle for deriving and interpreting results. The success of the project was underpinned by supporting software and documentation, incorporated into SPSS. For Gifi, categorical variables are the norm, and if originally numerical, they are often initially transformed to categorical form, so monotonic transformations are determined by the data. This is in contrast to generalized linear models (GLIMs) where the link function is determined at the outset.

Three-mode analysis is an area associated with the Gifi school. Three-mode analysis includes generalized canonical correlation, already mentioned, but also STATIS, variant forms of individual scaling analysis, generalized Procrustes analysis, the analysis of several dissimilarity matrices, the analysis of conventional three-way tables, and many more. Attempts have been made to unify these methods (see Kroonenberg, 2008a; Chapter 6 in this book), but in my opinion the jury is still out on their success.

2.4 Epilogue

I end where I should have begun. What is data analysis? My answer is that data analysis is just another name for statistics (see Gower, 2006b). Confusion began because mathematical statistics hijacked the name *statistics*. John Nelder put it well when he remarked that mathematical statistics should be termed statistical mathematics, with only peripheral relevance to statistics. In this scenario, data analysis is another name for statistics, while mathematical statistics is not statistics. I suspect that much of the problem stems from a perception in some circles that statistics is not a respectable academic study. The success of correspondence analysis may be associated with making statistics more palatable by renaming it data analysis—*analyse des données* in French.

Of course, the techniques of statistics range from sophisticated models (whose parameters need estimation and assessment by using inferential considerations) to descriptive methods (with little appeal to any underlying model). Most of the methods discussed in this book belong to the latter group. Even then there are subdivisions that are difficult to disentangle, we can distinguish exploratory data analysis (EDA) (as defined by Tukey, 1977) from *analyse des données* (as defined by Benzécri), with the Gifi (1990) school defining a more extensive overlapping group. Even EDA has subdivisions into initial data analysis (IDA) and confirmatory data analysis (CDA)—not to mention symbolic data analysis (SDA) (Billard and Diday, 2006b; see Chapter 16 of Verde and Diday in this book). But it is all statistics.

3

Historical Elements of Correspondence Analysis and Multiple Correspondence Analysis

Ludovic Lebart and Gilbert Saporta

CONTENTS
3.1 Simple Correspondence Analysis 32
    3.1.1 Optimal Scaling or Quantification of Categorical Variables 34
    3.1.2 Reciprocal Averaging 34
    3.1.3 Geometric Approach 35
    3.1.4 CA as a Discretization of a Continuous Bivariate Distribution 36
3.2 Multiple Correspondence Analysis 37
    3.2.1 First Traces of a Technology 37
    3.2.2 Preliminary Bases 39
    3.2.3 Multiple Correspondence Analysis: Formulas and Methodology 40
    3.2.4 First Implementations 41
    3.2.5 Dissemination of MCA 42
    3.2.6 MCA and Multiway Analyses 42
    3.2.7 Stability, Validation, Resampling 43
    3.2.8 Some Related Methods 44
    3.2.9 Conclusion 44

The use of visual displays as an exploratory tool in the 1960s is probably one of the distinctive features of correspondence analysis (CA) compared to other techniques, and the explanation of its success. Once the usefulness of a technique has been established, it is always easy to find afterwards pioneering works, and it is now well known that the equations of CA had been found many years before, based on quite different motivations. Multiple correspondence analysis (MCA) can be considered a mere variant of CA, since its history is inseparable from that of CA. Its present name dates back to the beginning of the 1970s, but its history also goes back much further in the past. The dissemination of these techniques is contemporary, and an active component of the upsurge of exploratory multivariate data analysis that followed the year 1965, notably under the name *analyse des données* in some francophone countries. We focus here on works prior to 1980. The history of both CA and MCA together with the simultaneous revival of induction in statistics will be a pretext to evoke 'the giants on whose shoulders we are standing'.

3.1 Simple Correspondence Analysis

The simple form of CA is a particular way of analysing contingency tables by means of visualizing jointly rows and columns in a low-dimensional space. Let us exemplify that aspect of CA by considering the cross-classification of eye and hair colours of Scottish children used by Fisher (1940) in his pioneering paper about the precision of discriminant functions (Table 3.1). This example is given as a mere illustration and is by no means a technical reminder.

Think about the four eye colour profiles, i.e., vectors of proportions summing to 1, as four points in a three-dimensional space (since four points lie in a three-dimensional space) and the five hair colour profiles as five points also in a three-dimensional space (they have four coordinates that sum to 1). CA consists of two simultaneous weighted principal components analyses (PCAs) of row and column profiles, using the chi-squared distances in each space, the weights being the marginal frequencies (see Chapter 9 from Cazes in this book).

Both PCAs give the same eigenvalues. Each principal component of the PCA of row profiles is proportional to the vector associated with the same eigenvalue, containing the loadings derived from the PCA of the column

TABLE 3.1

Fisher–Maung (Historical) Data about Hair and Eye Colours

	Hair Colour					
Eye colour	H_Fair	H_Red	H_Medium	H_Dark	H_Black	Total
E_Blue	326	38	241	110	3	718
E_Light	688	116	584	188	4	1,580
E_Medium	343	84	909	412	26	1,774
E_Dark	98	48	403	681	85	1,315
Total	1,455	286	2,137	1,391	118	5,387

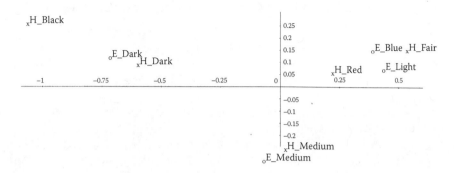

FIGURE 3.1
Sketch of the superimposed graphical displays of the row point and the column points of Table 3.1.

profiles, and vice versa. Like in usual PCA, components have variances, or weighted mean squares, equal to the eigenvalue, while the loadings have a weighted mean square equal to 1. Figure 3.1 is one of the possible displays of simultaneous representation of the rows and columns of Table 3.1 through simple CA, called the symmetric map, where both row and column points have inertias equal to the respective eigenvalues. As shown by the eigenvalues (Table 3.2), the first (horizontal) axis is markedly dominant. We may discard the third eigenvalue; the planar representation will be reliable.

The first axis ranks the categories from the darkest to the lightest colours. Each point (row or column) is the centroid of individuals possessing the corresponding category. Hair-medium and eye-medium are close, and children with dark eyes are close to children with dark hair (see Table 3.1). It does not mean that all children with dark hair have dark eyes. The second axis, which isolates the two 'medium' categories, shows the so-called arch or Guttman effect.

Note that this use of CA as a tool of descriptive statistics is very far from the concerns of Fisher and his contemporary. We may distinguish two tracks or sources in the history of CA, both developed in the 1930s: one is the quantification point of view, and the other the reciprocal averaging approach.

TABLE 3.2
The Corresponding Eigenvalues

Axis Number	Eigenvalue	Percentage	Cumulative Percentage
1	0.1992	86.56	86.56
2	0.0301	13.07	99.63
3	0.0009	0.37	100.00

3.1.1 Optimal Scaling or Quantification of Categorical Variables

Assigning optimal scores to rows and columns categories of a contingency table has been studied very early. De Leeuw (1983) traces back to Karl Pearson (1906) the idea of looking at such scoring as transformations that linearize regressions. This problem was solved later by Hermann Otto Hirschfeld (1935), better known under his subsequently anglicized name of H. O. Hartley. He obtained the eigenequations of CA when solving the following question (Hirschfeld, 1935, p. 520): 'Given a discontinuous bivariate distribution, is it always possible to introduce new values for the variates x, y such that both regressions are linear?' Hirschfeld showed that for the two-variable case with n and m categories, there are $\min(n, m) - 1$ solutions (the dimensions of CA). The new values, or scores, are nothing else than row and column coordinates in CA up to some standardization. Hirschfeld proved that the sum of eigenvalues was equal to Pearson's mean-square contingency coefficient, usually denoted by Φ^2, which is identical to the familiar Pearson's chi-squared statistic for a two-way contingency table, divided by the total of the table.

Fisher (1940) obtained CA equations when generalizing discriminant analysis to a single categorical predictor. His question was how to give scores to columns in order to discriminate as best as possible between row categories. He used an iterative technique similar to reciprocal averaging in order to find the optimal scores and applied it to the cross-classification of eye and hair colours of Table 3.1. Note that Fisher was interested only in the optimum, i.e., the first dimension of CA. Apart from a sign inversion, his solution is equal to the standardized coordinates of the seven categories on the first axis (Figure 3.1).

Maung (1941b) related Fisher's result to Hotelling's canonical correlation analysis by proving that Fisher's system of scores maximizes the correlation coefficient between the quantified variables. A reconstruction formula, attributed to Fisher, is presented without proof. Surprisingly, neither Fisher nor Maung refers to Hirschfeld. One can find later developments in Williams (1952) with chi-squared tests for the significance of eigenvalues. Chapter 33 of the famous treatise by Kendall and Stuart (1962) was devoted to 'canonical analysis of contingency tables', again CA in different terminology. Note that in this chapter the derivation of the distribution of the eigenvalues is flawed, as proved by Lancaster (1963).

The theory of optimal scaling culminated with several papers published in *Psychometrika* by Young, Takane, and de Leeuw in the 1970s and de Leeuw's dissertation (1973, reprinted in 1984). The contributions of the Dutch school of data analysis were gathered in the book of Gifi (1990).

3.1.2 Reciprocal Averaging

In a short article, Horst (1935) coined the name of *method of reciprocal averaging* for an algorithm proposed two years earlier by Richardson and Kuder (1933).

Richardson and Kuder aimed to achieve a better selection of salesmen for the company Procter and Gamble, and empirically discovered the method of reciprocal averaging. Basically, this algorithm consists in assigning arbitrary scores to the columns of a contingency table (or to the rows, if preferred), then computing the weighted averages, or barycentres, for the rows (row scores), then computing new column scores as weighted averages, etc. After each cycle the scores need to be standardized to avoid collapsing toward zero. After convergence and standardization, rows and columns scores are identical to rows and columns coordinates along the first principal axis of CA, the axis associated with the largest eigenvalue. In his note, Horst, who did not use any mathematical formula, indicates an identity with Hotelling's equations, but actually those of principal components, not of canonical analysis.

The idea that on some scale rows are represented as weighted barycentres of columns (and vice versa, if possible) is familiar to ecologists under the name of ordination. Legendre and Legendre (1998, p. 387) say that ordination 'consists in plotting object-points along an axis representing an ordered relationship, or forming a scatter diagram with two or more axes'. It is worth noticing that ecologists were interested in graphical displays, which was not the common case in the optimal scoring framework. However, the arch or Guttman effect, which occurs when a single numeric latent variable is present or dominant, has been considered a drawback by many ecologists. In this case higher-order factors are polynomial functions of the first factor, which may not bring useful information. This led to the development of detrended correspondence analysis by Hill and Gauch (1980).

Among other works related to CA and ordination let us cite ter Braak's (1986) canonical correspondence analysis, where the ordination axes are constrained to be linear combinations of external variables (see also ter Braak's Chapter 5 in this book).

The reciprocal averaging property was extensively used by Nishisato under the name of dual scaling (1980, 2007). Nishisato proposed this name to avoid choosing between many others since CA had been rediscovered many times. The name is related to the duality property, characterized by symmetric relations between row and column scores. The principle consists in maximizing the correlation ratio between the quantification of the variable associated to row categories and the categorical variable associated to column categories, and vice versa: dual scaling of a contingency table is then equivalent to CA, and dual scaling of a disjunctive table is equivalent to MCA.

3.1.3 Geometric Approach

CA was presented and developed under the French name *analyse des correspondances* for the first time by Cordier (1965) and Benzécri (1969). The founders of CA stressed a geometric and algebraic point of view that enables a simultaneous visual display of rows and columns of a contingency table. Here they are both sets of weighted points in a Euclidean space, and CA consists in projecting these points on a low-dimensional subspace in the same spirit as the presentation of principal components analysis (PCA) by Pearson (1901), but not by Hotelling. The use of the chi-squared metric ensures the coherence between the PCA of weighted row profiles and the PCA of weighted column profiles. The concept of inertia is highlighted, and the total inertia is equal to Pearson's Φ^2 (see Chapter 9 by Cazes, this book).

CA was applied in Benzécri's laboratory to a great number of fields and was adopted by many French researchers. However, the dissemination out of the French-speaking community was limited until the publication of the books by Greenacre (1984) and Lebart et al. (1984) popularized the method. Nevertheless, it took some time for correspondence analysis to be accepted as the standard name for the method, which was facilitated by the availability of software in commercial packages such as SAS, SPSS, and Stata.

CA may also be presented as an example of Gabriel's (1971) biplot for contingency tables (Gower and Hand, 1996; Greenacre, 2010), where rows, for example, depict vectors of relative frequencies, called profiles in CA, in a space structured by the chi-squared metric, and columns describe biplot axes onto which the row points can be projected.

3.1.4 CA as a Discretization of a Continuous Bivariate Distribution

When a contingency table is obtained by discretization of an underlying bivariate continuous distribution $f(x, y)$, the problem arises to study the behaviour of the solution as the partitions of the two variables are refined, when the CA scores asymptotically tend to continuous functions of x and y. Hirschfeld (1935, p. 524) wrote that the squared canonical correlation between transformed discrete variables 'could be defined for continuous distributions as well. However the complexity of its construction makes its practical use nearly impossible'.

Lancaster (1957, 1969) solved the general problem of canonical analysis of a bivariate distribution establishing a general reconstruction formula for the ratio of the bivariate density divided by the product of the marginal densities. This ratio may be expanded as a sum of products of canonical variables that are two complete sets of orthonormal functions defined on the margins. He also proved the convergence of canonical analysis of contingency tables (i.e., CA) toward the theoretical continuous solution, but concluded that 'there is no general method for determining the canonical variates in the distributions on infinitely many points in both variables'. However, for some distributions the solution is known: hermite polynomials in the case of the bivariate normal distribution, and Laguerre polynomials for the bivariate gamma distribution. Naouri (1971) studied the Fredholm integral equations giving the canonical variates, and gave several examples. Since then, there have been few publications on continuous correspondence analysis (for example, Cuadras et al., 1999).

3.2 Multiple Correspondence Analysis

Among several possible presentations, multiple correspondence analysis (MCA) can be viewed as a simple extension of the area of applicability of correspondence analysis (CA), from the case of a contingency table to the case of a 'complete disjunctive binary table', also called an indicator matrix. The properties of such a table are interesting; the computational procedures and the rules of interpretation of the obtained representations are simple, albeit specific. MCA can emerge as both a particular case and a generalization of CA; it is not easy to disentangle its history from that of CA. We use the same historical data set to exemplify the link existing between CA and MCA, in the particular case of a two-way table.

In Table 3.3, the first column contains intuitive identifiers of each cell. Columns 2 and 3 contain the addresses of each cell, whereas the fourth column ('Weights') contains the frequencies of the corresponding cell. The rest of the table is the binary disjunctive matrix, made up of two blocks that constitute, respectively, a coding of both columns 2 and 3.

The CA of the contingency table (Table 3.1), the CA of the weighted binary disjunctive table (Table 3.3), and the CA of the Burt matrix (Table 3.4) produce the same eigenvectors with norm 1. However, the eigenvalues from these three CAs are distinct: if μ is the eigenvalue from the original CA of Table 3.1, λ from the second CA (Table 3.3), and ξ from the third CA (Table 3.4), we have the relationships between eigenvalues:

$$\lambda = \frac{1+\sqrt{\mu}}{2} \text{ and } \xi = \lambda^2.$$

The values of λ are displayed in Table 3.5.

Unlike the original Table 3.1, the tables **Z** and **B** can be easily generalized to more than two blocks (or questions, variables), leading then to the general framework of MCA. A vast domain of applications is then opened (socioeconomic surveys, and more generally, all kinds of research questions involving numerous categorical variables).

3.2.1 First Traces of a Technology

According to a famous paper of Healy (1978), statistics is more a *technology* than a *science*. This terminological choice aptly applies to MCA, a tool characterized more by its usefulness than by its theoretical content and context. The history of a particular technology, because of its involvement with real life, can be fascinating. From this point of view, we can compare, for instance, MCA with the minimum spanning tree, a technique as useful in statistics as

TABLE 3.3
Another Presentation of Table 3.1 Data, Leading to MCA

Hair X Eye	Hair	Eye	Weight	Binary Disjunctive Table Z											
				HF	HR	HM	HD	HB	EB	EL	EM	ED
HF_EB	1	1	326	1	0	0	0	0	1	0	0	0
HF_EL	1	2	688	1	0	0	0	0	0	1	0	0
HF_EM	1	3	343	1	0	0	0	0	0	0	1	0
HF_ED	1	4	98	1	0	0	0	0	0	0	0	1
HR_EB	2	1	38	0	1	0	0	0	1	0	0	0
HR_EL	2	2	116	0	1	0	0	0	0	1	0	0
HR_EM	2	3	84	0	1	0	0	0	0	0	1	0
HR_ED	2	4	48	0	1	0	0	0	0	0	0	1
HM_EB	3	1	241	0	0	1	0	0	1	0	0	0
HM_EL	3	2	584	0	0	1	0	0	0	1	0	0
HM_EM	3	3	909	0	0	1	0	0	0	0	1	0
HM_ED	3	4	403	0	0	1	0	0	0	0	0	1
HD_EB	4	1	110	0	0	0	1	0	1	0	0	0
HD_EL	4	2	188	0	0	0	1	0	0	1	0	0
HD_EM	4	3	412	0	0	0	1	0	0	0	1	0
HD_ED	4	4	681	0	0	0	1	0	0	0	0	1
HB_EB	5	1	3	0	0	0	0	1	1	0	0	0
HB_EL	5	2	4	0	0	0	0	1	0	1	0	0
HB_EM	5	3	26	0	0	0	0	1	0	0	1	0
HB_ED	5	4	85	0	0	0	0	1	0	0	0	1

Note: The cell frequencies are now the weights of (artificial) individual profiles.

TABLE 3.4

Burt Matrix **B**, Product of the Above Binary Table **Z** by Its Transpose

1455	0	0	0	0	326	688	343	98
0	286	0	0	0	38	116	84	48
0	0	2137	0	0	241	584	909	403
0	0	0	1391	0	110	188	412	681
0	0	0	0	118	3	4	26	85
326	38	241	110	3	718	0	0	0
688	116	584	188	4	0	1580	0	0
343	84	909	412	26	0	0	1774	0
98	48	403	681	85	0	0	0	1315

Note: The original contingency table and its transpose are the nondiagonal blocks.

TABLE 3.5

Eigenvalues from the MCA of the Weighted Binary Disjunctive Table **Z**

Number	Eigenvalue	Percentage	Cumulative Percentage
1	0.723	20.7	20.7
2	0.587	16.8	37.4
3	0.514	14.7	52.1
4	0.500	14.3	66.4
5	0.485	13.9	80.3
6	0.413	11.8	92.1
7	0.277	7.9	100.0

in operations research. A prominent example is provided by the history of the minimum spanning tree (almost a thriller; see Graham and Hell (1985)).

3.2.2 Preliminary Bases

At the foundation of the contemporary presentation of principal axis analyses is a theorem based on the singular values decomposition, that was presented by Eckart and Young (1936) for rectangular matrices and which generalized the works of Cauchy (1830) and Sylvester (1889b) (to quote only two noteworthy milestones) concerning square matrices. The problem that is at hand is one of pure numerical reduction, i.e., of data compression: how to fit, in the least-squares sense, a matrix by another matrix of inferior rank. Among the first articles that were published on the algebraic and geometric methods of principal axis methods, we like to note Gower (1966; see also his chapter in this book) and Gabriel (1971).

MCA, like PCA, deals with tables involving individual-level data. PCA is the oldest and most established of the methods of principal axes visualization. Conceived for the first time in a limited (albeit quite modern) setting by

Pearson in 1901, and integrated into mathematical statistics by Hotelling in 1933, PCA has not really been used before the arrival and diffusion of computational aids. For the traditional statistician, it is about searching for the principal axes of a multivariate normal distribution from a sample. This is the initial presentation of Hotelling (1933), and later that of classic manuals of multivariate analysis, such as the fundamental treatise by Anderson (1958). It is also a special case of factor analysis (case of null or equal specificities; cf. Horst (1965)). A presentation that is closer to current thinking can be found in the pioneering and synthetic article by Rao (1964).

3.2.3 Multiple Correspondence Analysis: Formulas and Methodology

The basic formulas underlying MCA can be traced back to Guttman (1941), who devised it as a method of scaling, but also to Burt (1950), in a wider scope. The paper by Guttman contains, with modern matrix notations, all the formulas that concern and characterize MCA as we know it. It includes a careful description of 'binary disjunctive data sets' and mentions the 'chi-squared metric'.

In his (legitimate) claim (published in the *British Journal of Mathematical and Statistical Psychology*, directed by Burt himself) for an acknowledgement of priority addressed to Burt, Guttman (1953) insists on the limited scope that should be assigned to the method: to derive a *unique* scale. After a courteous introduction (Guttman, 1953, p. 1) ('It is gratifying to see how Professor Burt has independently arrived at much the same formulation. This convergence of thinking lends credence to the suitability of the approach'.), Guttman (1953, p. 2) states, 'While the principal components here are formally similar to those for quantitative variables, nevertheless, their interpretation may be quite different', and further on (Guttman, 1953, p. 2), 'If a correct, but non-linear prediction technique is used, the whole variation can sometimes be accounted for by but the single component. In such a case, the existence of more than one component arises merely from the fact that a linear system is being used to approximate a non linear one'. Guttman (1953, p. 2) then explains that the other dimensions will be polynomial functions of the first one, being similar to the solutions 'of a second order differential equation classical in physics'.

Note that Benzécri et al. (1973, chapter entitled 'Sur l'analyse de la correspondance définie par un graphe') has defined a 'generalized Guttman effect', linked to the eigenvectors of the Laplace operator, where components of higher order are polynomial functions of the first two components. In fact, Guttman has discovered all the formulas underlying MCA, but was uniquely and perhaps obstinately interested by the prediction of a perfect scale. In his paper of 1953 entitled 'Scale Analysis and Factor Analysis (Comments of Dr. Guttmann's Paper)', Burt, while recognizing the anteriority of Guttman with regard to all the details of the formulas, promotes a more exploratory role for the method, in the vein of the recommendations of Thurstone (1947) in the context of factor analysis. 'It is encouraging to find that Dr. Guttman has discerned points of resemblance between the methods we have independently reached for analysing qualitative data; And I willingly agree that this convergence in our lines of approach lends additional plausibility to the results ... my aim was to factorize such data, his to construct a scale' (Burt, 1953, p. 5). The most amazing in these interesting discussions, which involved painstaking calculations by hand on very small numerical examples, is precisely the absence of computers and real-sized experiments.

Note that the response of Burt is cleverly articulated—he was obviously perfectly mastering his own mind at that time. The fact that his reputation has been stained 20 years later about some alleged fraud (see Hearnshaw, 1979; Gould, 1983; Joynson, 1989; Fletcher, 1991) can be summarized by a sentence from the *Encyclopaedia Britannica*: 'From the late 1970s it was generally accepted that he had fabricated some of the data, though some of his earlier work remained unaffected by this revelation'.

The first real applications of MCA as an exploratory tool probably date back to Hayashi (1956). The availability of computing facilities entailed a wealth of new developments and applications in the 1960s and 1970s, notably around Benzécri (1964); Benzécri et al. (1973). The term *multiple correspondence analysis* was coined at that time.

MCA has been developed in another theoretical framework (closer to the first approach of Guttman) under the name of *homogeneity analysis* by the research team around Jan de Leeuw since 1973 (cf. Gifi, 1981, 1990) and under the name of *dual scaling* by Nishisato (1980), the latter more inspired by Hayashi.

3.2.4 First Implementations

A first internal note by Benzécri proved the equivalence between the analysis proposed by Burt (1950) of the Burt contingency table (now called the Burt matrix in MCA; see Table 3.4) relating to two variables and the same CA of the single two-way contingency table. Another note by Benzécri (1972) provided a review of the properties that hold in the case of more than two variables. Lebart and Tabard (1973) presented all these properties, and a complete source code of a Fortran software of MCA, including the diagonalization of a $(J - Q) \times (J - Q)$ matrix, instead of a $J \times J$ matrix, where J = total number of categories and Q = number of categorical variables. These papers and reports are referred to in Lebart (1974), in which the MCA solution is obtained through a direct diagonalization algorithm involving uniquely the response pattern matrix, i.e., the $N \times Q$ matrix, where N is the number of respondents. This paper is perhaps the first paper in English that contains the name MCA. In Lebart (1975), MCA is presented as a technology to drive the methodology of survey data processing: in a first phase, a visualization through MCA of all the basic socioeconomic characteristics of the respondents provides a 'socioeconomic grid' (aka the 'basic frame of the survey'). In a second phase, all the variables constituting the content of the survey are projected (as supplementary variables) onto this grid. In so doing, the researcher obtains at a glance an overview of the main correlations that are usually obtained through a painstaking inspection of numerous cross-tabulations. The book by Lebart et al. (1977) (translated in English seven years later with the kind support of Ken Warwick: Lebart et al. (1984)) dealt with all these results and computer programs. MCA was one of the most salient components of the software SPAD that was devised at that time.

Note that like the bourgeois gentleman of Molière, who has been 'speaking prose all his life, and didn't even know it', some statisticians, such as Nakache (1973), have applied CA to complete disjunctive tables, and didn't even know that they were using MCA.

3.2.5 Dissemination of MCA

Descriptive statistical analysis allows us to represent statistical information in a graphical form by simplifying and schematizing it. Multidimensional descriptive statistical analysis generalizes this idea in a natural way when the information concerns several variables or dimensions.

Sample surveys are popular tools in the social sciences, and their questionnaires generally contain a wealth of categorical data. Replacing or complementing the dull arrays of cross-tabulations with suggestive graphical displays has been an undeniable improvement in survey data processing. It is then not a surprise that MCA has been welcome, immediately used, and probably overused during the first years of its dissemination. Emblematic and large-scale applications of CA and MCA in social science are to be found in the works of the sociologist Pierre Bourdieu: *La Distinction* (Bourdieu, 1979) and *Homo Academicus* (Bourdieu, 1984) (see also Chapter 13 by Blasius and Schmitz in this book).

One of the innovations in statistics after 1960 was the appearance of techniques in the form of 'products': software developed with financial and commercial constraints on its conception, production, and distribution. Like any finalized product, the advantage of the software was its ability to diffuse, and its inconvenience that it entailed certain rigidity. Like any product intended to be used by specialists, it induced new divisions of labour that were not very desirable in a knowledge process. Software that is accessible and easy to use allows methods to be widely spread but lead to a careless use in areas where much caution would be called for.

3.2.6 MCA and Multiway Analyses

In the pioneering papers of Hirschfeld, Fisher, and Maung mentioned previously, simple CA was, implicitly or explicitly, considered a particular case of canonical correlations analysis between two sets of dummy variables (categorical variables coded under binary disjunctive form). It was then natural to devise the extensions of simple correspondence analysis to more than two dimensions through generalized canonical analysis as proposed by various authors, among them Horst (1961), Carroll (1968), and Kettenring (1971). In this wealth of methods that cannot be described in this chapter, the generalization proposed by Carroll is the closest to MCA as we know it. For p groups of variables (or p blocks $\mathbf{X}$) this generalization consists in obtaining an auxiliary variable $\mathbf{z}$ that maximizes

$$\sum_{j=1}^{p} R^2(\mathbf{z}, \mathbf{X}_j).$$

When blocks are made of indicator variables of categorical variables X_j, the multiple correlation is equal to the correlation ratio: $R^2(\mathbf{y}, \mathbf{X}_j) = \eta^2(\mathbf{y}, X_j)$.

There is no unique generalization of the theorem of Eckart and Young to the case of three-way tables. This can be expressed in the following way: the hierarchical decomposition of an element of the tensor product of two Euclidean spaces into a sum of tensor products of pairs of vectors belonging to each space is unique. But such decomposition is not unique in the case of an element of the tensor product of more than two Euclidean spaces (cf. Benzécri et al., 1973; Tome 2B, no. 6). Therefore, in this case, there cannot be an exploratory approach that is as well established as in the case of two-way tables.

We find a synthesis and a classification of the main approaches to multiway analysis in the work of Kroonenberg (1983a). The first works on this theme were those of Tucker (1964, 1966) followed by those of Harshman (1970) in the context of classic factor analysis. Let us also mention the works of Pagès et al. (1976) based on the operators defined by Robert and Escoufier (1976). Kroonenberg (Chapter 6 in this book) gives a complete account of the history of three-way methods.

3.2.7 Stability, Validation, Resampling

An important issue in MCA is certainly the stability of the results, together with the confidence attached to the location of points in the principal subspaces. From a purely numerical point of view, Escofier and Le Roux (1976) have treated the stability of the axes in principal axes analysis (PCA and CA). These authors study the maximal variations of the eigenvectors and the eigenvalues when well-defined modifications alter the data: removal or addition of elements to the data tables, the influence of regrouping several elements or small modifications of the values in the table, and the influence of the chosen distances and weightings. The study of the validity of the results of dimension-reducing techniques has led to much research, but which has since the 1980s taken a different direction with 'computer-intensive methods' such as bootstrapping.

The bootstrap, a particular simulation technique introduced by Efron (1979), consists in simulating samples of the same size as the initial sample. With the exception of the works of Gifi (1981, 1990) that specifically concern correspondence analysis (in fact, the bootstrap principle differs significantly according to the various principal axes techniques), one of the first works where bootstrap is applied to validate results of PCA is probably that of Diaconis and Efron (1983). Meulman (1982) introduced bootstrapping in the homogeneity analysis framework, which is another presentation of MCA. She performed a new CA for each bootstrap sample. Independently, Greenacre (1984) used what we later called the partial bootstrap; this is when the initial configuration is fixed and the bootstrap samples are added as supplementary points.

3.2.8 Some Related Methods

During the main period under consideration (before 1980), several improvements were brought to the practice of MCA. Saporta (1977) proposed a technique of discrimination from a set of qualitative variables (DISQUAL) based on the principal axes extracted from a preliminary MCA. Cazes (1977b, 1980) has studied the properties of the CA of subtables of a Burt contingency table. The technique of supplementary elements, proposed earlier by Gower in another framework (1968), has been thoroughly investigated with all its possible variants by Cazes (1977). A note by Benzécri (1979) gave a procedure to remedy the poorly significant percentages of variances provided by MCA. Escofier (1979a, 1979b) proposed variants of MCA allowing for processing (or representing as supplementary elements) both categorical and numerical variables.

3.2.9 Conclusion

It is certainly presumptuous to attempt to write the history of a technology without a certain perspective. We have tried as much as possible to restrict ourselves to works prior to 1980, but both the context and the complexity of discoveries are such that we had to mention some more recent contributions that may either reveal some past results or summarize scattered methods. We have encountered a series of ideas, formulas, and techniques, translated later into methodologies, textbooks, and software. The process of discovery is by no means sequential, and to specify, for example, a date of birth is always difficult, sometimes almost meaningless. The endeavour to briefly clarify the genesis of CA and MCA will remain, as expected, a personal point of view. However, the fascination toward our methodological ancestors deserved to be communicated. For other historical references about the history of CA, see Benzécri (1982b), de Leeuw (1983), Tenenhaus and Young (1985), Nishisato (2007, Chapter 3), Armatte (2008), Gower (2008), Heiser (2008), and Beh and Lombardo (2012).

4

History of Nonlinear Principal Component Analysis

Jan de Leeuw

CONTENTS

4.1 Linear PCA 46
4.2 Simple and Multiple Correspondence Analysis 46
    4.2.1 Correspondence Analysis 47
    4.2.2 Multiple Correspondence Analysis 47
4.3 Forms of Nonlinear Principal Component Analysis 48
4.4 NLPCA with Optimal Scaling 48
    4.4.1 Software 49
4.5 NLPCA in the Gifi Project 49
4.6 Example 51
4.7 NLPCA Using Pavings 54
4.8 NLPCA Using Aspects 56
4.9 Logit and Probit PCA of Binary Data: Gifi Goes Logistic 59
4.10 Conclusion 60

In this chapter we discuss several forms of nonlinear principal components analysis (NLPCA) that have been proposed over the years. Our starting point is that ordinary or classical principal components analysis (PCA) is a well-established technique that has been used in multivariate data analysis for well over 100 years. But PCA is intended for numerical and complete data matrices, and cannot be used directly on data that contain missing, character, or logical values. At the very least, in such cases the interpretation of PCA results has to be adapted, but often we also require a modification of the loss functions or the algorithms.

Multiple correspondence analysis (MCA) is a method similar to PCA that can be and is routinely applied to categorical data. We discuss MCA as a form of NLPCA, and explain the relationship with classical PCA. In addition, we discuss various other forms of NLPCA, such as linear PCA with optimal scaling, aspect analysis of correlations, Guttman's multidimensional scalogram analysis (MSA), logit and probit PCA of binary data, and logistic homogeneity analysis.

4.1 Linear PCA

Principal components analysis (PCA) is often attributed to Hotelling (1933), but that is surely incorrect. The equations for the principal axes of quadratic forms and surfaces, in various forms, were known from classical analytic geometry. There are some modest PCA beginnings in Galton (1889, pp. 100–102, and Appendix B), where the principal axes are connected for the first time with the 'correlation ellipsoid'.

There is a full-fledged (although tedious) discussion of the technique in Pearson (1901), and there is a complete application (seven physical traits of 3,000 criminals) by a Pearson coworker in MacDonell (1902). The early history of PCA in data analysis, with proper attributions, is reviewed in Burt (1949).

Hotelling's introduction of PCA follows the now familiar route of making successive orthogonal linear combinations of the variables with maximum variance. He does this by using Von Mises (power) iterations, discussed in Von Mises and Pollackzek-Geiringer (1929).

Pearson, following Galton, used the correlation ellipsoid throughout. He casts the problem in terms of finding low-dimensional subspaces (lines and planes) of best (least-squares) fit to a cloud of points, and connects the solution to the principal axes of the correlation ellipsoid.

The data for the problem are n points in $\mathbb{R}^m$, collected in an $n \times m$ matrix $\mathbf{Y}$. We want to find n points in a p-dimensional subspace of $\mathbb{R}^m$ that are close to the n data points. We measure closeness using squared Euclidean distances, which implies we want to minimize SSQ $(\mathbf{Y} - \mathbf{XB}^T)$ over $n \times p$ matrices $\mathbf{X}$ and $m \times p$ matrices $\mathbf{B}$ with normalization conditions on $\mathbf{X}$ or $\mathbf{B}$. Throughout this chapter we use SSQ( ) for the sum of squares. For $p = 1$ we find the best line, for $p = 2$ the best plane, and so on.

4.2 Simple and Multiple Correspondence Analysis

The history of simple and multiple correspondence analysis is reviewed expertly in Chapter 3 by Lebart and Saporta in this book. We merely give some additional references that serve to connect multiple correspondence analysis with nonlinear principal components analysis.

4.2.1 Correspondence Analysis

Simple correspondence analysis (CA) of a bivariate frequency table was first discussed, in a rather rudimentary form, by Pearson (1906), by looking at transformations linearizing regressions—see de Leeuw (1983). This was taken up by Hirschfeld (1935), where the technique was presented in a more complete form to maximize correlation and decompose contingency. This approach was later adopted by Gebelein (1941) and by Renyi (1959) and his students in their study of maximal correlation.

Fisher (1938) scores a categorical variable to maximize a ratio of variances (quadratic forms). This is not quite the same as CA, because it is presented in an (asymmetric) analysis of variance context. Both CA and the reciprocal averaging algorithm are discussed, however, in Fisher (1940, Section 3), and applied by his coworker Maung (1941a, 1941b).

Then in the early 1960s the chi-square distance-based form of CA, relating CA to metric multidimensional scaling (MDS), with an emphasis on geometry and plotting, was introduced by Benzécri, and published (with FORTRAN code) in the thesis of Cordier (1965).

4.2.2 Multiple Correspondence Analysis

Different weighting schemes to combine quantitative variables into an index that optimizes some variance-based discrimination or homogeneity criterion were proposed in the late 1930s by Horst (1936), Edgerton and Kolbe (1936), and Wilks (1938). Their proposals all lead to the equations for linear PCA.

The same idea of weighting (or quantifying) was applied to qualitative variables in a seminal paper by Guttman (1941), who was analysing qualitative data for the war department. He presented, for the first time, the equations defining multiple correspondence analysis (MCA). The equations were presented in the form of a row-eigen (scores), a column-eigen (weights), and a singular value (joint) problem. The paper introduced, without introducing explicit names, complete disjunctive coding (*codage disjonctif complet* in French), the Burt table (*tableau de Burt* in French), and pointed out the connections with the chi-square metric. There was no geometry, and the emphasis was on constructing a single scale. In fact, Guttman warned explicitly against extracting and using additional eigenpairs.

Guttman (1946) extended scale or index construction to paired comparisons and ranks. Then in Guttman (1950) it was extended to scalable binary items. In the 1950s and 1960s Hayashi introduced the quantification techniques of Guttman in Japan, where they were widely disseminated through the work of Nishisato. Various extensions and variations were added by the Japanese school; see Chapter 3 by Lebart and Saporta in this book for references. Starting in 1968, MCA was studied as a simple form of metric MDS by de Leeuw (1968, 1973).

Although the equations defining MCA were basically the same as those defining PCA, the relationship between the two remained problematic. These problems were compounded by 'arches', or the Guttman effect (*l'effet Guttman* in French), i.e., by artificial curvilinear relationships between successive dimensions (eigenvectors).

4.3 Forms of Nonlinear Principal Component Analysis

There are various ways in which we can introduce nonlinearity into PCA to obtain what is abbreviated as NLPCA. First, we could seek indices that are nonlinear combinations of variables that discriminate maximally in some sense. For example, we could look for a multivariate polynomial P of the observed variables y_j, with some normalization constraints on the polynomial coefficients, such that $P(y_1,\cdots,y_m)$ has maximum variance. This generalizes the weighting approach of Hotelling. Second, we could find nonlinear combinations of unobserved components that are close to the observed variables. In a polynomial context this means we want to approximate the observed y_j by polynomial functions $P(x_1,\cdots,x_p)$ of the components. This generalizes the reduced rank approach of Pearson. Third, we could look for transformations of the variables that optimize the linear PCA fit. We approximate $T(y_j)$ by the linear combination $\mathbf{X}\mathbf{b}_j$, by fitting both transformations and low-rank approximation. This is known as the *optimal scaling* (OS) approach, a term of Bock (1960).

The first approach has not been studied much, although there may be some relations with item response theory. The second approach is currently popular in computer science, as nonlinear dimension reduction—see, for example, Lee and Verleysen (2007). There is no unified theory, and the papers are usually of the 'well, we could also do this' type familiar from cluster analysis. The third approach preserves many of the properties of linear PCA and can be connected with MCA as well. We shall follow the history of PCA-OS and discuss the main results.

4.4 NLPCA with Optimal Scaling

Guttman (1959) observed that if we require that the regressions between monotonically transformed variables are linear, then these transformations are uniquely defined. In general, however, we need approximations. The loss function for PCA-OS is SSQ $(Y - XB^T)$, as before, but now we minimize over components X, loadings B, and also over transformations Y, thereby obtaining NLPCA-OS. Transformations are defined column-wise (over variables) and belong to some restricted class (monotone, step, polynomial, spline). Algorithms often are of the alternating least-squares (ALS) type, where optimal transformation and low-rank matrix approximation are alternated until convergence.

4.4.1 Software

PCA-OS only became interesting after it became computationally feasible. Consequently, the development and availability of software was critical for the acceptance of NLPCA. Shepard and Kruskal used the monotone regression machinery of the nonmetric breakthrough to construct the first PCA-OS programs around 1962. Their paper describing the technique was not published until much later (Kruskal and Shepard, 1974). Around 1970 versions of PCA-OS (sometimes based on Guttman's rank image principle) were developed by Lingoes and Roskam—see Roskam (1968), Lingoes and Guttman (1967), and Lingoes (1973). The rank image principle is an alternative to monotone regression, without clear optimality properties.

In 1973 de Leeuw, Young, and Takane started the alternating least squares with optimal scaling (ALSOS) project, which resulted in PRINCIPALS (Young et al., 1978) and PRINQUAL in SAS (SAS, 1992). In 1980 de Leeuw (with Heiser, Meulman, Van Rijckevorsel, and many others) started the Gifi project (Gifi, 1990), which resulted in PRINCALS (de Leeuw and Van Rijckevorsel, 1980), CATPCA in SPSS (SPSS, 1989), and the R package `homals` (de Leeuw and Mair, 2009a).

Winsberg and Ramsay (1983) published a PCA-OS version using monotone spline transformations. The loss function is the same, but the class of admissible transformations is different. The forms of PCA-OS we have discussed so far use polynomials or step functions, which may or may not be monotonic. Koyak (1987), using the ACE smoothing methodology of Breiman and Friedman (1985), introduced `mdrace`. Again, the class of transformations is different. Transformations in ACE do not necessarily decrease an overall loss function, but they approximate the conditional expectations by using smoothers. Indeed, ACE stands for alternating conditional expectation.

4.5 NLPCA in the Gifi Project

The Gifi project followed the ALSOS project. Its explicit goal was to introduce a system of multivariate analysis methods, and corresponding computer software, on the basis of minimizing a single loss function by ALS algorithms. The techniques that fit into the system were nonlinear regression, canonical analysis, and PCA.

The Gifi loss function is different from the previous PCA loss functions. The emphasis is not on transformation or quantifications of variables and low rank approximation, as in NLPCA. All variables are thought of as categorical, and the emphasis is on the reciprocal averaging and the corresponding geometry of centroids (*principe barycentrique* in French). The loss function is

$$\sigma(\mathbf{X}, \mathbf{Y}) = \sum_{j=1}^{m} \text{SSQ}(\mathbf{X} - \mathbf{Z}_j \mathbf{Y}_j),$$

which must be minimized over $n \times p$ scores $\mathbf{X}$ for the n objects satisfying $\mathbf{X}^T \mathbf{X} = \mathbf{I}$, and over $k_j \times p$ category quantifications $\mathbf{Y}_j$ of the m variables. The $\mathbf{Z}_j$ are the indicator matrices, coding category membership of the objects (*codage disjonctif complet*), where variable j has k_j categories—see Chapter 11 on MCA by Husson and Josse in this book. Thus, we require that the score of an object or individual in $\mathbf{X}$ is as close as possible, in squared Euclidean distance, to the quantifications of the categories $\mathbf{Y}_j$ that the object falls in.

In the context of generalized canonical analysis, the Gifi loss function is identical to the loss function proposed earlier by Carroll (1968). By using indicator matrices we make the technique identical to MCA, called homogeneity analysis by Gifi, while the various other techniques are special cases resulting from imposing restrictions on the quantifications $\mathbf{Y}_j$. One of the main contributions of Gifi is to show that the transformation approach of NLPCA can actually be fitted into the homogeneity analysis loss function. Interpreting the loss function in terms of Euclidean distance relates homogeneity analysis to nonmetric scaling and nonmetric unfolding.

The basic result that the unconstrained minimization of the Gifi loss function gives MCA follows from the fact that minimizing $\sigma(\mathbf{X}, \mathbf{Y})$ is equivalent to finding the largest eigenvalues of $\mathbf{P}^*$, where $\mathbf{P}^*$ is the average of the projectors $\mathbf{P}_j = \mathbf{Z}_j \mathbf{D}_j^{-1} \mathbf{Z}_j^T$, with $\mathbf{D}_j = \mathbf{Z}_j' \mathbf{Z}_j$ the diagonal matrix of marginals. This is also equivalent to finding the largest eigenvalues of the generalized eigenvalue problem $\mathbf{BY} = m \mathbf{DY} \Lambda$, where $\mathbf{B} = \mathbf{Z}^T \mathbf{Z}$, with $\mathbf{Z} = (\mathbf{Z}_1 \, \mathbf{Z}_2 \cdots \mathbf{Z}_m)$, is the Burt matrix of bivariate marginals and $\mathbf{D}$ is its diagonal.

NLPCA results by imposing the restriction that $\mathbf{Y}_j = \mathbf{z}_j \mathbf{a}_j^T$; i.e., the category quantifications are of rank 1. To see what the effect of rank 1 restrictions is, substitute $\mathbf{Y}_j = \mathbf{z}_j^T \mathbf{a}_j^T$ into the Gifi loss function. Define $\mathbf{q}_j = \mathbf{Z}_j \mathbf{z}_j$ and normalize so that $\mathbf{q}_j^T \mathbf{q}_j = \mathbf{z}_j^T \mathbf{D}_j \mathbf{z}_j^T = 1$. Then $\sigma(\mathbf{X}, \mathbf{Y}) = m(p - 1) + \text{SSQ}(\mathbf{Q} - \mathbf{X}\mathbf{A}^T)$, which is the loss function for linear PCA applied to $\mathbf{Q}$. Thus, the ALS algorithm alternates optimal transformations of the variables in $\mathbf{Q}$ and doing a linear PCA. Further restrictions can require the single quantifications $\mathbf{z}_j$ to be either linear, polynomial, or monotonic functions of the original measurements. The monotonic case gives nonmetric PCA in the classical Kruskal-Shepard sense, and the linear case gives classical linear PCA. In the Gifi approach, and the homals program, we can combine different types of transformations for all variables, as well as multiple (unconstrained) and single (constrained) category quantifications.

The homals program makes it possible to impose additional so-called additivity constraints on the category quantifications. This allows us to write $Z_j Y_j = Z_{j1} Y_{j1} + \cdots + Z_{jv} Y_{jv}$, and we can interpret each j as a set of variables, not just a single variable. In this way we can incorporate regression, discriminant analysis, canonical analysis, and multiset canonical analysis in the Gifi loss function. Combined with the possibility of scaling each variable linearly, polynomially, or monotonically, and with the possibility of treating each variable as rank constrained or not, this gives a very general system of multivariate analysis methods.

The relation between MCA and NLPCA was further investigated in a series of papers by de Leeuw and his students (de Leeuw, 1982, 1988b, 2006b; Bekker and de Leeuw, 1988; de Leeuw et al., 1999). The research is centred on assuming simultaneous linearizability of the regressions. We assume that separate transformations of the variables exist that make all bivariate regressions linear. The transformations are not necessarily monotone or continuous, and simultaneous linearizability does not say anything about the higher-order multivariate regressions.

Simultaneous linearizability generalizes the result of Pearson (1906) to $m > 2$. It also generalizes the notion (Yule, 1912) of a strained multivariate normal, i.e., a multivariate distribution obtained by applying monotone and invertible transformations to each of the variables in a multivariate normal.

If simultaneous linearizability is satisfied (as it is in the case of two variables, in the case of m binary variables, and in the case of a strained multivariate normal distribution), then MCA can be interpreted as performing a sequence of NLPCAs on a sequence of related correlation matrices. All solutions to the MCA equations are also solutions to the NLPCA equations. This also elucidates the arch or Guttman effect and the role of rank 1 constraints.

4.6 Example

We use the YouthGratitude data from the R package psychotools (Zeileis et al., 2012). They are described in detail in the article by Froh et al. (2011). The six 7-point Likert scale variables of the GQ-6 scale are used, with responses from 1,405 students aged 10–19 years. Froh et al. (2011) indicate that classical linear factor analysis shows a one-dimensional structure for the first five items, while they removed the sixth item from further analysis because of a very low factor loading. We keep all six items.

We use the homals package for the first analysis. Allowing multiple nominal quantifications for all six variables, i.e., doing an MCA, gives the object scores in Figure 4.1.

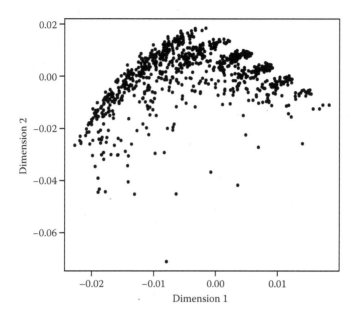

FIGURE 4.1
Object scores multiple nominal.

The figure shows the shape of the point cloud, which is an arch with various subclouds corresponding to closely related profiles.

Figure 4.2 shows the category quantifications, centroids of the corresponding object scores, for variables 1 and 6. The poor discriminatory power of variable 6 is obvious.

In the next analysis we use single numerical quantifications; i.e., we do a linear PCA. The object scores are in Figure 4.3. The arch has disappeared and the two dimensions appear much less related.

Figure 4.4 gives projection plots for variables 1 and 6. These are joint plots in which component loadings define a direction in space, with category quantifications being lines perpendicular to the loading direction. Object scores are projected on the category they belong to for the variable. The total sum of squares of the projection distances is the loss for this solution.

In this case the seven category quantifications are equally spaced perpendicular lines.

In the single ordinal case we allow for monotonic transformations of the category quantifications. This leads to object score plot and projection plots in Figures 4.5 and 4.6.

We see that variable 6 now defines the second dimension, with the loadings almost perpendicular to the loading of the first variable (in fact, the first five variables). Monotone regression creates ties, so we do not see seven category quantification lines any more, and they certainly are not equally spaced.

History of Nonlinear Principal Component Analysis

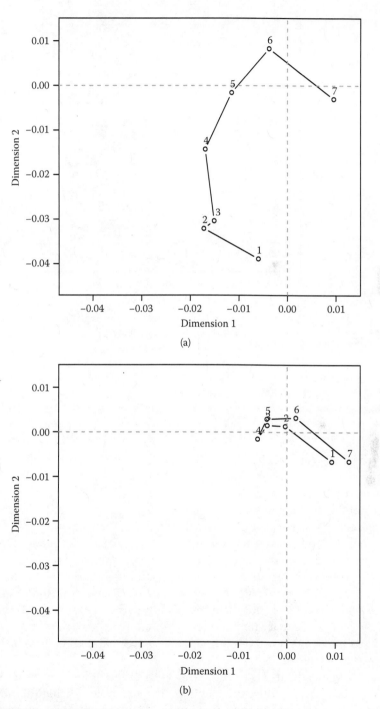

FIGURE 4.2
Category quantifications multiple nominal. (a) Variable 1. (b) Variable 6.

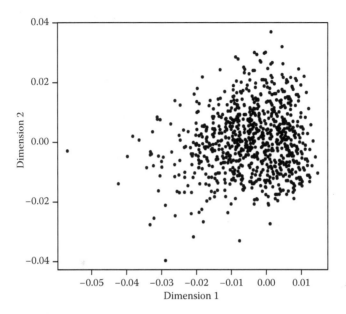

FIGURE 4.3
Object scores single numerical.

4.7 NLPCA Using Pavings

The geometrical approach to NLPCA and MCA has given rise to other related techniques. Suppose we map n objects into low-dimensional Euclidean space, and use a categorical variable to label the points. Each variable defines a partitioning of the points into subsets, which we call a paving. In NLPCA we want these category subsets to be either small (relative to the whole set) or separated well from each other. And we want this for all variables simultaneously. The two objectives are not the same, although they are obviously related.

In MCA we want small subsets, where smallness is defined in terms of total squared Euclidean distance from the category centroids. In PCA-OS we want separation of the subsets by parallel hyperplanes, and loss is defined as squared Euclidean distance to approximate the separating hyperplanes. Total loss measures how well our smallness or separation criteria are satisfied over all variables.

There have been various experiments with different loss functions, based on different ways to measure pavings. For example, Michailidis and de Leeuw (2005) used multivariate medians and sums of absolute deviations to measure homogeneity. Alternatively, de Leeuw (2003) used the length of the

History of Nonlinear Principal Component Analysis

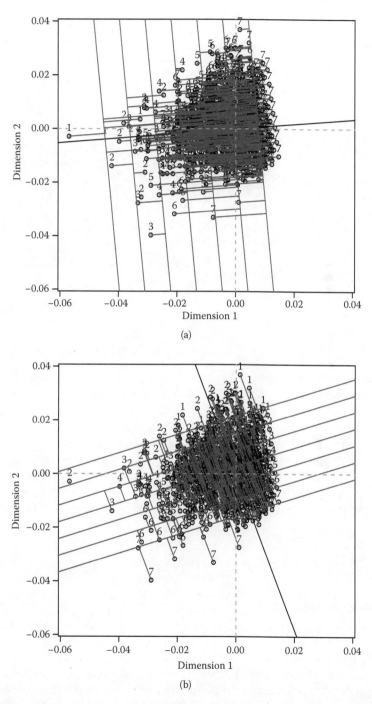

FIGURE 4.4
Projection plots single numerical. (a) Variable 1. (b) Variable 6.

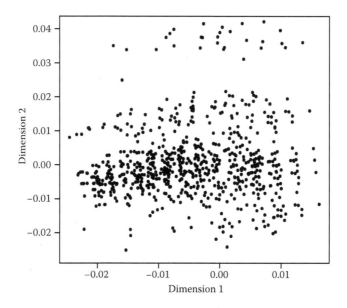

FIGURE 4.5
Object scores single ordinal.

minimum spanning tree defined by each category. Generally these attempts to try alternative loss functions and alternative definitions of homogeneity have not worked out well, seemingly because the sparse indicators do not provide enough constraints to find nontrivial solutions.

This points to other ways to define separation and homogeneity, which have been explored mostly by Guttman, in connection with his facet theory (cf. Lingoes (1968); see also Chapter 7 by Groenen and Borg in this book). In particular, Guttman's MSA-I can be thought of as a form of NLPCA that has a way of measuring separation by using a pseudotopological definition of inner and outer points of the subsets defined by the categories. There is an illustration in Guttman (1985).

4.8 NLPCA Using Aspects

In Mair and de Leeuw (2010) the R package aspect is described. This implements theory from de Leeuw (1988a, 1988b), and gives yet another way to arrive at NLPCA. An aspect is defined as any real-valued function of the matrix of correlation coefficients of the variables. The correlation matrix itself is a function of the quantifications or transformations of the variables.

History of Nonlinear Principal Component Analysis

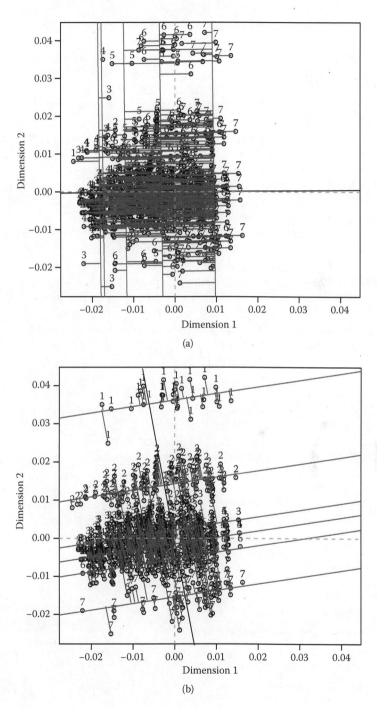

FIGURE 4.6
Projection plots single ordinal. (a) Variable 1. (b) Variable 6.

Many different aspects can be handled. We mention the determinant of the correlation matrix, the squared multiple correlation of one variable with the others, the maximum of the multinormal log-likelihood over a parametric correlation structure, and so on. The `aspect` software maximizes the aspect over transformations, by using majorization methods (de Leeuw, 1994), which are guaranteed to converge if the aspect is a convex function of the correlation matrix.

We briefly indicate why this works. As in other forms of optimal scaling, the correlation matrix $\mathbf{R}(\mathbf{Y}) = \mathbf{Y}^T\mathbf{Y}$ is a function of the standardized transformed or quantified variables. If f is a differentiable convex aspect, we write $\mathbf{D}f(\mathbf{R}(\mathbf{Y}))$ for the $m \times m$ matrix of partial derivatives of f with respect to $\mathbf{R}$, evaluated at $\mathbf{R}(\mathbf{Y})$. Then $f(\mathbf{R}(\mathbf{Y})) \geq f(\mathbf{R}(\tilde{\mathbf{Y}})) + \text{trace}[\mathbf{D}f(\mathbf{R}(\tilde{\mathbf{Y}}))(\mathbf{R}(\mathbf{Y}) - \mathbf{R}(\tilde{\mathbf{Y}}))]$, because a convex function is above its tangents. This means that if $\tilde{\mathbf{Y}}$ is the current solution, and $\hat{\mathbf{Y}}$ maximizes the quadratic $\text{trace}[\mathbf{Y}\mathbf{D}f(\mathbf{R}(\tilde{\mathbf{Y}}))\mathbf{Y}^T]$ over normalized transformations $\mathbf{Y}$, then $f(\mathbf{R}(\hat{\mathbf{Y}})) \geq f(\mathbf{R}(\tilde{\mathbf{Y}}))$, and we have increased the aspect. Iterate these steps and we have a convergent algorithm.

In MCA the aspect is the largest eigenvalue, and $\mathbf{D}f(\mathbf{R}(\mathbf{Y})) = \mathbf{v}\mathbf{v}$, with $\mathbf{v}$ the corresponding normalized eigenvector. Each MCA dimension provides a stationary value of the aspect. In PCA-OS the aspect is the sum of the largest p eigenvalues, and $\mathbf{D}f(\mathbf{R}(\mathbf{Y})) = \mathbf{V}\mathbf{V}^T$, with the columns of $\mathbf{V}$ containing the eigenvectors corresponding to the p largest eigenvalues. We can also easily define regression and canonical analysis in terms of the aspects they optimize.

For our example of the six gratitude variables we applied four different aspects: the largest eigenvalue, the sum of the two largest eigenvalues, the determinant, and the sum of the correlations. We give the eigenvalues of the correlation matrices corresponding to these aspects in Table 4.1.

The differences between the four solutions are obviously very small, which gives us confidence in the optimal scaling of the categories that are computed.

In addition, the R package also has a loss function defined as the sum of the differences between the ½m(m–1) correlation ratios and squared correlation coefficients. Minimizing this loss function quantifies the variables to optimally linearize all bivariate regressions, close to the original objective of Pearson (1906) and Guttman (1959).

TABLE 4.1

Eigenvalues for Different Aspects

	λ_1	λ_2	λ_3	λ_4	λ_5	λ_6
Largest eigenvalue	3.39	0.83	0.60	0.48	0.43	0.27
Largest two eigenvalues	3.19	1.07	0.56	0.48	0.43	0.26
Determinant	3.38	0.84	0.61	0.48	0.43	0.26
Sum of correlations	3.39	0.83	0.60	0.48	0.43	0.27

4.9 Logit and Probit PCA of Binary Data: Gifi Goes Logistic

The idea of using separation as a basis for developing NLPCA has been popular in social science. Let's consider binary data first, using some old ideas of Coombs and Kao (1955). Think of politicians voting on a number of issues. We want to map the politicians as points in low-dimensional space in such a way that, for all issues, those voting in favour can be linearly separated by those voting against. Techniques based on this idea have been developed by political scientists such as Poole and Rosenthal (1985) and Clinton et al. (2004).

A general class of NLPCA techniques for binary data, using logit or probit likelihood functions, in combination with majorization algorithms was initiated by de Leeuw (2006a). The basic idea for defining the loss function is simple. Again, we use the idea of an indicator matrix. Suppose variable j has an $n \times k_j$ indicator matrix $\mathbf{Z}_j$. Let us assume the probability that individual i chooses alternative ℓ for variable j is proportional to $\beta_{j\ell}\exp\{\phi(\mathbf{x}_i, \mathbf{y}_{j\ell})\}$, where ϕ is either the inner product, the negative Euclidean distance, or the negative squared Euclidean distance between vectors $\mathbf{x}_i$ and $\mathbf{y}_{j\ell}$. The $\beta_{j\ell}$ are bias parameters, corresponding to the basic choice probabilities in the Luce or Rasch models. Assuming independent residuals, we can now write down the negative log-likelihood and minimize it over object scores and category quantifications. The negative log-likelihood is

$$\mathcal{L} = \sum_{i=1}^{n}\sum_{j=1}^{m}\sum_{\ell=1}^{k_j} z_{ij\ell} \log\left\{\frac{\beta_{j\ell}\exp\{\phi(\mathbf{x}_i,\mathbf{y}_{j\ell})\}}{\sum_{v=1}^{k_j}\beta_{jv}\exp\{\phi(\mathbf{x}_i,\mathbf{y}_{jv})\}}\right\}.$$

Majorization (see Chapter 7 by Groenen and Borg in this book) allows us to reduce each step to a principal component (if ϕ is the inner product) or multidimensional scaling (if ϕ is negative distance or squared distance) problem.

This formulation allows for all the restrictions on the category quantifications used in the Gifi project, replacing least squares by maximum likelihood and ALS by majorization (de Leeuw, 2005). Thus, we can have multiple and single quantifications, polynomial and monotone constraints, as well as additive constraints for sets of variables. This class of techniques unifies and extends ideas from ideal point discriminant analysis, maximum likelihood correspondence analysis, choice models, item response theory, social network models, mobility tables, and many other data analysis areas.

4.10 Conclusion

NLPCA can be defined in various ways, but we have chosen to stay close to MCA, mostly by using the rank constraints on the category quantifications in the Gifi framework. The transformation approach to NLPCA, which was developed in the nonmetric scaling revolution, generalizes naturally to the aspect approach. The MDS approach to scaling categorical variables, which inspired the Gifi loss function, can be generalized to various geometric definitions of homogeneity and separation, implemented in both the pavings approach and the logit approach.

The logit and probit approach to categorical data analysis is a promising new development. It can incorporate the various constraints of the Gifi framework, but it also allows us to unify many previous approaches to categorical data proposed in statistics and the social sciences. Both the aspect framework and the logit and probit framework show the power of majorization algorithms for minimizing loss functions that cannot be tackled directly by the alternating least squares methods of Gifi.

5

History of Canonical Correspondence Analysis

Cajo J. F. ter Braak

CONTENTS

5.1 Overview of Canonical Correspondence Analysis 62
5.2 History of Correspondence Analysis in Ecology 63
5.3 History of Canonical Correspondence Analysis (CCA) 67
5.4 Triplets and Biplots in CCA 70
5.5 Relation of CCA with Canonical Variate Analysis (CVA) 71
5.6 Later Landmarks 73
Appendix: Canonical Correspondence Analysis via Singular Value Decomposition 74

Canonical correspondence analysis (CCA) was introduced in ecology by ter Braak (1986) as a new multivariate method to relate species communities to known variation in the environment. Since then, four CCA papers (ter Braak, 1986, 1987, 1988c; ter Braak and Verdonschot, 1995) have been cited more than 3,000 times in the Web of Science, approximately 10% of which were in 2010 and 6% (= 180) outside ecology. Independently, CCA was invented by Jean-Dominique Lebreton and Daniel Chessel (Chessel et al., 1987; Lebreton et al., 1988a, 1988b).

This chapter presents the history of CA and CCA in ecology, the definition of CCA in terms of statistical triplets and associated biplots, the relation of CCA to canonical variate analysis (multigroup linear discriminant analysis), and concludes with extensions and ramifications.

5.1 Overview of Canonical Correspondence Analysis

CCA extends correspondence analysis (CA) with predictor variables. If CA is applied to the $n \times m$ matrix $\mathbf{Y}$ ($y_{ij} \geq 0$), CCA treats this matrix as a matrix of multivariate responses and requires a second $n \times p$ matrix $\mathbf{X}$ with predictor variables (columns of $\mathbf{X}$). As in CA, it is convenient notationally to start with the response matrix $\mathbf{P}$ equal to the original matrix $\mathbf{Y}$ divided by its grand total, so that the sum of all elements of $\mathbf{P}$ is 1. The row and column totals of $\mathbf{P}$ are denoted by $\mathbf{r}$ and $\mathbf{c}$, respectively, contained in the diagonal of the diagonal matrices $\mathbf{D}_r$ and $\mathbf{D}_c$. CCA integrates CA and regression analysis. As in regression analysis, response and predictors must be measured at the same set of n sites, and predictors can be continuous or categorical, whereby categorical ones are converted into sets of indicator (1/0) variables. In ecology, $\mathbf{Y}$ typically contains species data with y_{ij} the presence-absence (1/0) or abundance (or another related nonnegative index such as biomass) of species j in site i, and $\mathbf{X}$ contains environmental variables with x_{ik} the measurement of environmental variable k in site i. The common procedure for 'indirect gradient analysis' (Prodon and Lebreton, 1981; ter Braak and Prentice, 1988) consists of (1) applying CA or detrended CA (Hill and Gauch, 1980) to the species data and (2) interpreting the factorial axes in terms of (external) variables, particularly environmental variables. By contrast, CCA integrates these two steps into one, treating the external variables as predictors. Prodon and Lebreton (1994) show that CCA is more efficient than the two-step approach. A major product of CCA is an ordination diagram (factorial plane) that displays the pattern of community variation that can be explained best by the known environment (Figure 5.1).

The shortest summary is perhaps that CCA is (multiple) CA with external linear restrictions on the row points (Gifi, 1990). As such, CCA is the counterpart of redundancy analysis, which is principal components analysis with external linear restrictions on the row points (Rao, 1964; Sabatier et al., 1989; Takane and Hunter, 2001). The solutions of CCA are obtained from the two-sided eigenequation:

$$(\mathbf{X}^T \mathbf{P} \mathbf{D}_c^{-1} \mathbf{P}^T \mathbf{X})\, \mathbf{b} = \lambda\, (\mathbf{X}^T \mathbf{D}_r \mathbf{X})\, \mathbf{b} \qquad (5.1)$$

with λ the eigenvalue. For a particular ordination dimension, the associated eigenvector $\mathbf{b}$ (of length p, the number of predictors) gives the CCA site scores $\mathbf{s} = \mathbf{X}\mathbf{b}/\lambda$, which are thus a linear combination of the predictor variables, and the CCA species scores $\mathbf{t}$, which are at the centroid (barycentre) of the site scores where they occur: $\mathbf{t} = \mathbf{D}_c^{-1} \mathbf{P}^T \mathbf{s}$. A more general term for species is *taxon* (plural: taxa), which gives a nice mnemonic for $\mathbf{s}$ and $\mathbf{t}$, referred to as site and taxon scores from now on. Other scalings of $\mathbf{s}$ and $\mathbf{t}$ are possible, as shown in Table 5.1 and discussed in Section 5.5. Equation (5.1) looks like the eigenequation of canonical correlation analysis for the matrix pair $\mathbf{X}$ and $\mathbf{P}$,

History of Canonical Correspondence Analysis

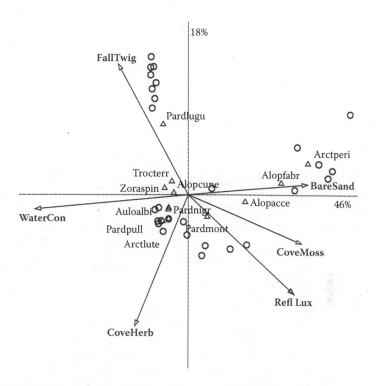

FIGURE 5.1
Ordination diagram of CCA (factorial plane) for 12 taxa of hunting spiders (taxon scores: triangles) in pitfalls (site scores: circles) with respect to five environmental variables (arrows: environmental biplot scores). The sites and taxa form a joint plot of **Y**, or in relative form **P**, as in CA; it is a column-metric preserving biplot of **P**, so that taxa are at the centroid of the site points (Table 5.1). The taxa and environmental variables also form a weighted least-squares biplot of the matrix of weighted averages of taxa with respect to the environmental variables (Table 5.2). In this biplot, the environmental scores are correlations; the site and taxon scores are all multiplied by 0.3 to fit in. The horizontal axis explains 46% of the total inertia, and the vertical axis 18%.

except for $\mathbf{D}_r$ and $\mathbf{D}_c$ not being $\mathbf{I}$ and $\mathbf{P}^T\mathbf{P}$, respectively. What is the virtue of this equation and the associated (generalized) singular value decomposition (see the appendix at the end of this chapter), and how was it discovered?

5.2 History of Correspondence Analysis in Ecology

The first application of CA to ecological data was presumably by Roux and Roux (1967) and the second by Hatheway (1971), but CA became popular by its introduction by Hill (1973) under the name of reciprocal averaging. Hill (1973) acknowledges John Gower for pointing out that the method was

TABLE 5.1

Transition Formulae of Canonical Correspondence Analysis (CCA) of a Response Matrix **Y**, in Relative Form **P**, with Respect to a Predictor Matrix **X** with Biplot Scaling Constant α ($0 \leq \alpha \leq 1$)

CCA Term	Formula in Words (with Ecological Term)	Formula
Constrained site (row) scores	Row points (sites) obtained as a linear combination of **X**	$\lambda^{1-\alpha}\mathbf{s} = \mathbf{Xb}$ ($\mathbf{s}^T\mathbf{D}_r\mathbf{s} = \lambda^{\alpha}$)
Taxon (column) scores	Weighted averages of constrained site scores	$\lambda^{\alpha}\mathbf{t} = \mathbf{D}_c^{-1}\mathbf{P}^T\mathbf{s}$ ($\mathbf{t}^T\mathbf{D}_c\mathbf{t} = \lambda^{1-\alpha}$)
Site (row) scores	Weighted averages of the taxon scores	$\mathbf{s}^* = \mathbf{D}_r^{-1}\mathbf{Pt}$
Canonical weights	Coefficients of regression of site scores $\mathbf{s}^*$ on **X** with site weights **r**	$\mathbf{b} = (\mathbf{X}^T\mathbf{D}_r\mathbf{X})^{-1}\mathbf{X}^T\mathbf{D}_r\mathbf{s}^*$

Note: The matrices $\mathbf{D}_r$ and $\mathbf{D}_c$ are diagonal and contain the row and column relative totals of **P**, respectively, called *masses* in CA. For interpretation purposes and the biplot, the columns of **X** are often standardized, so that diag($\mathbf{X}^T\mathbf{D}r\mathbf{X}$) = **I**; this does not influence λ, **s**, and **t**. For $\alpha = 0$, the resulting biplot is column-metric preserving; for $\alpha = 1$ it is row-metric preserving.

indeed an *analyse factorielle des correspondances* and refers to Benzécri (1969) and Escofier-Cordier (1969), with more references to older work in Hill (1974). Tenenhaus and Young (1985) point to Richardson and Kuder (1933) as the inventors of the reciprocal averaging approach (in psychology). The averages for rows (individuals) in Richardson and Kuder (1933) had the same denominator (the number of categories). Hill (1973) was arguably more general, as both his row and column averages had unequal denominators.

The popularity of CA in ecology is, in my view, due to the unimodal response that CA can discover in data. The possibility of analysing unimodal relationships with CA was first noticed by Mosteller (1948, in Torgerson, 1958, p. 338). Unimodal response is common in ecology and derives from two well-known laws. The first law, Liebig's law, states that each taxon requires a minimum amount of resource (for example, nitrogen)—fertilizer usage in agriculture builds on this law. The second law, Shelford's (1919) law of tolerance (Allaby, 1998), states that, in addition, each taxon tolerates no more than a certain maximum. These two laws yield the niche of a taxon, that is, the region in resource space where the taxon can actually grow and reproduce. Niches vary among taxa because the required minima and tolerated maxima differ among taxa. Taxa also tend to prosper best around the centre of the ecological niche, yielding unimodal (single-peaked) response (Figure 5.2). Unimodal response is also an important feature in psychology with the ideal point model (Coombs and Avrunin, 1977) and unfolding method (Coombs and Kao, 1960; Heiser, 1981, 1987). An example in personal preference is 'I like coffee, but only at the right temperature, not too cold and not too hot', and persons differ in their ideal temperature.

Ecologists (Gause, 1930; Ellenberg, 1948; Whittaker, 1956; Zelinka and Marvan, 1961; Ellenberg et al., 1991) developed the method of weighted

History of Canonical Correspondence Analysis

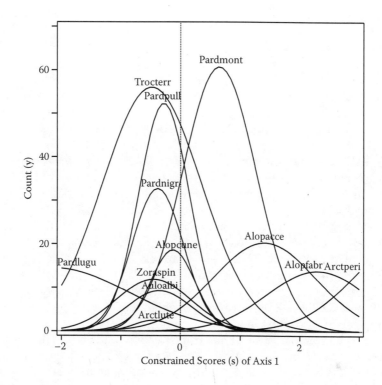

FIGURE 5.2
The niche model: Unimodal response curves of taxa of hunting spiders with respect to the first constrained axis (s_1) of the CCA of Figure 5.1. The scores run up to about 3 and are plotted in Figure 5.1 after multiplication by 0.3.

averaging in this niche context. The preference or indicator value a_{jk} of a taxon j with respect to a physical gradient k, for example, acidity (pH), can be estimated by the average of the pH values of the sites where the taxon occurs. If abundance is taken into account, a weighted average is taken with weight proportional to abundance. In formula,

$$a_{jk} = \sum_i y_{ij} x_{ik} / \sum_i y_{ij} = \sum_i (p_{ij} / c_j) x_{ik} \tag{5.2}$$

with x_{ik} the measured pH at the site i. The values p_{ij}/c_j, $i = 1, \ldots, n$, constitute the jth taxon's *profile* across the sites. Conversely, when indicator values a_{jk} are known, but the acidity of a site is not, an estimate $\hat{x}_{ik}$ of the acidity at the site can be estimated on the basis of the taxa that are recorded there. The estimate is simply the average of the indicator values of the taxa recorded there. If abundance is taken into account, a weighted average is taken with weight proportional to abundance. In formula,

$$\hat{x}_{ik} = \sum_j y_{ij} a_{jk} / \sum_j y_{ij} = \sum_j (p_{ij}/r_i) a_{jk} \qquad (5.3)$$

where the values p_{ij}/r_i, $j = 1, \ldots, m$, constitute the ith site's profile across the taxa. For presence-absence data the weighted average formulae hold true with $y_{ik} = 0$ or 1. Note that absences ($y_{ik} = 0$) do not count in weighted averaging. This is a distinctive and useful feature of the method, as a taxon may not occur at a site just because of another unfavourable factor.

Hill (1973) extended the method of weighted averaging to CA by proposing to iterate between the weighted averaging equations (5.2) and (5.3) until convergence (for a fixed range of x, for example). By this process, the meaning of the measured variable, in our example pH, is lost and the variable obtained after convergence is a latent one; i.e., it is the best *hypothetical* environmental variable. It is well known to be the first nontrivial axis of CA. The well-known transition formulae of CA are essentially Equations (5.2) and (5.3) with the first nontrivial eigenvalue λ ($0 < \lambda \leq 1$). Weighted averaging is, of course, nothing more than the centroid principle (*principe barycentrique* in French), and CA takes it both ways, although not exactly if $\lambda < 1$.

In an ecological context (and many more; see de Leeuw's Chapter 4) the second axis has often an almost quadratic relation to the first. This is known as the *arch effect* (Gauch, 1982) or *Guttman effect* (*effet Guttman* in French— notice that it is not a 'horseshoe' as the ends do not bend in). An easy, non-mathematical explanation of this effect is presented in Jongman et al. (1987, 1995, p. 105). Hill and Gauch (1980) proposed to remove this relationship by detrending within the iteration algorithm, yielding *detrended correspondence analysis*. Their efficient computer program DECORANA (Hill, 1979a) implementing this has allowed major usage of both CA and detrended CA in ecology. Detrended CA is currently available in the CANOCO software package (ter Braak and Šmilauer, 2012) and as function decorana in the R package vegan (Oksanen et al., 2011). The arch effect, which is a mathematical artefact of CA, is in my view the main reason why Guttman never wanted to go beyond the first CA axis. Detrending is somewhat controversial (Wartenberg et al., 1987), but is not the only way to avoid the arch effect. External linear constraints as in CCA are another (Palmer, 1993).

But, what does CA optimize to make it suited for ecological applications and the unimodal response model in particular? CA has many nice optimality properties when applied to contingency tables (Greenacre, 1984), but is the taxon data matrix **Y** a contingency table? Often it is a presence/absence matrix with just zeros and ones (but not a (super) indicator matrix) or, if **Y** contains counts, these counts arise from sampling a number of individuals from each of a fixed number of sites. **Y** is thus not a contingency table, unless sites are treated as a (random) nominal variable with n categories. What then is the ecological interpretation of the fact that CA finds the row and column quantification that maximizes the correlation between rows and columns (Hill, 1974; Tenenhaus and Young, 1985)? What else could motivate CA for use in ecology, rather than being just principal components analysis (PCA) with some fancy pre- and posttransformations? First, CA is useful for seriation of presence/absence data, i.e., incidence matrices (Torgerson, 1958): 'If the rows and columns of an incidence matrix can be permuted so that all the ones in every row and column come together (a Petrie matrix), then there is (apart from degeneracies) a unique correct ordering of rows and columns which is generated by the first non-trivial axis of CA of the matrix' (freely quoted after Hill, 1974). Heiser (1981) extended this result to so-called complete matrices with consecutive ones in the rows (or columns) only. Second, CA is able to find disjoint blocks in two-way tables (Hill, 1974), and this feature is used in the cluster analysis program TWINSPAN (Hill, 1979b; Jongman et al., 1995). The features of CA that make it less suited for ecological applications (Jongman et al., 1995) are its sensitivity to rare taxa and outlier sites (a feature related to block structure detection) and the arch effect (see above). The ecological meaning of the analysis of variance approach to CA with its correlation ratio escaped me until ter Braak (1987). For the relation of CA to the Gaussian response model see the next section.

5.3 History of Canonical Correspondence Analysis (CCA)

In this section I present my route to CCA followed by notes on what I know of Jean-Dominique Lebreton and Daniel Chessel's route. The background of my invention of CCA is as follows. From 1978 I worked for an ecology group, while being based in a statistics group, where regression analysis and analysis of variance (ANOVA) were the workhorses of statistics. I studied the two Benzécri volumes (Benzécri et al., 1973) in some detail. The Satellite Program in Statistical Ecology (Patil, 1995) schooled me in statistical ecology. Kooijman (1977) and Kooijman and Hengeveld (1979) introduced me to the Gaussian model for ordination and pointed out the numerical problems with the maximum likelihood approach when applied in several dimensions (see also Goodall and Johnson, 1982). While on leave for doing an MSc in statistics at Newcastle upon Tyne (UK), I met Colin Prentice and Mark Hill, without whom I would never have invented CCA. Colin Prentice introduced me to ordination methods current at the time (Prentice, 1977, 1980a, 1980b). Mark Hill kindly pointed out that he disliked the draft of my paper on PCA biplots and diversity (ter Braak, 1983), for two reasons: (1) ecology is not linear and (2) differences in niches' locations create diversity, as in DECORANA (Hill, 1979a). There I also learned from Professor R. L. Placket about the maximum likelihood equations in the exponential family; these are important in my derivation of CCA. Back from the UK, I attended the second Gifi course (Gifi, 1981, 1990) and was asked for statistical advice on the usage of species indicator values in nature management and acidification research (van Dam et al., 1981; ter Braak and Gremmen, 1987; ter Braak and van Dam, 1989).

This led me to the study of the properties of the method of weighted averaging, outlined above. I put weighted averaging in the context of response curve modelling and asked myself: For which shape of species response curves and under which conditions would weighted averaging be close to maximum likelihood (ter Braak and Barendregt, 1986; ter Braak and Looman, 1986)? These papers show the link of weighted averaging with the ecological niche model and the Gaussian response model in particular. The paper on CA of incidence and abundance models in terms of a unimodal response model (Gaussian ordination) followed naturally (ter Braak, 1985), despite being published earlier.

Independently, Ihm and van Groenewoud (1984) compared correspondence analysis and a variant of Gaussian ordination that is attractive for compositional data, that is, when samples at sites vary in size and only relative abundance is meaningful. They show that their model is identical to Goodman's row–column (RC) model (Goodman, 1979) and, citing Escoufier (1982), that a first-order Taylor expansion yields the reconstitution formulae of CA. This result, discovered earlier by Goodman (1981), applies to small λ (close to 0; that is, for data close to row-column independence), whereas the result of ter Braak (1985) holds true for large λ (close to 1).

With this background, how was CCA derived? In October 1984 Colin Prentice and I made an outline in Uppsala, Sweden, for 'A theory of gradient analysis', which was eventually published (ter Braak and Prentice, 1988). We searched for something like canonical correlation analysis for niche models. The night before leaving Uppsala I got the idea to linearly constrain the scores in Gaussian ordination and approximate the maximum likelihood equations, just as I did for CA (ter Braak, 1985). This derivation, given in the appendix of ter Braak (1986), yielded the precise row and column weights that are particular for CCA (and CA), the transition formulae (Table 5.1), and from these, the eigenequation (5.1). This explains the adjective *canonical* in the name of the method. A better, later motivation of the term is that CCA transforms a quadratic into its canonical form (ter Braak, 1988c). The close relationship to canonical variate analysis provides additional motivation.

In an attempt to find a derivation without explicit use of the Gaussian model, I derived CCA as a linear combination of predictors $\mathbf{X}$ that best separates taxon niches (ter Braak, 1987). The maximized criterion is the dispersion of the taxon scores (weighted averaged site scores) with respect to standardized site scores, which is precisely the correlation ratio η^2 in the analysis of variance approach to CA (Nishisato, 1980; Tenenhaus and Young, 1985). The difference is that the criterion is now maximized subject to linearly constrained sites scores ($\mathbf{s} = \mathbf{Xb}$), which is achieved by optimizing with respect to $\mathbf{b}$.

I presented partial CCA (a CCA with the effects of nuisance variables partialled out) in 1987 at the first International Federation of Classification Societies

(IFCS) conference in Aachen, Germany, after which Yves Escoufier kindly invited me to Montpellier, France, to get to know statistical triplets, duality diagrams, and his French colleagues who independently invented CCA. Early work related to partial CCA (ter Braak, 1988c) is Yoccoz and Chessel (1988), Cazes et al. (1988), Sabatier et al. (1989), and Lebreton et al. (1991).

Jean-Dominique Lebreton worked at the Université Lyon I (Laboratoire de Biométrie), where he met Daniel Chessel and his PhD student Nigel Yoccoz, and then moved to CEPE/CNRS at Montpellier, where he developed further work with Yves Escoufier and Robert Sabatier. Lebreton was interested not only in multivariate data analysis, but also in survival analysis of animals by capture-mark-recapture sampling and had been exposed to generalized linear models from the onset of these models in the 1970s. This sampling method usually yields estimates of capture and survival probabilities with extremely wide confidence intervals. In that context he invented the trick to borrow strength across years and sampling occasions by constraining the probabilities of the model by logistic functions with environmental covariates (Clobert and Lebreton, 1985; Lebreton et al., 1992). This gave Lebreton the idea of applying the same trick in CA (December 1984). Daniel Chessel contributed the general perspective brought by the duality diagram approach to multivariate analysis that was commonly used in France at that time (Cailliez and Pagès, 1976). They then developed CCA under the point of view of linearly constrained CA in several papers (Chessel et al., 1987; Lebreton et al., 1988a, 1988b; Sabatier et al., 1989). In the process they discovered a preprint to ter Braak (1986). The connection between the unimodal model and CA was well known to Chessel and Lebreton (see Chessel et al., 1982). Chessel was well aware of orthogonal and oblique projectors (Afriat, 1957) and Rao's (1964) 'principal components of instrumental variables' (Nigel Yoccoz, personal communication), later reinvented as redundancy analysis (van den Wollenberg, 1977). Neither paper was known to me when I invented CCA. For biplots in redundancy analysis see ter Braak and Looman (1994).

CCA was first implemented in the computer program CANOCO (ter Braak, 1988a,b), which is now in version 5 (ter Braak and Šmilauer, 2012). CANOCO version 1 was an extension of DECORANA (Hill, 1979a) that was already quite popular in ecology because of the large data sets it could handle. DECORANA conveniently used sparse matrix algebra and an efficient eigenvector routine (ter Braak and de Jong, 1998). CCA has also been included in the R packages vegan (Oksanen et al., 2011), ade4 (Dray and Dufour, 2007), and anacor (de Leeuw and Mair, 2009). CCA in vegan allows a powerful formula interface: for example, the statement cca(Y ~ A*B + Condition(C)) specifies a partial CCA with covariables C and with predictors A*B, that is, the main effects of factors (or variables) A and B and their interaction.

5.4 Triplets and Biplots in CCA

This section points to the connections between statistical triplets (Cailliez and Pagès, 1976; Tenenhaus and Young, 1985; Escoufier, 1987), reduced rank least-squares approximation (Greenacre, 1984), and the biplot (Gabriel, 1971) and uses these to provide triplets and biplots for CCA. Triplets belong to the French multivariate data analysis tradition, and reduced rank approximations, associated singular value decompositions, and biplots to the English-speaking one.

A statistical triplet (Y, C, R) of an $n \times m$ data matrix Y and positive semidefinite metrics (weight matrices) C ($m \times m$) and R ($n \times n$) corresponds to a rank r approximation of $Y \approx FG^T$ with F ($n \times r$) and G ($m \times r$), where $r < \min(n, m)$, by minimizing the Frobenius norm (Greenacre, 1984):

$$||R^{\frac{1}{2}}(Y - FG^T) C^{\frac{1}{2}}||^2 = \text{trace}[(Y - FG^T)^T R(Y - FG^T) C] \quad (5.4)$$

where $||\ldots||^2$ denotes the sum of squares of the elements of the matrix argument. The solution matrices can be obtained from the generalized singular value problem (Greenacre, 1984). This result is related to the famous Eckart–Young theorem (see Gower's Chapter 2 in this book). The biplot graphic display is nothing more than a vector representation of the rows of both F and G with the mutual inner products providing the rank r least-squares approximation to Y (Gabriel, 1971). The key statistical triplets of CCA are given in Table 5.2. The eigenequation (5.1) follows directly from the third triplet; the associated singular value decomposition is given in the appendix at the end of the chapter. A duality diagram (Escoufier, 1987) of CCA with the corresponding transition formulae is given in Figure 5.3—see Dray's Chapter 18 in this book and also Chessel et al. (1987) for related duality diagrams.

TABLE 5.2

Key Statistical Triplets and Corresponding Biplots of CCA

Data Matrix Y	Triplet	Biplot Scores Row	Biplot Scores Column
Fitted contingency ratios	$(\Pi_X D_r^{-1} P D_c^{-1}, D_c, D_r)$	s (rows)	t (columns)
Table of weighted averages of taxa with respect to X	$(D_c^{-1} P^T X, (X^T D_r X)^{-1}, D_c)$	t (columns)	c (variables)
Regression coefficients of contingency ratios with respect to X	$((X^T D_r X)^{-1} X^T D_r (D_r^{-1} P D_c^{-1}), D_c, (X^T D_r X))$	b (variables)	t (columns)

Note: The matrices P, D_r, and D_c are defined previously (see, for example, Table 5.1) and $\Pi_X = X(X^T D_r X)^{-1} X^T D_r$, the r-weighted projection operator on X. Between parentheses is the entity in terms of the original data P (rows, columns) and X (rows, variables).

History of Canonical Correspondence Analysis

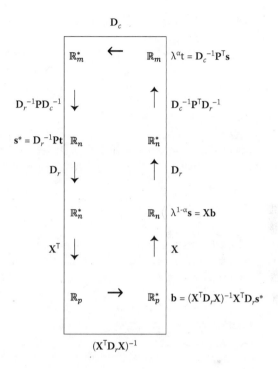

FIGURE 5.3
A duality diagram of CCA of response matrix **Y**, in relative form **P**, and predictor matrix **X** with the corresponding transition formulae of Table 5.1. Note that **s** is the projection of **s*** on **X**, in short, $\mathbf{s} = \Pi_X \mathbf{s}^*$ (see the note to Table 5.2).

5.5 Relation of CCA with Canonical Variate Analysis (CVA)

Chessel et al. (1987) and Lebreton et al. (1988a) were the first to recognize the formal equivalence between CCA and linear discriminant analysis (CVA) with m groups (taxa) on reformatted (inflated) data. With the knowledge that CCA is a form of CVA, CCA has an early precursor in the ecological literature in the form of Green's (1971, 1974) multigroup linear discriminant analysis, as noted by ter Braak and Verdonschot (1995). Interest in Green's method was lost, ironically in the same period as the popularity of CA surged. The main difference between the two methods is that the unit of statistical analysis in CVA is the individual (a row in the inflated data matrices), whereas it is the site (row of **Y**) in CCA. Statistical tests designed for CVA as used by Green (1971, 1974) are thus invalid in the context of CCA (ter Braak and Verdonschot, 1995). Valid tests can be based on Monte Carlo permutation of sites (instead of individuals), as implemented in CANOCO. This part of the history is similar to that of CA, as CA is also a particular form of CVA and canonical correlation analysis on data inflated to indicator matrices.

The analysis of variance approach to CA is essentially one of discriminant analysis/CVA. CVA extracts orthogonal components of the form $\mathbf{s} = \mathbf{Xb}$ such that the between- (B) to within- (W) group sum of squares is maximized. As the total sum of squares is the sum of the two ($T = B + W$), maximizing B/W is identical to maximizing B/T, which is the correlation ratio (Nishisato, 1980), and identical to minimizing W/T. The W/T criterion puts CCA (and CA) in the context of unfolding models (Heiser, 1981, 1987). With a rescaling so that $T = 1$, W can be rewritten, for a one-dimensional model, as

$$W = \sum_i \sum_j p_{ij}(s_i - t_j)^2 \tag{5.5}$$

because for any fixed choice of $\mathbf{s}$, Equation (5.5) is minimized for $\mathbf{t} = \mathbf{D}_c^{-1}\mathbf{P}^T\mathbf{s}$, so that the right-hand side of Equation (5.5) becomes the usual within-group sum of squares (Heiser, 1981). In CA, W is minimized subject to the constraints $\mathbf{1}^T\mathbf{D}_r\mathbf{s} = \mathbf{0}^T$ and $\mathbf{s}^T\mathbf{D}_r\mathbf{s} = 1$. CCA imposes the additional constraint $\mathbf{s} = \mathbf{Xb}$. Note that the data enter (5.5) as weights. A second component can be extracted by requiring it to be orthogonal to the first; that is, if $\mathbf{s}_1$ is determined first, then the next $\mathbf{s}$ satisfies $\mathbf{s}^T\mathbf{D}_r\mathbf{s}_1 = 0$. Heiser (1987) noted that the orthogonality constraint is 'not free from arbitrariness *under the present rationale* of the method' (CA)—hence the arch effect artefact and the occasional need for detrending in CA. In CVA, and thus CCA, the orthogonality is not so arbitrary, for example, if we take as a starting point the least-squares approximation of the group averages $\mathbf{A}$ (second triplet in Table 5.2, detailed below). Zhu et al. (2005) showed the equivalence between CCA and CVA via a weighted sample model.

The usual CCA ordination diagram (Figure 5.1) is similar to that of CVA. The scores for the variables in $\mathbf{X}$ are usually not based on the eigenvector $\mathbf{b}$ (a vector with regression coefficients), as these are very unstable if predictors show high mutual correlation. Instead, scores for predictor variables are based on the weighted least-squares approximation of the weighted averages (the $\{a_{jk}\}$ of Equation (5.2)) of the m taxa with respect to the p predictor variables, represented by the $m \times p$ matrix $\mathbf{A} = \mathbf{D}_c^{-1}\mathbf{P}^T\mathbf{X}$. The matrix $\mathbf{A}$ corresponds to the group means when CCA is viewed as a CVA on inflated data. The corresponding triplet for CCA in Table 5.2 is $(\mathbf{A}, (\mathbf{X}^T\mathbf{D}_r\mathbf{X})^{-1}, \mathbf{D}_c)$, whereas that of CVA would use $(\mathbf{A}, \mathbf{W}^{-1}, \mathbf{D}_c)$ with $\mathbf{W}$ the pooled within-group covariance matrix. This difference corresponds to the fact that CCA maximizes B/T, whereas CVA maximizes B/W. If the eigenvalue of CCA is λ, the corresponding eigenvalue of CVA is therefore $\lambda/(1 - \lambda)$, a relationship absent in Chessel et al. (1987). This difference does not influence the biplot of $\mathbf{A}$ in either method, as the biplot does not change when we rescale the pair ($\mathbf{t}, \mathbf{c}$) in Table 5.2 to ($r\mathbf{t}, \mathbf{c}/r$) for any scalar r. However, it does influence the interpretation of, for example, the distances between the taxa (see Gower's Chapter 2 in this book). In a default CCA diagram in the column-preserving metric, the intertaxon distances approximate chi-square distances (Meulman, 1986), whereas in CVA, the intergroup distances then approximate Mahalanobis distances. The CCA diagram can be transformed into the CVA form by using the so-called Hill's scaling, which equalizes the within-taxon variance W across different components. In this scaling the taxon scores are scaled to $t^T D_c t = \lambda/(1-\lambda)$, with λ the eigenvalue of the corresponding component (Hill, 1979a; ter Braak and Verdonschot, 1995). The Mahalanobis distance is then a natural measure of distance between niches of taxa (Green, 1971, 1974).

What is then correspondence discriminant analysis (Perriere et al., 1996)? It is simply CCA with a nominal predictor variable and can be computed with any of the available implementations of CCA.

5.6 Later Landmarks

With many predictors CCA no longer constrains the CA and is then identical to CA. To avoid this, Dolédec and Chessel (1994) developed co-inertia analysis, which ignores the correlations among predictors. As an alternative, ter Braak and Verdonschot (1995) developed a partial least-squares (PLS) variant of CCA. The regression setting of CCA allows easy variance (inertia) partitioning (Borcard et al., 1992; Okland and Eilertsen, 1994). Doubly linearly constrained CA, or double CCA for short, was developed by Böckenholt and Böckenholt (1990), whereas the co-inertia version is by Dolédec et al. (1996). These variants are endpoints (usage or no usage of the within-set covariance matrix) in the framework of Tenenhaus and Tenenhaus (2011). Lavorel et al. (1999) noted that double CCA is essentially a weighted canonical correlation analysis. Later assessments of CCA are Palmer (1993), Johnson and Altman (1999), Graffelman (2000, 2001), Graffelman and Tuft (2004), and Zhang and Thas (2012). A modern view on CCA is Zhu et al. (2005).

A related method is co-correspondence analysis, a method designed to relate two taxon communities (ter Braak and Schaffers, 2004; Schaffers et al., 2008). This method can also be used to relate two sets of (many) nominal variables in small samples. It avoids the multicollinearity problems in such data and is, as such, an alternative to two-set OVERALS (Gifi, 1990), available in SPSS, or of CANALS (van der Burg and de Leeuw, 1983, 1990), which provide (different) optimal scaling approaches to canonical correlations analysis of two sets of nominal variables. It appears that all these CA-related methods can be obtained from standard (unweighted) linear methods by inflating the data matrices to (super)indicator matrices as in the dual scaling approach (Nishisato, 1980; Gifi, 1990; see de Leeuw's Chapter 4 in this book). After inflation, the unit of statistical analysis is the individual instead of being the site. Alternatively, they can be obtained from unweighted linear methods by pre- and posttransformation (ter Braak and Verdonschot, 1995), as in the principal components approach to CA (Tenenhaus and Young, 1985; Chessel et al., 1987).

In this chapter I emphasized the unimodal properties of CA and CCA, but the linear ones are explicit in the first triplet of Table 5.2 that forms the basis of the reconstitution formulae of CA (Greenacre, 1984; van der Heijden et al., 1994). So, CA and CCA are chameleons: sometimes they show up as linear methods, as in the reconstitution formulae and associated biplot of the contingency ratios, and sometimes as unimodal methods, as in their relationship to CVA and unfolding (Equation (5.5)). How is that possible? How can this be explained? In the RC model this duality is fully mathematically understood (Ihm and van Groenewoud, 1984; ter Braak, 1988c; de Rooij, 2007), but its understanding in (C)CA is still limited. When there is a strong arch effect (not very likely in CCA with few predictors), the rank 1 reconstitution appears bad. Perhaps power transformation (Greenacre, 2009) can shed light (and remove the arch)? The relation between CA, CCA, and the (constrained) RC model, ideal point discriminant analysis (Takane et al., 1987), and Anderson's stereotype model (Anderson, 1984) has been further developed in de Rooij and Heiser (2005) and van der Heijden et al. (1994).

Appendix: Canonical Correspondence Analysis via Singular Value Decomposition

The two-sided eigenequation (5.1) of canonical correspondence analysis can equally well be obtained via a singular value decomposition (SVD). On using the notation of the main text, define

$$\mathbf{Y}^* \equiv (\mathbf{X}^T \mathbf{D}_r \mathbf{X})^{-1/2} \mathbf{X}^T \mathbf{P} \mathbf{D}_c^{-1/2} \qquad (5.6)$$

and obtain the SVD of $\mathbf{Y}^*$,

$$\mathbf{Y}^* = \mathbf{U} \mathbf{\Sigma} \mathbf{V}^T \qquad (5.7)$$

with $\mathbf{U}$ and $\mathbf{V}$ orthonormal matrices and $\mathbf{\Sigma}$ a diagonal matrix with singular values that are the square root of the CCA eigenvalues ($\mathbf{\Sigma} = \mathbf{\Lambda}^{1/2}$). The canonical weights (**b**) and the taxon scores (**t**) for the various dimensions are then in the columns of the matrices

$$\mathbf{B} = (\mathbf{X}^T \mathbf{D}_r \mathbf{X})^{-1/2} \mathbf{U} \mathbf{\Sigma}^\alpha \qquad (5.8)$$

and

$$T = D_c^{-1/2} V \Sigma^{1-\alpha} \quad (5.9)$$

respectively, where α is a biplot scaling constant ($0 \le \alpha \le 1$); see Table 5.1. The third triplet in Table 5.2 and associated biplot follows directly from these equations.

If X contains a column with ones for the intercept, as usual in multiple regression, the first singular value is 1, and the first column of B is $b_1 = (1, 0, 0, \ldots, 0)^T$ and that of T is $(1, 1, 1, \ldots, 1)^T$, similar to the trivial first solution in CA. It can be avoided by first D_r centring the columns of X such that $X^T D_r 1_n = 0$, as can be seen as follows. As U and V are orthonormal, $B^T(X^T D_r X)B = (XB)^T D_r(XB) = \Sigma^{2\alpha}$ and $T^T D_c T = \Sigma^{2(1-\alpha)}$, showing that the site scores $S = XB$ and taxon scores T are D_r- and D_c- orthogonal, respectively. With the first site score vector $s_1 = Xb_1 = 1$, where 1 is a vector of n 1s. We thus have $s^T D_r 1 = b^T X^T D_r 1 = 0$ for any higher-numbered canonical weight vector b and its associated site score vector s. Similarly, for any higher-numbered taxon score vector t, $t^T D_c 1 = 0$, where 1 here denotes a vector of m 1s (the order of 1 is clear from its context). These null equations hold trivially true when we apply D_r centring to X, and thus entail no additional constraint to the eigenvalue problem. If $X = I$, the SVD (5.6) of CCA reduces to that of CA and centring of the identity matrix amounts to the usual centring of the matrix P in CA. The trivial solution and the way to avoid it come as no surprise when you understand CA.

Equations (5.6)–(5.9) reduce to those of principal components analysis with respect to instrumental variables (Rao, 1964), alias redundancy analysis, by redefining $D_r = I$ and $D_c = I$ with corresponding unweighted column centring of X. By contrast, in CCA D_r and D_c contain the row and column totals of P, without which there is no weighted averaging (Equations (5.2) and (5.3)) and no link to the unimodal model (Figure 5.2). This shows that CCA is more than just a minor generalization of redundancy analysis. Section 5.5 serves to show that CCA is much more closely related to multigroup linear discriminant analysis than to redundancy analysis.

6

History of Multiway Component Analysis and Three-Way Correspondence Analysis

Pieter M. Kroonenberg

CONTENTS

6.1 Multiway Continuous Data 78
    6.1.1 Overview of Four Basic Models 78
        6.1.1.1 Tucker3 Model 79
        6.1.1.2 Tucker2 Model 79
        6.1.1.3 PARAFAC Model 80
        6.1.1.4 INDSCAL Model 81
    6.1.2 Relationships between the Models 81
    6.1.3 Dawn of Multiway Analysis 82
    6.1.4 Motivations 82
    6.1.5 Early Contributors 85
    6.1.6 Computational Methods 86
    6.1.7 Software 87
    6.1.8 Applications 87
    6.1.9 Current Status 88
6.2 Multiway Categorical Data 89
    6.2.1 Three-Way Correspondence Analysis 90
6.3 Conclusion 93

This historical overview will deal with the development of the field of three-mode/three-way/multiway component-based models in a conceptual rather than a technical sense. It covers primarily the period until the early years of the 21st century. A special section is devoted to the history of three-way correspondence analysis.

    The history of techniques and models for three-mode data analysis started in 1963 with the work of Tucker, who first described three-mode extensions of principal component analysis and factor analysis (Tucker, 1963, 1964, 1966, 1972). A second founding father was Harshman, who proposed the parallel factors model (PARAFAC) for extending principal component analysis from two to three ways (Harshman, 1970, 1972a, 1972b, 1984; Harshman and Lundy, 1984a, 1984b, 1994). Carroll is the third pioneer, and his contributions are especially in the fields of individual differences, multidimensional scaling, and three-way clustering. He proposed a host of models and algorithms for three-way analysis, such as the INDSCAL model for individual differences scaling and the three-way canonical decomposition model (CANDECOMP), which is identical to Harshman's model (Carroll and Chang, 1970, 1971). The fourth major early contributor was Kruskal, whose work dealt especially with mathematical problems, in particular the uniqueness properties of multiway models and the rank of multiway arrays (Kruskal, 1977).

Obviously many other individuals have made important contributions to the field, but these four psychometricians set three-way methods on the map. They have had a deep influence on many fields of science, on a par with the influence of the originators and propagators of factor analysis and principal component analysis, techniques that also have solid roots in the psychometric tradition. Surprisingly, the areas where they were most influential were not the social and behavioural sciences, but analytical chemistry and signal processing.

In part, the impact of these pioneers can be gauged from the number of citations their work has received over the years. The outcomes of Harzing's program 'Publish or Perish' (dd. July 2, 2012) are used to illustrate this—for details of this program see her website, http://www.harzing.com/pop.htm. Tucker received 1,438 citations for his three seminal papers on three-mode factor analysis (1963, 1964, 1966), of which the 1966 paper has been cited at a rate of 21 times per year. Harshman's unpublished 1970 paper on the PARAFAC model was cited 1,082 at a rate of 25 times per year. Carroll and Chang's (1970) seminal paper on INDSCAL received 2,183 citations, amounting to 51 citations per year. Finally, Kruskal's (1977) paper on three-way rank and uniqueness was cited 550 times at a rate of 15 times per year.

6.1 Multiway Continuous Data

6.1.1 Overview of Four Basic Models

Before discussing the emergence of multiway analysis, and three-way analysis in particular, it is useful to provide a brief summary of the four major three-way component-based models, the Tucker3, Tucker2, PARAFAC or CANDECOMP/PARAFAC (CP), and INDSCAL models. Several three-way cluster and other types of models have also been proposed, but these are not part of this overview.

The starting point for the component models is the $(I \times J \times K)$ three-way data array $\mathbf{X} = \{x_{ijk}\}$, which may also be conceived as a collection of frontal slices $\mathbf{X}_k = \{x_{ij}\}^k$, with $k = 1, \ldots, K$.

6.1.1.1 Tucker3 Model

The Tucker3 model is the factorization of such a three-way array in which all three ways are represented by their components, i.e.,

$$x_{ijk} = \sum_{p=1}^{P} \sum_{q=1}^{Q} \sum_{r=1}^{R} g_{pqr} a_{ip} b_{jq} c_{kr} + e_{ijk}$$

in which the coefficients a_{ip}, b_{jq}, and c_{kr} are the elements of the component matrices $\mathbf{A}$, $\mathbf{B}$, and $\mathbf{C}$ respectively. The g_{pqr} are the elements of the three-way core array $\mathbf{G}$ linking the components of the three ways, and the e_{ijk} are the errors of approximation collected in the three-way array $\mathbf{E}$. For identification purposes the component matrices are generally taken to be orthonormal, but as in standard principal component analysis (PCA) they may be nonsingularly transformed afterwards without loss of fit provided the core array is counterrotated.

In a typical social science application the data array consists of subject scores (first mode) on variables (second mode) and conditions (third mode). In this case $\mathbf{A}$ is the $(I \times P)$ matrix with the coefficients for the subject components, $\mathbf{B}$ is the $(J \times Q)$ coefficient matrix for the variables, and $\mathbf{C}$ is the $(K \times R)$ coefficient matrix for the conditions. In the original data array $\mathbf{X}$ every element x_{ijk} represents the value of the combination of the i, j, k levels of the original modes, and in a similar manner each element of the core array g_{pqr} represents the size of the link of the pth component of $\mathbf{A}$, the qth component of $\mathbf{B}$, and the rth component of $\mathbf{C}$.

Even though Tucker's basic model is now commonly known as the Tucker3 model, a term coined by Kroonenberg and de Leeuw (1980), several researchers have used other names for the same model. Kolda and Bader (2009, p. 474) list as alternative names: three-mode factor analysis (3MFA/Tucker3), three-mode principal component analysis (3MPCA), N-mode principal component analysis, higher-order SVD (HOSVD), and N-mode SVD.

6.1.1.2 Tucker2 Model

The Tucker2 model is also a factorization of the three-way data array $\mathbf{X} = \{x_{ijk}\}$, but in this case only two of the three ways are represented by their components, while the third way is uncondensed, i.e., it has the same number of levels as the third way of the data array:

$$x_{ijk} = \sum_{p=1}^{P} \sum_{q=1}^{Q} \breve{g}_{pqk} a_{ip} b_{jq} + e_{ijk}$$

Here, **A**, **B**, and **E** have the same meaning as above, while $\breve{\mathbf{G}} = (\breve{g}_{pqk})$ is a three-way extended ($P \times Q \times K$) core array. Each element of the extended core array $\breve{g}_{pqk}$ represents for each subject k the size of the link of the pth **A** component and the qth **B** component. Because the third way is not condensed, this model can also be seen as a model for the frontal slices $\mathbf{X}_k$ with common components for the first and for the second way.

6.1.1.3 PARAFAC Model

The parallel factors model, PARAFAC or CANDECOMP/PARAFAC (CP) model, is the factorization of the three-way data matrix $\mathbf{X} = \{x_{ijk}\}$, which is conceptually closest to the standard principal component analysis, as it has only an additional term for the coefficients of the third way.

$$x_{ijk} = \sum_{s=1}^{S} g_s^* a_{is} b_{js} c_{ks} + e_{ijk}$$

The scalars g_s^* fulfil the same role as the singular values or square roots of the eigenvalues in ordinary principal component analysis, and the meaning of the other coefficients is the same as in the models above. Note that we now have only one summation sign compared to three in the Tucker3 model, and that $P = Q = R = S$; i.e., the ways have the same number of components and the core array $\mathbf{G}^*$ is a cube with only nonzero elements (g_s^*) on its diagonal.

Harshman (1970, 1972a, 1972b) conceived the model as an extension of regular component analysis using the parallel proportional profiles principle proposed by Cattell (1944; Cattell and Cattell, 1955). His motivation was to solve the rotational indeterminacy of ordinary two-mode principal component analysis. In his papers he was the first to derive uniqueness properties, in particular that there was only one solution for the parameters and that under very general conditions the components in the model could not be rotated without loss of fit. Thus, no orthonormality restrictions are necessary for identification of the model. In various other contexts the PARAFAC model was independently proposed, for instance, in the context of individual differences multidimensional scaling by Carroll and Chang (1970) under the name *canonical decomposition* (CANDECOMP) and in EEG research under the name *topographic component model* (Möcks, 1988a, 1988b; Pham and Möcks, 1992). Following Hitchcock (1927a, 1927b) the model is also referred to as the *canonical polyadic decomposition* (CP decomposition), giving the abbreviation CP a double meaning, namely, CANDECOMP/PARAFAC and canonical polyadic, thus honouring several origins.

A full-blown exposé of the model and some extensions are contained in Harshman and Lundy (1984a, 1984b); a more applied survey can be found in Harshman and Lundy (1994), and a tutorial with a chemical slant in Bro (1997). Some of the earliest applications by Harshman showing the potential of the PARAFAC model are in phonetics (Harshman et al., 1977) and psychology (Harshman and Berenbaum, 1981).

6.1.1.4 INDSCAL Model

The INDSCAL model is a special case of the PARAFAC model in the sense that the first two ways contain the same elements. Moreover, it is especially employed for data that have symmetric frontal slices X_k. Therefore, the model has the form

$$x_{ii'k} = \sum_{s=1}^{S} g_s^* a_{is} a_{i's} c_{ks} + e_{ii'k}$$

INDSCAL has been the model of choice for individual differences multidimensional scaling. It is typically applied to sets of (dis)similarity matrices, each of which contains the (dis)similarity judgements between the row and column entities given by a single subject.

6.1.2 Relationships between the Models

Figure 6.1 displays the relations between the four models. Indicated next to the arrows is how one can get from one to the other by imposing various restrictions.

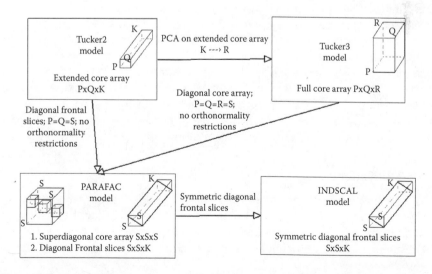

FIGURE 6.1
Three-way model chart.

A clarification of the relations between various other models proposed both in the French literature for three-mode analysis and in the English language literature was provided by Kiers in two papers (Kiers, 1988, 1991); other discussions of model comparisons are contained in Smilde et al. (2004, chap. 5) and Kroonenberg (2008a, chap. 4).

In several areas multiway developments have taken place, such as block models, clustering, conjoint analysis, hierarchical clustering, longitudinal data analysis, partial least-squares (PLS) modelling, partitioning procedures, structural equation modelling, and unfolding, but a historical overview of these specialized proposals will not be given here.

6.1.3 Dawn of Multiway Analysis

Very early in the 20th century mathematicians, especially those working in linear algebra, were interested in handling more than one matrix at a time and in examining the properties and eigendecompositions of sets of matrices and multiway arrays. In fact, one can even trace such an interest as far back as a paper by Jordan (1874), who was working on the singular value decomposition but was also interested in simultaneously diagonalizing two matrices at a time. The models now known as the PARAFAC and Tucker3 were already known to Hitchcock, who also discussed rank problems related to multiway structures (Hitchcock, 1927a, 1927b). Oldenburger (1934) also wrote about properties of multiway arrays. However, these exercises in linear algebra were not linked to real data analysis but stayed within the realm of mathematics.

In 1946 Cattell was probably the first to present three-way data in the form of a three-way data box, which he called the *covariation chart* (Figure 6.2). His presentation was not so much aimed at analysing a three-way data array in its entirety, but more an attempt to systematize thinking about data designs. He wanted to show how the different entities involved in the designs could be related to one other. On the basis of this covariation chart he defined six different types of factor analyses, two for each face of the cube, i.e., one type for the arrangement as shown in Figure 6.2 and one for its transpose. In 1966 Cattell extended his ideas to 10-way data by introducing the basic data relation matrix (Cattell, 1966, p. 69). In this paper he also discussed the ideas of centring, normalization, and their combinations (pp. 115ff).

6.1.4 Motivations

The early mathematical papers of Hitchcock and Oldenburger were probably unknown within psychology, and therefore one can safely say that multiway *data* analysis started in psychometrics. In an interview with Neil Dorans in 2004, Tucker had the following to say on the origin of three-mode data analysis:

> Three-mode factor analysis grew out of this multidimensional scaling work. While I was at ETS, I had observed that Charles Osgood of the

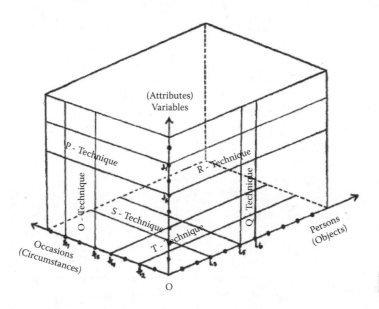

FIGURE 6.2
Cattell's covariation chart. This chart shows how one can define six types of factor analyses (O, P, Q, R, S, T techniques) depending on which face of the cube one intends to analyse. In principle, the idea is to form and analyse a covariation matrix of the entities in the columns by taking inner products over the rows. (From Cattell, R. B., *Psychological Bulletin*, 49, 501, 1952. An earlier version of this chart appeared in Cattell, R. B., *Description and Measurement of Personality*, World Book Company, Oxford, 1946, and the version shown here has appeared in several of Cattell's publications.)

> University of Illinois had collected data from three modes—concepts, scales, and subjects—in his semantic differential research. I thought that the data should be analysed differently than it was. He collapsed over people and threw away individual differences data [i.e., he flattened his data box, PMK]. So I developed the 3-mode factor analysis approach, a very general model for evaluating individual differences data. It was able to deal with the variety of people. (Dorans, 2004, p. 8)

Tucker illustrated his objection to 'flattening' with the example of the car manufacturer Chrysler, who designed a car for the average customer without appreciating the importance of individual differences. Unfortunately, the car did not sell, because there was no such thing as the average customer, and Chrysler almost went bust on this enterprise. The necessity of paying attention to individual differences is still the guiding principle for almost all work in the area of three-mode analysis.

The interest in chemistry resulted from several fundamental problems, but primarily from the desire in analytical chemistry to be able to quantify chemicals in the presence of other possibly undesirable chemicals or interferents. This quantifying was first made practical by rank annihilation factor analysis (RAFA), which was developed by Ho and coworkers (1978).

The approach was advocated first and foremost by Lorber and Kowalski (e.g., Lorber, 1984), and only later it merged with the psychometric models discussed here when some of the variants of RAFA were seen to be special cases of PARAFAC (Sanchez and Kowalski, 1986, 1990); for more technical details see Bro (1998, p. 30ff). One of the chemical problems in question was the determination of the identity and quantity of compounds present in a mixture from their fluorescence spectra. To solve this problem for mixtures, it was necessary to have unique solutions of the component models used for this purpose. By using not one but several mixtures, it was shown that the PARAFAC model, which is generally unique, provided exactly the type of information necessary. The PARAFAC model directly corresponds with already available physical models so that the existing algorithms for the PARAFAC model could be used for their estimation. In analytical chemistry often measurement instruments are used in tandem; for instance, a gas chromatographic system is often coupled with a mass spectrometer. This is but one example of a hyphenated procedure that routinely produces three-way and multiway data (Hirschfeld, 1985). Furthermore, important commercial applications were developed on the basis of the multiway models, for example, in batch processing (see Nomikos and McGregor, 1994).

Signal processing joined the multiway movement at the end of the 20th century, closely followed by a resurgence of interest by the mathematicians. Within signal processing the main motivation is what is termed *blind source separation*. Given a received compound signal by some measuring instrument like an antenna, it typically is a mixture of a number of independently generated signals, such as the signals from cell phones from different people. Multiway techniques can assist in unravelling the compound signal into its constituents. Such compound signals arise in many scientific areas, including magnet resonance, EEG, and other brain studies. As a consequence, many applications have evolved in this area (see, e.g., Sidiropoulos et al., 2000).

In mathematics multiway models are generally referred to as higher-order tensors. On the site of the 2010 Workshop on Tensor Decompositions and Applications some further insight is given by De Lathauwer on the motivation for multiway methods in a variety of disciplines (http://homes.esat.kuleuven.be/~sistawww/TDA2010/):

> In the '90s non-Gaussian signal processing and higher-order statistics became popular, where the basic quantities are higher-order tensors. Around 2000 it was understood that the concept of 'diversity' in telecommunication corresponds to the order of a tensor. Exponential signals, which can be considered as the atoms of signal processing, can be represented by rank-1 tensors. Tensors have led to new efficient and accurate computation techniques. Semantic graphs, multilayer networks and hyperlink documents are represented by higher-order tensors.... The development of tensor methods is relevant to countless applications.

6.1.5 Early Contributors

After its conception by Tucker, a further step in the development of three-way component analysis was taken when Kroonenberg and de Leeuw (1980; Ten Berge et al., 1987) presented an improved, alternating least-squares solution for Tucker's component model as well as computer programs to carry out the analysis, Tuckals2 and Tuckals3. The published paper had a precursor in Kroonenberg's master thesis (Kroonenberg and de Leeuw, 1977). A very similar algorithm, called COMSTAT, was independently proposed by Röhmel et al. (1983; see also Herrmann et al., 1983), but these authors do not seem to have followed up their initial algorithmic work. A different kind of algorithm was developed by Weesie and Van Houwelingen (1983) using regression analyses rather than eigen-techniques; however, their approach has only seen limited application (De Ligny et al., 1984). Also in 1983, Kroonenberg (1983a) produced his thesis on three-mode component analysis, which contains an overview of the then state of the art with respect to Tucker's three-mode component model, as well as an annotated bibliography that was independently published by Kroonenberg (1983b).

The psychometrician Kiers has been an important contributor to the algorithmic and model development of three-mode and multiway analysis, as well as a proposer of procedures for standardizing terminology (Kiers, 2000a), graphical procedures for three-mode analysis (Kiers, 2000b), and procedures to handle rotations in three-mode analysis (Kiers, 1998a, 1998b). Several of his publications on a variety of multiway methods are together with Bro (e.g., Bro and Kiers, 2003) and Ten Berge (Kiers et al., 1999; Ten Berge and Kiers, 1999). In addition, Ten Berge has been especially active on the subject of the rank of multiway problems and their uniqueness. In this area he was building on earlier partially published work of Kruskal, e.g., Ten Berge (2000), Kruskal (1977); see also a summary given by Kolda and Bader (2009, pp. 10ff).

In chemistry, the real upsurge of multiway data analysis, initially in its three-way form, started with two amazingly concise papers by Appellof and Davidson (1981, 1983). They discussed the merits of the PARAFAC model for chemistry and in passing independently reinvented the Tucker3 model.

Key authors for the early period in chemistry are Geladi (1989), Sanchez and Kowalski (1990), and Smilde (1990, 1992). Smilde was also the thesis supervisor of Bro (1998), who is probably at present the most prolific author and coauthor in the area. Smilde remained a major contributor to the development of multiway methods, and together with Bro and Geladi, he produced a major book on multiway methods in the chemical sciences: Smilde et al. (2004). The book arose from their work in the area of chemistry, in particular, chemometrics. Their contributions range far and wide and deal with mathematical and algorithmic improvements, combined additive and multiplicative models for three-way data, tests for numbers of components, applications in the chemical sciences, and many other aspects. Further important authors within the chemical domain are Nomikos and McGregor (e.g., 1994; multiway component analysis) and Tauler (e.g., 1995; multivariate curve resolution).

By the year 2000 three- and multiway methods had drawn the attention of mathematicians such as De Lathauwer, Kolda, Lim, Vasilescu, and many others (for excellent and extensive overviews and literature references, see Kolda and Bader, 2009; Açar and Yener, 2009). Even earlier, Wansbeek and Verhees (1989) developed new notational systems for the generalization from three-mode to multiway techniques, but this notation has not really caught on. Harshman (2001) also put forward a proposal for new notation, but his proposal has also not had much success to date.

One of the factors that has contributed to unifying the work on multiway analysis across sciences was the series of three-yearly specialist conferences under the name of Three-Way Methods in Chemistry and Psychology (Tricap), which gradually also drew the attendance of key researchers from many other disciplines. The series started in 1993 (Epe, Netherlands), followed by conferences in Lake Chelan (United States, 1997), Faaborg (Denmark, 2000), Lexington (United States, 2003), Xania (Greece, 2006), Vall de Núria (Catalonia, Spain, 2009), and Bruges (Belgium, 2012) (see http://ppw.kuleuven.be/okp/tricap2012/, which also contains links to the earlier conferences). Another series of specialist workshops that helped to unify the area were organized by multiway mathematicians under the name 'Workshop on Tensor Decompositions and Applications'. These took place at irregular intervals in Palo Alto (United States, 2004), Luminy (France, 2005), and Bari (Italy, 2010), with several researchers participating in both series of meetings. At present many conferences have specialist sessions on theoretical and practical aspects of multiway analysis.

6.1.6 Computational Methods

Even though Cattell (1966, p. 120) proposed in his basic data relation matrix a 10-way data arrangement, in his time 10-way algorithms were of course out of the question. Carroll and Chang (1970) were the first to propose and develop actual (least-squares) algorithms for multiway data analysis. In particular, they presented a seven-way version of their CANDECOMP computer program. Their program was written in FORTRAN, which could not handle more than seven nested loops. Lastovicka (1981) proposed a four-way generalization for Tucker's three-way method, and applied it to an example based upon viewer perceptions of repetitive advertising marketing. Kapteyn et al. (1986) generalized Kroonenberg and de Leeuw's (1980) alternating least-squares solution to the Tucker$n$ models and compared its results with Lastovicka's on the same data. A detailed paper comparing several of the available algorithms for the PARAFAC model was published by Faber et al. (2003). The book by Smilde et al. (2004) contains the references to the developments that have taken place, especially in chemometrics, with respect to multiway algorithms.

6.1.7 Software

With respect to software, the central package, developed within MATLAB®, is the *N*-way Toolbox of Andersson and Bro (2000; http://www.models.life.ku.dk/nwaytoolbox/). Multiway procedures are also contained in the PLS_ Toolbox (http://www.eigenvector.com/software/pls_toolbox.htm). A graphical interface for the *N*-way Toolbox, CuBatch, was developed by Gourvénec et al. (2005; http://www.models.life.ku.dk/CuBatch). Bader et al. (2012) developed a toolbox for multiway analysis called the Tensor Toolbox (http://www.sandia.gov/~tgkolda/TensorToolbox/), as did Comon et al. (2009) (http://www.i3s.unice.fr/~pcomon/TensorPackage.html). Within R, Leibovici (2010) has developed the PTAk package (cran.r-project.org/web/packages/PTAk/index.html) and Kroonenberg has a completely menu-driven suite of programs under Windows, 3WayPack (Kroonenberg and De Roo, 2010; http://three-mode.leidenuniv.nl). A sophisticated developer's tool is the multilinear engine, a program developed by Paatero (1999; www.helsinki.fi/~paatero/PMF/), which allows extremely sophisticated multiway modelling. Apart from these programs, many researchers have developed their own software for both general purpose and specific methods.

6.1.8 Applications

Apart from his theoretical papers, Tucker also wrote or collaborated on about 10 applications, not all of them published. One of his PhD students wrote maybe the first expository published paper on applying the technique in psychology (Levin, 1965). Thereafter, the number of psychological applications and theoretical papers increased gradually, but slowly.

A few isolated applied papers appeared in the early years of three-way analysis outside the social, behavioural, and related sciences, such as Hohn (1979) in geology and De Ligny et al. (1984), Arcelay et al. (1988), and Blyshak et al. (1989) in chemistry. For an overview up to 1983, see Kroonenberg (1983a, 1983b, pp. 355ff.).

Some other areas of application are agriculture including plant breeding, analytical chemistry such as spectroscopy, batch processing, computer vision, data mining, environmental studies, graph analysis, growth curve analysis, linguistics, longitudinal analysis, medicine, music perception, neuroscience such as functional magnetic resonance imaging (fMRI) and EEG studies, numerical analysis, numerical linear algebra, process analysis, receptor modelling for airborne particles, signal processing, social network analysis, and text mining. At the same time, there is a certain stagnation of applications within the social and behavioural sciences, as indicated by Kroonenberg (2012).

An extensive bibliography on the website of the Three-Mode Company (http://three-mode.leidenuniv.nl) contains references to most of the papers dealing with multiway issues, especially to those before 2004, while Kroonenberg (1983b) provides a virtually complete listing of publications, both statistical and applied, up until 1983. Further references can be found in the books by Law et al. (1984), Coppi and Bolasco (1989), Smilde et al. (2004), and Kroonenberg (2008a), as well as the review papers by Henrion (1994), Bro et al. (1997), Kiers and Van Mechelen (2001), Bro (2006), Comon (2001), Kolda and Bader (2009), and Açar and Yener (2009).

6.1.9 Current Status

In order to provide some idea of the fortunes of multiway analysis over the years I have constructed Figure 6.3, which shows the roughly calculated citation curves for the main protagonists and their research groups. The implicit assumption is thus that citations are an indication of the popularity of both the research area and its application. The citation counts have been accumulated per group, consisting of psychometricians (Carroll, Harshman, Kiers, Kroonenberg, Kruskal, Ten Berge, Tucker), chemometricians (Appellof, Bro, Hopke, Smilde, Nomikos, Tauler, Westerhuis), mathematicians (Kolda, Paatero), and signal detectors (Comon, De Lathauwer, Sidiropoulos, Vasilescu) using the authors' primary work as a basis for the subdivision. It should be realized, however, that nearly all mentioned authors have made contributions beyond the realm of their mentioned disciplines. For the authors the numbers of citations per year were based on those publications listed in the Web of Science that concerned multiway analysis.

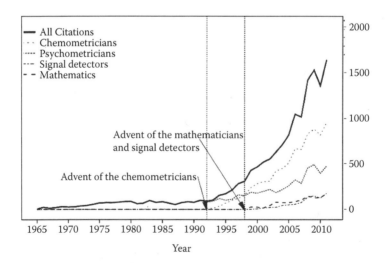

FIGURE 6.3
Citation to papers dealing with three/multiway analysis (1964–2011) by authors from all disciplines. The lines indicate the numbers of citations to papers by authors subdivided into psychometricians, chemometricians, mathematicians, and signal detectors, as well as the overall curve for all cited authors together.

What is most striking in this diagram is the steep increase in the number of citations during the 1990s, and the fact that this escalation is almost exclusively due to the chemometricians. Moreover, the rising number of citations for the psychometricians is actually also due to the stormy developments in chemometrics. As soon as chemometricians realized the power of multiway analysis, especially the PARAFAC models, there was no stopping them; the causes for this were already indicated in Section 6.1.5.

An interesting aspect of this development in chemistry is that psychometricians also started to publish in chemometrics journals, because that was where their biggest customers were located. However, as can be seen in Figure 6.3, there is an increasing interest in mathematics and signal processing for multiway methods and their development (see the review papers by Kolda and Bader, 2009; Açar and Yener, 2009).

Finally, a word should be said about the exclusivity of the English language. The work of non-English speakers has only been able to gain attention by their publications in English language journals. It is because of this that two of the most prominent techniques developed in France have not gained the attention in the literature they deserve. In particular, it concerns the techniques *structuration des tableaux à trois indices de la statistique* (STATIS) or *analyse conjointe de tableaux* (ACT) (Lavit, 1988; Lavit et al., 1994) and *analyse factorielle multiple* (AFMULT, AFM) (Escofier and Pagès, 1984a, 1989, 1990, 1994, 1998). Even though the four editions of Escofier and Pagès's book (1988, 1990, 1998, 2008) have gained wide acceptance, a cursory inspection of its citations seems to indicate that the citing authors had predominantly a French and certainly European background, and that even their English 1994 paper is cited primarily by French authors. Both recent books on multiway analysis (Smilde et al., 2004; Kroonenberg, 2008a) fail to mention the work of Escofier and Pagès, while only Kroonenberg (2008a) mentions the work of Lavit (1998) and Lavit et al. (1994).

6.2 Multiway Categorical Data

One of the special characteristics of data in the social and behavioural sciences is that they are often categorical rather than continuous, unlike data in many other scientific disciplines. Developments have taken place to tackle such data within the multiway context. Primarily for binary data a series of hierarchical classes (HICLAS) models have been developed by Van Mechelen, Ceulemans, and coworkers, which models have also been extended to rating scale data (e.g., Ceulemans and Van Mechelen, 2004, 2005; Ceulemans et al., 2003; Van Mechelen et al., 2007). Their approach is particularly novel because they directly deal with binary data via binary models based on Boolean algebra.

Another type of technique for categorical data is based on correspondence analysis (see next section). There has only been one attempt to develop a three-way model for fully crossed data with mixtures of continuous and categorical variables using optimal scaling (Sands and Young, 1980), but this has never been followed up.

6.2.1 Three-Way Correspondence Analysis

Given the focus of the present volume on correspondence analysis, a special section of this historical account of multiway data analysis is dedicated to three-way correspondence analysis. However, the treatment of the history of three-way and multiway correspondence analysis is rather elementary and sketchy. The main reason is that the technique is not really a mainstream one, and that the number of the publications in this area depends very heavily on what one considers multiway data.

Several streams exist that generalize standard correspondence analysis from two- to three-way and multiway tables. One may examine sets of tables, or one may investigate data sets with many categorical variables and treat them similarly to numerical variables by constructing variance-covariance or Burt tables. Characteristic of such approaches is that only the two-way interactions between the variables come into play. However, in this paper I will restrict myself to models that include the three-way interaction. Yoshizawa (1976, 1988) and Choulakian (1988a, 1988b) have presented the mathematics for higher-way generalizations, but no serious follow-up in the realm of data analysis has been presented as far as I know.

Conspicuously absent from our historical overview are several techniques that are akin to canonical correlation analysis, which investigate sets of two-way tables with special features for adjusting the marginals so that the two-way interactions can be considered simultaneously. A prime example is the aforementioned *multiple factor analysis for contingency tables* (AFMULT, AFM).

The basic idea behind three-way correspondence analysis is the modelling of the global dependence in a three-way contingency table with a multiplicative three-way model. The global dependence of a cell (i,j,k) Π_{ijk} is defined as the deviation from the three-way independence model: $\Pi_{ijk} = (p_{ijk} - p_{i\bullet\bullet}p_{\bullet j\bullet}p_{\bullet\bullet k})/p_{i\bullet\bullet}p_{\bullet j\bullet}p_{\bullet\bullet k}$. The modelling of this dependence is generally carried out with the Tucker3 model, i.e.,

$$\Pi_{ijk} = \sum_{p=1}^{P}\sum_{q=1}^{Q}\sum_{r=1}^{R} g_{pqr}a_{ip}b_{jq}c_{kr} + e_{ijk}$$

Again, the component matrices are taken to be orthonormal, but now orthonormal with respect to the marginal probabilities, e.g., $\mathbf{A}^T\mathbf{A} = \mathbf{D}_I$, where diagonal matrix $\mathbf{D}_I$ has the marginal proportions $p_{i..}$ on the diagonal.

In applications three-way correspondence analysis consists of three essential parts:

- Additive decomposition of the dependence; i.e., the deviation from three-way independence is decomposed into separate terms for the two-way and three-way interactions (see Lancaster, 1951)
- Modelling the dependence via a generalization of a three-way singular value decomposition, as shown above
- Plotting the resulting coordinates

On the basis of these three criteria one can evaluate authors' contributions to multiway correspondence analysis.

The analysis of interactions in three-way contingency tables is a natural extension of the two-way case and has a long history. Initially, the main emphasis was on multiplicative modelling via loglinear models (see, for instance, Agresti, 2007). However, loglinear analysis was primarily used for testing the relevance of interaction terms rather than analysing the interaction terms themselves. This situation was remedied by the advent of association models for multiway contingency tables, introduced by Clogg (1982), who posed specific models for the interactions. Van der Heijden et al. (1989) produced an important further development by combining interaction selection with modelling interactions via two-way correspondence analysis. Various other authors have developed methods to deal with three-way (and sometimes multiway) tables via two-way correspondence analysis. Unfortunately, one of them referred to their approach as three-way correspondence analysis (Romney et al., 1998; Batchelder et al., 2010).

The modelling of three-way contingency tables with true three-way models arose independently in various countries. In Japan Yoshizawa and Iwatsubo worked on such models and published a number of papers in Japanese (Iwatsubo, 1974; Yoshizawa, 1975, 1976), and only much later Yoshizawa published his work in English (Yoshizawa, 1988). Initially their work concentrated on generalizations of the singular value decomposition and one-dimensional modelling of the interactions, but Yoshizawa's 1988 paper also considered solutions more like correspondence analysis.

In the meantime in France, the work of Benzécri on correspondence analysis naturally created interest in multiway tables, and the French thesis of Dequier (1973) is probably the first publication properly extending correspondence analysis to three-way tables using three-way models, including the PARAFAC decomposition. He not only looked at modelling global dependence, but also paid attention to partial and marginal dependence. The next contribution was by Bener (1981, 1982). In a very technical paper without any references, probably based on his thesis (1981), Bener (1982) treats the decomposition of interaction terms and the chi-square statistic. Within the French tradition Choulakian published two papers in 1988, one in French (1988a) and one in English (1988b), in which he presented the Lancaster additive decomposition of the chi-square statistic in detail. He also discussed decompositions for both two-way and three-way interactions at a theoretical level, mentioning in passing the PARAFAC model for this purpose.

Kroonenberg presented in his thesis/book a three-way principal component analysis of a three-way contingency table without much explicit consideration to the standard measures and practices of correspondence analysis (Kroonenberg, 1983a). Following this he presented in 1989 a paper in which he discussed whether one should look at slice profiles (single subscripted arrays or matrices) or fibre profiles (double subscripted arrays or vectors). His approach was based on a three-way principal component analysis (Kroonenberg, 1989) on a three-way table of standardized residuals. The proper scaling of coordinates to get chi-squared distances did not feature in this paper, so that his approach cannot properly be called three-way correspondence analysis. A comprehensive approach to three-way correspondence analysis was developed by Carlier and Kroonenberg, which includes the Lancaster partitioning of the chi-squared statistic into independent contributions of all interactions, a modelling of the global dependence in the table, i.e., the deviations from the three-way independence model, and graphing of the global dependence (Carlier and Kroonenberg, 1993, 1996, 1998). It was also shown that from the global modelling, models for partial and marginal dependences could all be derived within one framework and plotted in a single graph. One of the few independent applications of the technique was published by Van Herk and Van de Velden (2007). Another application was presented by Kroonenberg at the 2011 CARME conference in Rennes. At an earlier CARME conference in Barcelona, Kroonenberg and Anderson (2006; http://www.youtube.com/watch?v=lrCYwSRoIgQ) explored the relationships between three-way correspondence analysis and three-way association models.

Around the same time that Carlier and Kroonenberg started their exploration, in Naples correspondence analysis for tables with a dependence structure, called nonsymmetric correspondence analysis, was gradually extended to three-way tables. The first proposals ignoring the two-way interaction between the two predictors were made by D'Ambra and Lauro (1989). An analysis procedure based on the work of Carlier and Kroonenberg and supervised by D'Ambra and Carlier was contained in the Italian thesis of Lombardo (1994). Her work resulted in the published paper by Lombardo et al. (1996). Lombardo, in cooperation with Beh, extended the method to include ordinal restriction on the components (Lombardo et al., 2007), and this proposal is still being developed further.

Parallel to the development sketched above, other initiatives were made to extend correspondence analysis as a multidimensional scaling technique along the lines of the work of Carroll and Chang (1970) by Meulman et al. (1999).

6.3 Conclusion

The field of multiway analysis is coming of age, and the large lines of research seem to have been drawn. The books by Smilde et al. (2004) and Kroonenberg (2008a) show a certain consolidation in the field. This does not imply that the field is now in a stable state. On the contrary, as Figure 6.3 shows, it seems that multiway analysis is only just past its puberty and may be now in its adolescence. We are all waiting to see what adulthood may bring. This much cannot be said of multiway correspondence analysis, as very few applications actually exist. Moreover, most of them have appeared in technical or tutorial papers on the subject. Perhaps what is needed is a concerted effort by the initiators to market their wares.

7

Past, Present, and Future of Multidimensional Scaling

Patrick J. F. Groenen and Ingwer Borg

CONTENTS

7.1 Basic Ideas of Multidimensional Scaling 96
7.2 Motives for Multidimensional Scaling: A Historical Account 97
    7.2.1 Early MDS in Geography 98
        7.2.1.1 Distance Formula as a Psychological Model of Dissimilarity Judgements 99
        7.2.1.2 Ordinal MDS as a Response to the Premise That Measurement in Psychology Must Build on Nonmetric Data 100
        7.2.1.3 Ordinal MDS as a Method to Study Generalization Gradients in Learning 101
        7.2.1.4 MDS as a General Data Analytic Tool 102
    7.2.2 Utilization of MDS Today 103
7.3 Technical Aspects of MDS: The Past 104
    7.3.1 Classical MDS 104
    7.3.2 Stress 105
    7.3.3 Facet Theory and Regional Interpretation in MDS 106
    7.3.4 Three-Way MDS Models 107
    7.3.5 Majorization Algorithm 110
    7.3.6 Other Algorithms 112
7.4 Present 112
    7.4.1 Distance-Based Multivariate Analysis 112
    7.4.2 Constant Dissimilarities 113
    7.4.3 Local Minima 113
7.5 Future 114
    7.5.1 MDS with Weights 114
    7.5.2 Dynamic MDS 114
    7.5.3 Large-Scale MDS 114
    7.5.4 Symbolic MDS 115
    7.5.5 What Needs to Be Done? 116

In this chapter, we pay tribute to several important developers of multidimensional scaling (MDS) and give a subjective overview of milestones in MDS developments. We also discuss the present situation of MDS and give a brief outlook on its future.

MDS has become one of the core multivariate analysis techniques discussed in any standard data analysis, multivariate analysis, or computer science textbook. Its success is due to its simple and easily interpretable representation of potentially complex structural data. These data are typically embedded into a two-dimensional map, where the objects of interest (items, attributes, stimuli, respondents, etc.) correspond to points such that those that are near to each other are empirically similar, and those that are far apart are different. A search in the Thomson Reuters Web of Science on the topic 'multidimensional scaling' yielded 5,186 papers, with a total of 68,429 citations (as per January 2013). Several important milestones in the development of MDS can be distinguished, and the present chapter is a subjective interpretation of these. The emphasis here lies on algorithmic milestones as they have cleared the way for practical use.

The remainder of this chapter is organized both chronologically and per topic. We have roughly distinguished three periods: past (until 1980), present (1980–2000), and future (from 2000, because it always takes time for developments to be used by a wider audience, which explains the lag of about 15 years). Table 7.1 gives an overview of these subjective milestones of the authors, which are discussed in more detail in the subsequent sections. We do not intend to provide an exhaustive overview of the history of MDS, as that could easily require a book by itself (for further details, see Borg and Groenen, 2005).

7.1 Basic Ideas of Multidimensional Scaling

The core idea of MDS is explained by the first sentence in Borg and Groenen (2005, p. 3): 'Multidimensional scaling (MDS) is a method that represents measurements of similarity (or dissimilarity) among pairs of objects as distances among points of a low-dimensional multidimensional space'. Thus, the data of MDS consist of measurements of (dis)similarity among pairs of objects, collectively called proximities. Objects could be persons, attributes, stimuli, countries, etc., and the measurements may be correlations of test items, similarity of politicians, dissimilarity of mobile telephones, etc., where each object is compared with every other item. The overall goal is to represent these objects as points in a low-dimensional (usually two-dimensional) space such that the distances among the points represent the (dis)similarities as closely as possible. The rationale for doing this is to visualize the

TABLE 7.1

Subjective Overview of Milestones in MDS

Years	Main Author(s)	Topic
Past		
1958, 1966	Torgerson, Gower	Classical MDS
1962	Shepard	First MDS heuristic
1964	Kruskal	Least-squares MDS through stress with transformations
1964	Guttman	Facet theory and regional interpretations in MDS
1969, 1970	Horan, Carroll	Three-way MDS models (INDSCAL, IDIOSCAL)
1977–	De Leeuw and others	The majorization algorithm for MDS
Present		
1986–1998	Meulman	Distance-based MVA through MDS
1994	Buja	Constant dissimilarities
1978, 1995–	Various	Local minimum problem
1998	Buja	Smart use of weights in MDS
Future		
1999–	Heiser, Meulman, Busing	Modern MDS software: PROXSCAL in SPSS
2000	Tenenbaum, De Silva, Langford	Large-scale MDS ISOMAP heuristic
2002	Buja, Swayne, Cook	Dynamic MDS in GGvis (part of GGobi)
2003	Groenen	Dynamic MDS visualization through iMDS
2005–	Groenen, Trosset, Kagie	Large-scale MDS through stress
2002	Denoeux, Masson, Groenen, Winsberg, Diday	Symbolic MDS of interval dissimilarities
2006	Groenen, Winsberg	Symbolic MDS of histograms
2009	De Leeuw, Mair	Package smacof in R

data in a 'picture' that makes the data structure much more accessible to the researcher than the data matrix with its many numbers.

7.2 Motives for Multidimensional Scaling: A Historical Account

MDS was not invented by statisticians. It was first developed to solve specific scaling problems that arose in practical and scientific contexts. In the following, we outline some of these developments.

7.2.1 Early MDS in Geography

The first traces of MDS can be found in the 17th century. Figure 7.1 shows a small table of distances among several towns and villages in Durham County, England. The order of the row and column towns is reversed so that an unusual matrix of distances appears that is symmetric over the diagonal from lower left to upper right. This diagonal does not contain the zero distances of a town to itself, but contains the distances to London. Apart from this table, Figure 7.1 also shows the geographical map of Durham County. It is considered the first instance of showing both a table of distances and the map that corresponds to these distances in a single figure (Gower, personal communication). Therefore, this case can be seen as a predecessor of MDS. (Note that this map is one of a series covering the counties of England, made by the Dutch cartographer Jacob van Langren in 1635.)

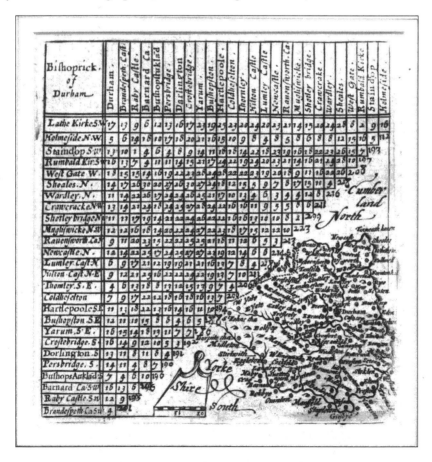

FIGURE 7.1
Map of Durham County by Jacob van Langren in 1635. (From John Gower, personal communication, 1992.)

Modern MDS is less concerned with cartography. Rather, similar to factor analysis, it evolved as a model for certain psychological phenomena, and only later became more and more popular as a general purpose data analytic tool. Historically, MDS can be related to at least four different purposes.

7.2.1.1 Distance Formula as a Psychological Model of Dissimilarity Judgements

The notion that dissimilarity judgements (or, inversely, similarity judgements) can normally be modelled as distances has been around in psychology for quite some time. It seems obvious that persons generate dissimilarity judgements for pairs of objects in a process that closely mimics the natural distance function in a Cartesian space. That is, if a judgement is needed, the person forms a mental representation of the objects in psychological space, a space spanned by the objects' attributes and with the objects corresponding to points in this coordinate system. Dissimilarity judgements are then formed by first assessing, dimension by dimension, the differences of each pair of points, and then summing these intradimensional differences. This generates a global distance, the basis for an overall dissimilarity impression or rating of the respective objects. To the outside observer—and possibly also to the person himself or herself—the psychological space itself is unknown (underlying, latent), but MDS promises to uncover it—as Kruskal, one of the MDS pioneers, claimed—from the individual's overall dissimilarity judgements. To do this, classical MDS first assumes that the given judgements are ratings on a metric scale, and that the distance function of the psychological space is the Euclidean distance. It was not, however, until MDS algorithms were developed that allowed us to process not metric but ordinal data before this distance model received a lot of attention. This met with the zeitgeist of the late 1960s, which emphasized ordinal data.

Later, the Euclidean distance formula was also generalized to the more general family of Minkowski distances (Borg and Groenen, 2005). A series of studies were done to investigate what particular Minkowski metric was most suitable for what kind of context (e.g., judgements under time pressure, perception of analytic vs. integral stimuli). Then, it was argued that Minkowski spaces have local validity only because numerical dissimilarity judgements in psychology are subadditive; i.e., the dissimilarity of very different stimuli is systematically underestimated relative to the dissimilarity of more similar stimuli. To account for this empirical lawfulness, Schönemann and Lazarte (1987) proposed an MDS model with a nonlinear geometry where distances are increasingly contracted as they approximate an upper bound. However, this idea (which is popular in physics) was not pursued much further in psychological research.

Besides such refinements of the similarities-explained-by-distances model, attention was also turned to modelling individual differences in MDS. This produced a hugely popular model, often identified with a particular computer program called INDSCAL (Individual Differences Scaling). It assumes that different individuals differ in how they weight the same set of dimensions of a common space. Numerous applications in the social sciences used this model, since it promised to identify the dimensions uniquely, until step-by-step approaches (e.g., Procrustean Individual Difference Scaling (PINDIS)) showed that rotations of the group space often are only slightly less successful explaining the data. Moreover, the individual dimension weights can be deceptive in the sense that they scatter a lot without explaining much more variance than unit weights. The biggest mistake when using INDSCAL modelling is, however, comparing individual weights all too loosely. These weights depend on the (arbitrary!) norming of the group space, so that only the order of the weights of different persons for the same dimension can be compared, while market researchers, in particular, had hoped that INDSCAL would show them what dimensions are most important in product perception. When these restrictions became clearer, INDSCAL became less important in applied research.

A further line of research used the MDS method to solve Coombs's unfolding model, where the data are not (dis)similarities but preferences of different individuals for the same set of objects. In unfolding, persons are represented as points in space, and choice objects by other points, and the distances between these two types of points represent the preference data. Each person point is taken as this person's point of maximal preference (ideal point), and circles about each ideal point as isopreference contours. Unfolding was used a lot to model voting behaviour, for example, but it was soon discovered that theory-guided multiple unidimensional unfolding for different subgroups can be superior to exploratory multidimensional unfolding of the total sample. Hence, unfolding's popularity as a model of preferential choice dropped in importance.

This type of research where the MDS geometry and its distance function are taken as psychological models has considerably advanced the understanding of human perception, judgement, and preference. In particular, it has become clear under what conditions such models yield good descriptions of empirical phenomena, and when they do not. Today, research where general MDS plays a major role as a psychological model is rare.

7.2.1.2 Ordinal MDS as a Response to the Premise That Measurement in Psychology Must Build on Nonmetric Data

In the late 1960s, methodologists were much concerned with foundations of measurement (Krantz et al., 1971). The cardinal premise of this research initiative was that numerical judgements (mostly ratings on, say, a scale from 0 to 10) cannot automatically be assumed to be real numbers. Rather, it was argued that real-valued measurements must be constructed and, first of all, justified by testing typically large sets of pairwise ordinal judgements that, together with some technical assumptions, establish structure-preserving maps of relational into numerical systems (homomorphisms). To respond to this measurement philosophy, scaling methodologists felt driven to replace the classical MDS of Torgerson (1958) and Gower (1966)—which assumed metric data as input—by ordinal MDS (or, as it was called at that time, nonmetric MDS).

Kruskal (1964a, 1964b) and Guttman (1968) with Lingoes (1973) developed computer programs for ordinal MDS. They both used gradient-based minimization to optimize the point coordinates, but Kruskal did this in combination with ordinal regression of the data onto the distances, while Guttman invented 'rank images' as targets, a method that is less likely to yield degenerate solutions. There are other technical differences (later harmonized in a best of both world's program called MINISSA (Michigan–Israel–Nijmegen Integrated Smallest Space Analysis)), but the main difference between Kruskal and Guttman was how they approached an MDS solution. Kruskal (as most users of MDS at that time) first asked: 'What do the dimensions mean?' Guttman, in contrast, was content driven. For him content came first, and methods only served as tools to build substantive theory in a partnership with data. He called MDS SSA (smallest space analysis; later reinterpreted as similarity structure analysis) because he wanted to emphasize that the Cartesian dimensions of an MDS representation are but an algebraic scaffolding for solving a geometric problem. Hence, any geometric patterns (such as dimensions, directions, clusters, figures, and, in particular, regions and neighbourhoods) that correspond to substantive knowledge about the objects can be meaningful. This perspective later developed into facet theory (see Section 7.3) and led to other data analytic methods, such as partial order scalogram analysis.

Ordinal MDS stimulated a huge number of applications, but over the years, interval and even ratio MDS recovered considerably in terms of utilization. This had statistical reasons on the one hand (interval MDS solutions, for example, are often less cluttered, with fewer tight point clusters), and theoretical reasons on the other hand (the emphasis of measurement foundations had shifted toward cumulative theory construction over replications, away from an almost endless testing of single data sets, and metric MDS solutions often allow for simpler and more robust interpretations than overfitted ordinal solutions). Today ordinal MDS is but one of several MDS models. Advanced computer programs generate solutions for each of them in seconds, and so they can easily be tested against each one at virtually no costs.

7.2.1.3 Ordinal MDS as a Method to Study Generalization Gradients in Learning

One historical motivation for MDS is closely linked to a special issue in the psychology of learning. Its main focus is not the MDS space itself, but the shape of the regression function of MDS distances to the data they represent. The theory that generates this interest is generalization of conditioned responses: If a response R (e.g., a pigeon pecking at a coloured disk) is conditioned to a stimulus S_R (a yellow disk, say), then a stimulus similar to S_R (a disk that is more orange coloured or more greenish, say), S'_R, also tends to trigger response R with a certain probability. Such probabilities should be a monotonically decreasing function of the distance of S'_R to S_R. Yet, it is difficult to tell the exact shape of this function, because S'_R and S_R lie in perceptual, not physical, space. However, given a set of stimuli with conditioned responses (e.g., from different groups of pigeons, each group conditioned to one particular colour) and measurements of the probabilities of giving the S_i response to stimulus S_k, the issue can be turned into an MDS problem: if the data are converted into similarity measures of the stimuli, one can first scale them via ordinal MDS; then, the regression trend of the MDS distances onto the data shows the shape of the generalization gradient. So, rather than postulating that the gradient is an exponential or a linear decay function, Shepard (1958) wanted to let the data speak for themselves, finding the generalization gradient empirically through what is now known as the Shepard diagram. Shepard (1962a, 1962b) struggled with the problem without really solving it, but his work motivated Kruskal (1964a, 1964b) to develop an ordinal MDS algorithm (MDSCAL) as a statistician's answer to the scaling task.

Today, applications of MDS where the Shepard diagram is of primary interest are exceedingly rare. Rather, researchers almost always focus on the MDS space itself and its relationship to known or assumed properties of the objects represented in this space. Ordinal MDS can sometimes be useful to check certain model assumptions empirically. For example, in classical Thurstonian scaling, dominance probabilities are mapped into scale differences by a cumulative Gaussian function. Rather than assuming such a mapping function, one can use ordinal MDS to scale the data into scale distances, and then check empirically if the mapping function is indeed S-shaped.

7.2.1.4 MDS as a General Data Analytic Tool

As soon as ordinal MDS became possible, it was enthusiastically received by many disciplines outside of psychology, in particular by market researchers. Green and his coworkers (see, for example, Green and Carmone, 1970; Green and Rao, 1972) published scores of papers and books that showed how MDS can be used to uncover how consumers perceive products. Sociologists also used MDS to study social networks and, in particular, attitudes and values. Schwartz (1992), for example, used MDS to develop his theory of universals in values (TUV), an influential theory on social values that is well and alive today. The TUV is intimately related to a circumplex of regions (a wheel of regional sectors) in two-dimensional MDS space. It partitions the MDS space into neighbourhoods that each contain only points representing values of the same category (e.g., achievement values, security values, enjoyment values). While Green had used MDS in a purely exploratory way, Schwartz (as Guttman) was content driven, and hence looked for correspondences of content theory about the MDS objects and their representation in space.

Schwartz, however, never enforced such external constraints onto the MDS solutions using confirmatory MDS (CMDS), although CMDS had been around since the early 1980s. De Leeuw and Heiser (1980), among others, developed certain forms of confirmatory MDS, and programs such as PROXSCAL (in SPSS) or smacof (in R) are able to handle most of them. However, many forms of external constraints onto the MDS configuration (except those on dimensions) are not easy (or simply impossible) to set up in the present MDS software. Nor is it often clear how to assess the effects of those constraints statistically. Sometimes the present programs also yield incorrect solutions, which can be difficult to diagnose for the applied researcher.

Another line of research deals with one general argument against distance models, i.e., that distances are always symmetric but (dis)similarity data may not be symmetric. Various proposals were made on how to handle non-symmetric data. Most amount to first splitting the data information into a symmetric part (which can be modelled via MDS) and a skew-symmetric part (which can be added to the points in MDS space in the form of small arrows). These models are not psychological models in the sense that they explain how a person generates asymmetric (dis)similarities. Rather, they are statistical tools that can be useful showing systematic trends in asymmetric (dis)similarity data. However, no user-friendly programs exist so far for these models, and hence their potential for statistical diagnostics has not been exploited.

7.2.2 Utilization of MDS Today

Today, many of the original motives that led to the development of MDS have become unimportant. What has survived, in particular, is using MDS in Guttman's sense, in particular in attitude and value research, where intercorrelations of survey items are studied for correspondences of the conceptual facets of the items to regions of their spatial representation. Yet, most applications of MDS today actually serve a much wider purpose; i.e., they are done to visualize tables of indices that can be interpreted as (dis)similarity data. For that purpose, MDS is highly useful, as it:

Can handle a vast variety of data as long as they are (dis)similarities (e.g., correlations, covariances, co-occurrence data, profile distances)

Does not require interval-scaled data but also handles ordinal (and even nominal) data

Is robust against missing data and coarse data

Often serves as a data smoother, showing a structure that is replicable even under conditions of high error

- Is easily explained to nonexperts and allows them to explore the solutions without much risk (given that the Euclidean metric is employed!)
- Is easy to run for nonexperts even though its solution algorithms are rather difficult (but, driving a car also does not require knowing how the engine works)
- Does not impose a particular interpretation (dimensions, in particular) onto the user but allows the data to speak for themselves

7.3 Technical Aspects of MDS: The Past

7.3.1 Classical MDS

Classical MDS can be considered the first algebraic approach to MDS. It has been independently proposed by several authors: Torgerson (1958), Gower (1966), and Kloek and Theil (1965). Classical MDS rests on the following equation: Let $\mathbf{X}$ be the $n \times p$ matrix of point coordinates (assumed here to be column centred for simplicity); then, the matrix of squared Euclidean distances with elements

$$d_{ij}^2(\mathbf{X}) = \sum_{s=1}^{p} (x_{is} - x_{js})^2$$

is

$$\mathbf{D}^{(2)} = \mathbf{1}\boldsymbol{\alpha}^T + \boldsymbol{\alpha}\mathbf{1}^T - 2\mathbf{X}\mathbf{X}^T, \qquad (7.1)$$

where $\mathbf{1}$ is a vector of ones of appropriate length and $\boldsymbol{\alpha}$ the vector with diagonal elements of $\mathbf{X}\mathbf{X}^T$. Given $\mathbf{D}$, $\mathbf{X}$ is found as follows. Let $\mathbf{J} = \mathbf{I} - \mathbf{11}^T/\mathbf{1}^T\mathbf{1}$ be the centring matrix with $\mathbf{I}$ the identity matrix. Then, multiplying the left- and right-hand side of (7.1) with $\mathbf{J}$ makes the terms with $\boldsymbol{\alpha}$ disappear as $\mathbf{J1} = 0$. An additional multiplication by $-\tfrac{1}{2}$ yields

$$-\tfrac{1}{2}\mathbf{J}\mathbf{D}^{(2)}\mathbf{J} = \mathbf{X}\mathbf{X}^T. \qquad (7.2)$$

Then, the eigendecomposition of $-\tfrac{1}{2}\mathbf{J}\mathbf{D}^{(2)}\mathbf{J}$ is $\mathbf{Q}\boldsymbol{\Lambda}\mathbf{Q}^T$, and so $\mathbf{X} = \mathbf{Q}\boldsymbol{\Lambda}^{1/2}$. Classical MDS rests on the idea that if the matrix of dissimilarities $\boldsymbol{\Delta}$ is not a Euclidean distance matrix (which is almost always true with real data), Euclidean distances can be approximated by inserting $\boldsymbol{\Delta}^{(2)}$ for $\mathbf{D}^{(2)}$ in (7.2), and then retaining the first p positive eigenvalues of the eigendecomposition used for computing $\mathbf{X}$.

Classical MDS minimizes a loss function called strain:

$$\text{Strain}(\mathbf{X}) = \tfrac{1}{4}\text{trace}\left[\mathbf{J}(\Delta^{(2)} - \mathbf{D}^{(2)})\mathbf{J}(\Delta^{(2)} - \mathbf{D}^{(2)})\mathbf{J}\right]$$

$$= \left\|(-\tfrac{1}{2}\mathbf{J}\Delta^{(2)}\mathbf{J}) - \mathbf{X}\mathbf{X}^\mathsf{T}\right\|^2$$

Gower (1966) was the first to realize that the dimension reduction of principal component analysis (often seen as the eigendecomposition of a correlation matrix or the singular value decomposition of the data matrix $\mathbf{Z}$ itself) has a dual method that can be obtained by doing classical MDS on the Euclidean distances of the rows of the data matrix $\mathbf{Z}$. This method was coined principal coordinate analysis and emphasizes the representation of the rows (usually individuals or samples) of the data matrix $\mathbf{Z}$. For more on classical MDS, we refer to Chapter 12 of Borg and Groenen (2005).

7.3.2 Stress

Arguably the two most important breakthroughs in MDS were (1) modelling dissimilarities directly by distances in a loss function and (2) allowing transformations of the dissimilarities that in turn are estimated by distances. Shepard (1962a, 1962b) proposed heuristic methods to do both aspects, but he did not provide a loss function. Kruskal (1964a, 1964b) then suggested the stress loss function:

$$\text{Stress}(\mathbf{X},\hat{\mathbf{d}}) = \frac{\sum_{i<j}\left(\hat{d}_{ij} - d_{ij}(\mathbf{X})\right)^2}{\sum_{i<j} d_{ij}^2(\mathbf{X})} \tag{7.3}$$

where $\hat{d}_{ij}$ is a transformed dissimilarity. For the moment assume that $\hat{d}_{ij} = \delta_{ij}$. Then, this stress loss function fits the distance $d_{ij}(\mathbf{X})$ directly to the dissimilarity δ_{ij} and minimizes simply the squared errors over all combinations i, j. The minimization of (7.3) over $\mathbf{X}$ is not trivial, as no analytical solution exists. Kruskal proposed a gradient-based minimization method to get the coordinates. The second breakthrough is to allow for transformations of the dissimilarities. One such transformation is the linear transformation $\hat{d}_{ij} = a + b\delta_{ij}$ for unknown a and b. With a large positive intercept a and a negative slope b the dissimilarities may be replaced by similarity measures, thereby opening up a large variety of applications that are based on similarity measurements (e.g., correlations) among the objects. Kruskal proposed an even more flexible transformation, that is, the ordinal transformation. This implies that the $\hat{d}_{ij}$ should be chosen such that whenever $\delta_{ij} \leq \delta_{kl}$, it must also hold that $\hat{d}_{ij} \leq \hat{d}_{kl}$ for any combination of pairs ij and kl. For fixed $\mathbf{X}$, the minimization of (7.3) over $\hat{\mathbf{d}}$ amounts to a quadratic program with linear inequality constraints on $\hat{\mathbf{d}}$. Kruskal provided a solution called monotone regression that provides a global minimum to this optimization problem. These two contributions can be seen as crucial milestones in the development of MDS as a statistical technique. When optimizing both over $\mathbf{X}$ and $\hat{\mathbf{d}}$, some adaptation is needed to avoid the trivial solution $\mathbf{X} = \mathbf{0}$ and $\hat{\mathbf{d}} = \mathbf{0}$. In (7.3), this trivial solution is avoided by dividing by the sum of squares of the $d_{ij}(\mathbf{X})$s.

7.3.3 Facet Theory and Regional Interpretation in MDS

In facet theory, 'content' information is available for the objects in the form of external coding variables. These variables are called facets. The objects of observation are assigned to a certain level on each facet. Guttman (1964) proposed to use such facets to form regions in MDS space. That is, it is hypothesized that if the facets are scientifically useful at all, then the points should fall into certain (nonoverlapping and exhaustive) neighbourhoods that correspond to the levels of a particular facet, facet by facet. Ordered facets should lead to correspondingly ordered regions, and this order can be linear (stripes in space) but also circular (wedges), with the usual 'dimensions' as special cases of linearly ordered stripes. Three types of regional patterns are often observed with empirical data: axial, modular, and polar regions (see Figure 7.2).

In empirical research, regions are almost always found by hand (drawing and redrawing partitioning lines on printouts of MDS plots until the partitioning seems optimal), but Borg and Groenen (1998) were the first to minimize stress while imposing axial constraints when the number of axial facets equals the number of dimensions. Groenen and Van der Lans (2004) extend this to the case where the number of axial facets exceeds the number of dimensions.

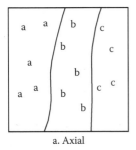
a. Axial

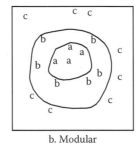
b. Modular

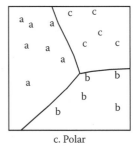
c. Polar

FIGURE 7.2
Three possible ways for regional partitioning by a facet: (a) an axial partitioning by parallel lines, (b) a modular partitioning by concentric circles, and (c) a polar partitioning by rays emanating from a common origin. (Reproduced with permission from Borg, I., and Groenen, P. J. F., *Modern Multidimensional Scaling*, 2nd ed., Springer, New York, 2005.)

Let us look at a classical example, the MDS of Morse signals. Content information is available on the Morse signals: each signal has a temporal length (from 0.05 to 0.95 seconds) and a certain composition of long and short beeps. We code the latter as 'only short beeps' (1), 'more short beeps than long beeps' (1 > 2), 'an equal number of short and long beeps' (1 = 2), 'more long than short beeps' (2 > 1), and 'only long beeps' (2). These two external variables are facets of the signals. An approximate axial partitioning of the unconstrained MDS representation of pairwise confusion probabilities of Morse signals is shown in Figure 7.3a, and the regionally constrained version in Figure 7.3b. The axially constrained (confirmatory) solution has a slightly higher stress (0.21) than the unconstrained solution (0.18), yet it gives a plot that is much easier to interpret in psychophysical terms. Moreover, the linearized structure is related to substantive laws of formation, and therefore it can be expected to be more robust over replications than the possibly overfitted exploratory MDS pattern with its partitioning lines that were inserted only afterwards.

In a recent application, Borg et al. (2011) impose two regional axial constraints with only two levels per facet (which effectively imposes a quadrant restriction). They then perform a permutation test on stress to test if the axial constraints perform better than random assignment of points to quadrants. For more information on facet theory, we refer to the book of Borg and Shye (1995), and for some other applications, see, for example, Borg and Groenen (1997, 1998).

7.3.4 Three-Way MDS Models

In many applications of MDS, there is not just one data matrix but K dissimilarity matrices, Δ_k, for $k = 1, \ldots, K$. To model such data in one MDS representation, Horan (1969) and Carroll and Chang (1970) proposed an important extension of the basic MDS model. Their weighted Euclidean model assumes that each Δ_k can be explained by distances of a single common space $\mathbf{G}$ transformed by the diagonal matrix $\mathbf{S}_k$ for each k. This means that each individual k simply stretches or compresses the common space along its dimensions as if each k attributes its own specific salience to each dimension. Figure 7.4a shows an artificial example of three individual spaces $\mathbf{X}_k = \mathbf{GS}_k$ that are derived from a single common space.

Carroll and Chang (1970) did not use stress but strain in their INDSCAL program. In this formulation, negative weights are a problem as they could lead to negative distances. The constrained MDS approach of de Leeuw and Heiser (1980) that uses stress avoids this problem, and the sign of the dimension weight is formally unimportant. To eliminate a basic indeterminacy in the $\mathbf{GS}_k$ model, $\mathbf{G}$ is normalized so that $\mathbf{G}^\mathsf{T}\mathbf{G} = \mathbf{I}$. As this restricts the sum of squares of all columns in $\mathbf{G}$ to 1, it becomes possible to compare the dimension weighting values among persons over dimensions, but only conditional on how $\mathbf{G}$ is normed (i.e., other norms lead to other dimension weights).

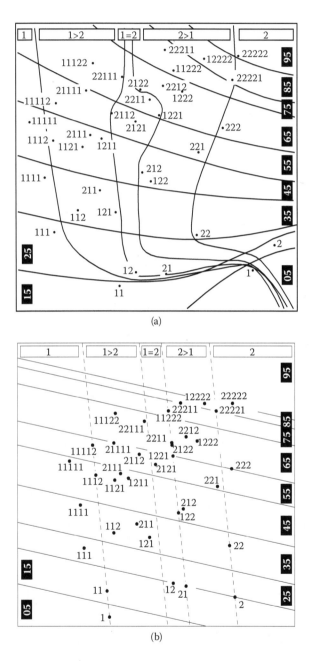

FIGURE 7.3
MDS solution of ordinal MDS on Rothkopf's Morse code confusion data with regions drawn by hand (a) and regionally constrained MDS (b). (Reproduced with permission from Borg, I., and Groenen, P. J. F., *Modern Multidimensional Scaling*, 2nd ed., Springer, New York, 2005.)

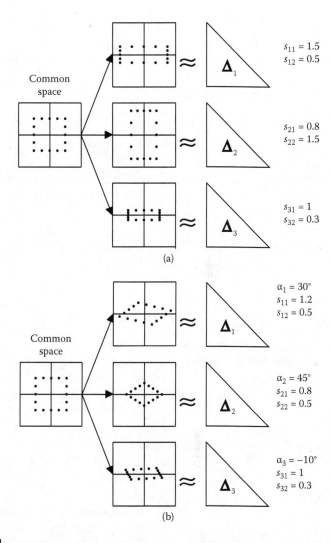

FIGURE 7.4
Illustration of the effect of dimension weighting for a common space **G** that forms a square of points and three examples for individual spaces **GS**$_k$. (a) The weighted Euclidean model with S_k diagonal, i.e., just stretching along the dimensions. (b) The generalized Euclidean model, with rotation and stretching.

Three extensions exist of this approach. The first one also allows for a rotation before stretching (see Figure 7.4b), which means that $\mathbf{S}_k$ is allowed to be any square matrix. Using the strain loss function, Carroll and Chang (1970) called this method IDIOSCAL (Individual Differences in Orientation Scaling). The second extension comes from the constrained MDS approach of de Leeuw and Heiser (1980) using stress that imposes a model that allows $\mathbf{S}_k$ to be of lower rank than the dimensionality of the common space **G**. For example, one could model the common space to be four-dimensional and the individual spaces as (rotated and stretched/compressed) subspaces of the four-dimensional common space. A third possibility is to also premultiply the common space coordinate matrix $\mathbf{G}$ by a weight matrix (Lingoes and Borg, 1978), but this leads to a model with few applications.

Note that these three-way models tend to be very restrictive when the dimensionality is low. The alternative of doing K separate MDS analyses allows the $\mathbf{X}_k$ to be estimated freely. On the other hand, the generalized Euclidean model or reduced rank model with very high dimensionality of the common space (close to n) will yield solutions close to K separate MDS analyses because the number of parameters in $\mathbf{S}_k$ becomes large. We believe the generalized Euclidean model or reduced rank model is most useful for common spaces whose dimensionality is not too small (say, larger than 3 and smaller than 6), and in the case of the reduced rank model, the rank of the individual spaces (thus the rank of $\mathbf{S}_k$) may be small (say, 2).

7.3.5 Majorization Algorithm

A key contribution to MDS was made by de Leeuw (1977) when he first used the idea of majorization, albeit in the context of convex analysis. Up to then, the minimization of stress was essentially done through gradient algorithms, such as the one proposed by Kruskal (1964b). The problem is that if only a single pair ij has a zero distance, then the gradient is not defined any more. De Leeuw (1977) proved that by using subgradients for those Euclidean distances that are zero, a convergent algorithm can be obtained. In de Leeuw and Heiser (1977), the idea of majorization was worked out further. The algorithm uses in each iteration an auxiliary function (called the majorizing function) that is simple (quadratic in $\mathbf{X}$), touches the original stress function at the current estimate, and is located above the original stress function anywhere else (or has the same value as the stress function). Consequently, the update of the majorizing function must have a smaller (or equal) value as the majorizing function and as the stress function at the current estimate (as these two functions touch there). Because the stress function either touches or is smaller than the majorization function by construction, it must be so that at the update of the majorizing function, the stress function also is smaller than (or equal to) the stress function of the current estimate. Hence, making the $\mathbf{X}$ that minimizes the majorization function the next current estimate reduces stress (or keeps it the same). In practice, the majorizing algorithm is fast and reduces stress until the reductions in stress become very small. This algorithm for minimizing stress was coined SMACOF (Scaling by MAjorizing a COmplicated Function).

The Smacof approach operates on a slightly different formulation of raw stress:

$$\sigma_r^2(\mathbf{X},\hat{\mathbf{d}}) = \sum_{i<j} w_{ij}\left(\hat{d}_{ij} - d_{ij}(\mathbf{X})\right)^2 \qquad (7.4)$$

where w_{ij} are nonnegative weights indicating the importance of misrepresentation of a particular pair of objects ij. An obvious choice is $w_{ij} = 1$ for all ij so that all pairs contribute equally to the loss of function. The w_{ij} can also be used for accommodating missing dissimilarities. In the SMACOF approach, degeneration to $\mathbf{X} = 0$ and $\hat{\mathbf{d}} = 0$ is avoided by imposing the explicit restriction that the sum of squared d-hats must equal some positive constant, for example,

$$\sum_{i<j} w_{ij}\hat{d}_{ij}^2 = n(n-1)/2.$$

The strength of the majorization approach lies in its generalizability and the desirable properties of the algorithm. One advantage is that it allows imposing constraints onto the configuration quite easily for those constraints for which least-squares solutions exist (de Leeuw and Heiser, 1980). The reason for this is that the majorizing function is quadratic in $\mathbf{X}$ in each iteration. For example, three-way MDS models (such as the weighted Euclidean model, the generalized Euclidean model, and the reduced rank model) are expressed in the stress framework as

$$\sigma_r^2(\mathbf{X}) = \sum_k \sum_{i<j} w_{ijk}\left(\delta_{ijk} - d_{ij}(\mathbf{GS}_k)\right)^2. \qquad (7.5)$$

Let Δ^*, $\mathbf{W}^*$, and $\mathbf{X}^*$ be defined as

$$\Delta^* = \begin{bmatrix} \Delta_1 & 0 & 0 \\ 0 & \Delta_2 & 0 \\ 0 & 0 & \Delta_3 \end{bmatrix}, \mathbf{W}^* = \begin{bmatrix} \mathbf{W}_1 & 0 & 0 \\ 0 & \mathbf{W}_1 & 0 \\ 0 & 0 & \mathbf{W}_3 \end{bmatrix}, \text{ and } \mathbf{X}^* = \begin{bmatrix} \mathbf{X}_1 \\ \mathbf{X}_2 \\ \mathbf{X}_3 \end{bmatrix}.$$

Now, three-way MDS can be viewed as doing a constrained MDS on Δ^*, $\mathbf{W}^*$, and $\mathbf{X}^*$, where $\mathbf{X}_k$ is constrained to be of the form $\mathbf{GS}_k$.

Other algorithmic properties on convergence of the algorithm were proved in de Leeuw (1988a). Note that such convergence properties are not available for other MDS algorithms. The SMACOF algorithm is available in the SPSS module PROXSCAL (Meulman et al., 1999) and as the smacof package in R (de Leeuw and Mair, 2009c).

Although Euclidean distances are the easiest to visually interpret, and therefore are predominantly used in MDS, there can be reasons to deviate from the Euclidean distance and use the more general Minkowski distance

$$d_{ij}(\mathbf{X}) = \left(\sum_s |x_{is} - x_{js}|^q\right)^{1/q}$$

for $q \geq 1$. The well-known special cases are the city block ($q = 1$), the Euclidean ($q = 2$), and the dominance ($q = \infty$) distances. The majorization approach to MDS was extended to deal with these cases in Groenen et al. (1995) for $1 \leq q \leq 2$ and also for $q \geq 2$ in Groenen et al. (1999).

7.3.6 Other Algorithms

The property of undefined gradients for stress led Takane et al. (1977) to propose the S-stress loss function that minimizes the sum over all pairs of the squared differences of squared Euclidean distances and squared dissimilarities as its gradient is defined for all distances, even if they are zero. The disadvantage of S-stress is that it will tend to overrepresent large dissimilarities, and that it can allow for relatively large errors for small dissimilarities.

Ramsay (1977) proposed the MULTISCALE loss function that equals the sum of squared differences of the logarithms of dissimilarities and distances, or, equivalently, the sum of squared logarithms of the ratio of a dissimilarity and its distance. If the ratio is 1, the log is zero and there is a perfect representation of the dissimilarity by its distance. Note that zero dissimilarities or zero distances cannot be handled by MULTISCALE. The advantage of this method is that it can be seen within a maximum likelihood framework. In particular, its three-way extension with replications by individuals allows inference and confidence ellipses for the points.

7.4 Present

7.4.1 Distance-Based Multivariate Analysis

In a series of publications in the period from 1986 to 1998, Meulman generalized the relation of principal coordinates analysis and classical MDS to a much wider range of multivariate analysis techniques such as (multiple) correspondence analysis, (generalized) canonical correlation analysis, and discriminant analysis. The emphasis in this approach is on the representation of the objects, and much less on the variables. It allows for optimal scaling of the variables in the Gifi (1990) approach approximating distances through stress or strain. A comprehensive overview was outlined in her thesis (Meulman, 1986), and a series of papers were written on that topic (see, for example, Meulman, 1992).

7.4.2 Constant Dissimilarities

Users of MDS should always check if their data have sufficient variation in the dissimilarities or d-hats before starting to interpret a solution. Even though the use of stress is dominant in MDS, little was known of the 'nullest of null models', that is, having constant dissimilarities without any variation (Buja et al., 1994). Many classical multivariate analysis techniques assume centred data so that constant data do not occur. Buja et al. proved what kind of solutions occur when stress with Euclidean distances is fed with constant dissimilarities. It turns out that in one dimension, the objects will be positioned equally spaced on a line, in two dimensions, the points will be in concentric circles, and in three dimensions, they are positioned on the surface of a hypersphere. The two-dimensional solution is used very often in MDS applications, and near-constant dissimilarity data can occur after transformations. The contribution of Buja et al. (1994) is that they focused the attention on such noninformative solutions.

7.4.3 Local Minima

The advantage of classical MDS is that there is an algebraic solution that yields a global minimum. Some of the disadvantages are that it cannot handle transformations of the dissimilarities, and that the resulting distances often underestimate the dissimilarities. The use of stress avoids these disadvantages but introduces the problem of local minima. From 1978 until recently, several contributions have been made. Three cases should be distinguished.

First, de Leeuw and Heiser (1977) and Defays (1978) noted that unidimensional scaling becomes a very large combinatorial problem. Hubert and Golledge (1981) and Hubert and Arabie (1986) used dynamic programming to globally optimize the combinatorial problem, and hence the unidimensional stress function, up to about 22 objects only at that time. The approach by Pliner (1996) that smoothes small distances in $\rho(\mathbf{X})$ by a quadratic function is very effective in finding global minima for even larger unidimensional scaling problems. Brusco (2001) applied simulated annealing.

The second case is for $2 \leq p < n - 1$. For small $p \geq 2$, Groenen and Heiser (1996) found that local minima occur frequently, more so in low dimensionalities than for higher dimensionalities. They proposed the tunnelling method that is indeed capable of finding a series of subsequent lower local minima that could end in a global minimum, although there is no guarantee of finding it. In Groenen et al. (1999), the smoothing approach of Pliner (1996) was adapted and extended to deal with any Minkowski distance and in any dimensionality. For city block distances and Euclidean distances, their smoothing approach was effective in locating a global optimum, but for Minkowski distances with exponents $q > 2$, in particular the dominance distance ($q = \infty$), the smoothing approach was not effective. For the special case of city block distances, combinatorial approaches have been proposed by Hubert et al. (1992). More recently, there has been a series of articles on simulated annealing approaches of several MDS variants that work well when appropriately tuned (Murillo et al., 2005; Vera et al., 2007).

The third case is full-dimensional scaling ($p = n - 1$). De Leeuw (1993) proved that there is a single local minimum for stress that is global.

7.5 Future

7.5.1 MDS with Weights

Apart from handling missings, the weights w_{ij} in the stress function (7.4) can be also be used to emphasize certain aspects of the dissimilarities. The first one to exploit this was Heiser (1988), who proposed to mimic certain other MDS loss functions by choosing appropriate weights. For example, the S-stress loss function can be mimicked by $w_{ij} = \delta_{ij}^2$. Buja and Swayne (2002) emphasized choosing $w_{ij} = \delta_{ij}^q$ allows a more refined weighting of the errors depending on the size of δ_{ij}. If the objective is to have the large dissimilarities well represented, then q should be chosen large (for example, $q = 5$ or 10). Conversely, if the interest is in the proper representation of small dissimilarities and the larger ones are less important in the representation, then this can be assured by choosing q small (for example, $q = -5$ or -10). Making the weights w_{ij} in such a way dependent on the dissimilarities allows emphasizing the proper representation of certain selection of the dissimilarities that is dependent of their values. This can be particularly useful in the context of large-scale MDS.

7.5.2 Dynamic MDS

For most of its lifetime, MDS has been a static method: dissimilarities are input to an MDS program producing a usually two-dimensional solution that is shown as a map. The GGvis software of Buja and Swayne (2002) (see also Buja et al., 2008) that is a part of GGobi was the first comprehensive interactive software. The advantage is that in real time MDS options can be changed, and their effects are immediately shown as the iterative process progresses. Such dynamics allow for a completely new, direct, and intuitive interaction with an MDS user on the interplay of the specific dissimilarities at hand and the possible MDS options.

7.5.3 Large-Scale MDS

Traditional MDS tends to have a small number of objects (say, between 10 and 200). Both computations and interpretation completely change when one is dealing with far more objects, e.g., 10.000 to 100.000 objects. With such large n the total number of pairs of objects $n(n-1)/2$ increases quadratically, which generally is prohibitive for standard MDS algorithms.

The ISOMAP (isometric feature mapping) algorithm of Tenenbaum et al. (2000) is based on classical MDS for large-scale MDS problems. In particular, it focuses on nonlinear manifolds in higher dimensionalities. By k-nearest neighbours a network of connected pairs of objects is created by declaring the bigger dissimilarities as missing. This structure of nonmissing dissimilarities can be seen as a weighted graph with weights at the vertices being the dissimilarities. As classical MDS cannot handle missing dissimilarities; they are replaced by the shortest path on the graph. This forms a fully filled matrix of pseudodissimilarities on which classical MDS is performed. In this way, ISOMAP is capable of recovering low-dimensional manifolds that exists in higher dimensionality.

Another approach was taken by Trosset and Groenen (unpublished; see also the dissertation of Kagie, 2010). Trosset and Groenen proposed to allow for many missing values, thereby creating a sparse dissimilarity matrix. In an adaptation of SMACOF, they provide an algorithm that can indeed handle large n provided there is sufficient sparseness. In joint work, the dissertation of Kagie expands on this approach. It was noticed that often large-scale MDS solutions are dominated by the mode of the distribution of dissimilarities and that, therefore, solutions comparable to the constant dissimilarity case occur often, such as in two dimensions, a circle filled with a blur of points. It was proposed to be solved by an a priori weighting of the dissimilarities to avoid a single mode becoming dominant and by appropriate transformations.

7.5.4 Symbolic MDS

The use of symbolic data and adapted multivariate analysis methods has been advocated in Bock and Diday (2000)—see Chapter 16 by Verde and Diday in this book. Symbolic data can be seen as richer forms of data values. Here, we discuss two such forms: (1) the case that for each pair ij not the dissimilarity is known, but the interval of the dissimilarity, and (2) the case that a distribution (histogram) of the dissimilarity for each pair ij is known. Often, such symbolic data are obtained by aggregation or summary statistics over larger units such as geographic areas, countries, etc.

For interval dissimilarities, Denoeux and Masson (2000) proposed to present the coordinates of object also as intervals yielding a rectangle to represent an object in two dimensions. Then, the smallest distance of rectangles i and j should match as closely as possible the lower boundary value of the interval for dissimilarity ij, and the largest distance of the rectangles should match as the upper boundary value of the interval.

Groenen and Winsberg (2006) proposed to model histogram dissimilarities. In this case, the distribution of a dissimilarity is summarized by several quantiles, for example, by the percentiles 20, 30, 40, 60, 70, and 80. In this model, the percentiles should be chosen in pairs around the median, so 20–80, 30–70, and 40–60. Therefore, each such pair consists of an interval dissimilarity. The current choice of percentiles can be perceived as a three-way interval dissimilarity matrix with three replications (as there are three percentile pairs). Just as in regular three-way MDS, there will be one common matrix **X** with the rectangle centres for all three replications. The heights and widths of the rectangles over the three replications should be such that each rectangle representing a percentile range closer to the median should fit within a rectangle representing a wider percentile range around the median. For more details, we refer to Groenen and Winsberg (2006).

7.5.5 What Needs to Be Done?

Of the developments described above in this section, what is most interesting for the typical MDS user is the possibility to interact more with MDS programs in a dynamic way (as in GGvis). Heady developed a powerful interactive MDS program called PERMAP, a stand-alone program that is available as free-ware on the Internet (http://cda.psych.uiuc.edu/mds 509 2013/permap/permap 11 8pdf.pdf). Unfortunately, this program is not supported anymore. For the user, it would be nice to see how MDS responds if he or she eliminates some objects/points from the solution, shifts some points in space, or draws in some partitioning lines that are then enforced (in some way such as linear axial constraints) onto the MDS solution. Such programs would be hard to write and test, of course, but often simpler programs are missing too. For example, programs that can handle asymmetric proximities and produce vector fields over MDS plots would be helpful to diagnose asymmetric data for systematic trends. Even Procrustean transformations that are needed when comparing different MDS solutions for similarities and differences are missing (or are difficult to find) in many of the statistics packages. It is hoped that such programs will soon be generated within the R environment where they should also survive longer than has been true for many of the old Fortran programs, such as KYST and PINDIS, for example.

Another area where we expect more developments is large-scale MDS. For this case, there are several technical and perceptual problems. When is large-scale MDS interesting? How many dimensions should be used? How can points in such plots informatively be labeled? How should dissimilarities and weights be adapted such that MDS yields informative solutions? How should we treat missing values that could lead to unconnected or only partially connected groups of objects? In this area, we expect that researchers from computer science and machine learning are and will contribute to new developments. One such development is the visualization of similarities (VOS) approach for bibliometrics; see Van Eck and Waltman (2010), and for software and more references, http://www.vosviewer.com/.

8
History of Cluster Analysis

Fionn Murtagh

CONTENTS

8.1 Early Clustering in the 1950s to 1970s and Legacy 118
    8.1.1 Clustering and Data Analysis in the 1960s, and the Establishment of the Classification Society 118
    8.1.2 Between Language Engineering and Language Understanding: Clustering and Data Analysis in the 1950s to 1970s 119
    8.1.3 Clustering in Computer Science in the 1950s and 1960s 120
    8.1.4 Clustering in Benzécri's and Hayashi's Vision of Science 121
8.2 Algorithms for Hierarchical Clustering: 1960s to 1980s and Their Legacy 123
    8.2.1 Hierarchical Clustering 123
    8.2.2 Ward's Agglomerative Hierarchical Clustering Method 124
    8.2.3 Lance–Williams Dissimilarity Update Formula 125
    8.2.4 Generalizing Lance–Williams 126
    8.2.5 Hierarchical Clustering Algorithms 126
8.3 Hierarchical Clustering as an Embedding of Observations in an Ultrametric Topology 127
    8.3.1 Ultrametric Space for Representing a Hierarchy 127
    8.3.2 Some Geometrical Properties of Ultrametric Spaces................ 128
    8.3.3 Ultrametric Matrices and Their Properties............................. 128
8.4 Then and Now: Continuing Importance of Clustering....................... 130
    8.4.1 Antecedents of Data Analytics 130
    8.4.2 Benzécri's Data Analysis Project 130
    8.4.3 Pattern Recognition 131
    8.4.4 A Data Analysis Platform 131
    8.4.5 Role of Cluster Analysis in Science 131
    8.4.6 Conclusion 132
Appendix............................... 132

Cluster analysis has a rich history, extending from the 1950s onward. It is noticeable how different disciplines were the primary areas of work in clustering at various times. Such applications were the drivers of methodology development and practical deployment. In turn, disciplines like computer science, library and information science, and statistics owe a great deal to the theory and practice of cluster analysis.

8.1 Early Clustering in the 1950s to 1970s and Legacy

8.1.1 Clustering and Data Analysis in the 1960s, and the Establishment of the Classification Society

In his description of early work with John Gower in the Statistics Department at Rothamsted Experimental Station in 1961, when Frank Yates was head of department, Gavin Ross reviewed data analysis as follows (Ross, 2007):

> We had several requests for classification jobs, mainly agricultural and biological at first, such as classification of nematode worms, bacterial strains, and soil profiles. On this machine and its faster successor, the Ferranti Orion, we performed numerous jobs, for archaeologists, linguists, medical research laboratories, the Natural History Museum, ecologists, and even the Civil Service Department. On the Orion we could handle 600 units and 400 properties per unit, and we programmed several alternative methods of classification, ordination and identification, and graphical displays of the minimum spanning tree, dendrograms and data plots. My colleague Roger Payne developed a suite of identification programs which was used to form a massive key to yeast strains. The world of conventional multivariate statistics did not at first know how to view cluster analysis. Classical discriminant analysis assumed random samples from multivariate normal populations. Cluster analysis mixed discrete and continuous variables, was clearly not randomly sampled, and formed non-overlapping groups where multivariate normal populations would always overlap. Nor was the choice of variables independent of the resulting classification, as Sneath had originally hoped, in the sense that if one performed enough tests on bacterial strains the proportion of matching results between two strains would reflect the proportion of common genetic information. But we and our collaborators learnt a lot from these early endeavours.

In establishing the Classification Society in 1964, the interdisciplinarity of the objectives was stressed: 'The foundation of the society follows the holding of a Symposium, organized by ASLIB on 6 April, 1962, entitled 'Classification: an interdisciplinary problem', at which it became clear that there are many aspects of classification common to such widely separated disciplines as biology, librarianship, soil science, and anthropology, and that opportunities for joint discussion of these aspects would be of value to all the disciplines concerned' (Classification Society, 1964). ASLIB, the Association for Information Management, began in 1924 as the Association of Special Libraries and Information Bureaux, in England.

The legacy of early work in clustering and classification remains in a key and central role to this day. In a retrospective view of a founder of the field of application-driven data analysis, Jean-Paul Benzécri (2007) sketched out target areas of application that range over analysis of voting and elections, jet algorithms for the Tevatron and Large Hadron Collider systems, gamma ray bursts, environment and climate management, sociology of religion, data mining in retail, speech recognition and analysis, sociology of natality—analysis of trends and rates of births, and economics and finance—industrial capital in Japan, financial data analysis in France, and monetary and exchange rate analysis in the United States. In all cases the underlying explanations are wanted, and not superficial displays or limited regression modelling.

8.1.2 Between Language Engineering and Language Understanding: Clustering and Data Analysis in the 1950s to 1970s

In this period, 1950s to 1970s, there was great interest in information retrieval, databases, and with that came linguistics too. In the UK, this led to mainstream computer science. In France and Japan, it led more towards quantitative social science.

Roger Needham and Karen Spärck Jones were two of the most influential figures in computing and the computational sciences in the UK and worldwide. The work of Roger Needham, who died in February 2003, ranged over a wide swathe of computer science. His early work at Cambridge in the 1950s included cluster analysis and information retrieval. In the 1960s, he carried out pioneering work on computer architecture and system software. In the 1970s, his work involved distributed computing. In later decades, he devoted considerable attention to security. In the 1960s he published on clustering and classification. Information retrieval was among the areas he contributed to. Among his early publications on clustering and classification were Needham (1965, 1967), Spärck Jones and Needham (1964, 1968).

Needham, who was the husband of Spärck Jones, set up and became the first director of Microsoft Research in Cambridge in 1997. Karen Spärck Jones died in April 2007. Among early and influential publications on her side, on the theory and practice of clustering, were the following: Spärck Jones (1965, 1970, 1971; with D. M. Jackson, 1970). (Note that we write Karen's published name as Sparck Jones in the references.)

Even in disciplines outside of formative or emergent computer science, the centrality of data analysis algorithms is very clear from a scan of publications in earlier times. A leader of classification and clustering research over many decades is James Rohlf (State University of New York). As one among many examples, we note his work on single-link clustering (Rohlf, 1973).

8.1.3 Clustering in Computer Science in the 1950s and 1960s

I will now turn attention to the early years of the *Computer Journal*. A leader in early clustering developments and in information retrieval, C. J. (Keith) van Rijsbergen (Glasgow University and now also Cambridge University) was editor in chief of the *Computer Journal* from 1993 to 2000. A few of his early papers include van Rijsbergen (1970, 1974, 1977; with Nick Jardine, 1971).

From 2000 to 2007, and again from 2012, I have been in this role as editor in chief of the *Computer Journal*. I wrote in an editorial for the 50th anniversary of the journal in 2007 the following (Wilks et al., 2007):

> When I pick up older issues of the *Computer Journal*, I am struck by how interesting many of the articles still are. Some articles are still very highly cited, such as Fletcher and Powell on gradient descent. Others ... on clustering, data analysis, and information retrieval, by Lance and Williams, Robin Sibson, Jim Rohlf, Karen Spärck Jones, Roger Needham, Keith van Rijsbergen, and others, to my mind established the foundations of theory and practice that remain hugely important to this day. It is a pity that journal impact factors, which mean so much for our day to day research work, are based on publications in just two previous years. It is clear that new work may, or perhaps should, strike out to new shores, and be unencumbered with past work. But there is of course another also important view, that the consolidated literature is both vital and a wellspring of current and future progress. Both aspects are crucial, the 'sleep walking' innovative element, to use Arthur Koestler's characterization, and the consolidation element that is part and parcel of understanding.

Business origins and influence can be noted in regard to research in the field. The very first issue of the *Computer Journal* in 1958 had articles by authors with industrial research lab affiliations for the most part: Ferranti Ltd., Leo Computers Ltd., Computing Machine Laboratory, University of Manchester, Business Group of the British Computer Society, English Electric Company Ltd., Norwich Corporation, Royal Aircraft Establishment (RAE) (Farnborough), and again RAE.

Then from later issues I will note some articles that have very clear links with data analysis: 'Matching Inquiries to an Index' (Wright, 1961), as well as several articles on character recognition in the 1961 edition, work by Gower (1958, 1962), and Lloyd (1964). In 'Classification of a Set of Elements', by Rose (1964), the abstract has this: 'The paper describes the use of a computer in some statistical experiments on weakly connected graphs. The work forms part of a statistical approach to some classification problems'. There were also several articles on regression analysis in issues at these times.

8.1.4 Clustering in Benzécri's and Hayashi's Vision of Science

In general the term *analyse des données* in the French tradition means data mining and unsupervised classification. The latter is the term used in the pattern recognition literature, and it can be counterposed to supervised classification, or machine learning, or discriminant analysis.

If Benzécri was enormously influential in France in drawing out the lessons of data analysis being brought into a computer-supported age, in Japan Chikio Hayashi played no less a role. Hayashi (1918–2002) led areas that included public opinion research and statistical mathematics, and was first president of the Behaviormetric Society of Japan. In Hayashi (1954), his data analysis approach is set out very clearly.

First, what Hayashi referred to as quantification was the scene setting or data encoding and representation forming the basis of subsequent decision making. He therefore introduced (Hayashi, 1954, p. 61) 'methods of quantification of qualitative data in multidimensional analysis and especially how to quantify qualitative patterns to secure the maximum success rate of prediction of phenomena from the statistical point of view'. So, first data, second method, and third decision making are inextricably linked.

Next comes the role of data selection, weighting, decorrelation, low dimensionality selection and related aspects of the analysis, and classification. 'The important problem in multidimensional analysis is to devise the methods of the quantification of complex phenomena (intercorrelated behaviour patterns of units in dynamic environments) and then the methods of classification. Quantification means that the patterns are categorized and given numerical values in order that the patterns may be able to be treated as several indices, and classification means prediction of phenomena'. In fact, the very aim of factor analysis type analyses, including correspondence analysis, is to prepare the way for classification: 'The aim of multidimensional quantification is to make numerical representation of intercorrelated patterns synthetically to maximize the efficiency of classification, i.e. the success rate of prediction'. Factorial methods are insufficient in their own right, maybe leading just to display of data: 'Quantification does not mean finding numerical values but giving them patterns on the operational point of view in a proper sense. In this sense, quantification has not absolute meaning but relative meaning to our purpose'. The foregoing citations are also from Hayashi (1954, p. 61).

This became very much the approach of Benzécri too. Note that Hayashi's perspectives as described above date from 1954. In Benzécri (1983, p. 11), a contribution to the journal *Behaviormetrika* that was invited by Hayashi, Benzécri draws the following conclusions on data analysis: 'In data analysis numerous disciplines have to collaborate. The role of mathematics, although essential, remains modest in the sense that classical theorems are used almost exclusively, or elementary demonstration techniques. But it is necessary that certain abstract conceptions penetrate the spirit of the users, who are the specialists collecting the data and having to orientate the analysis in accordance with the problems that are fundamental to their particular science'.

Benzécri (1983, p. 11) develops the implications of this. The advance of computer capability (remember that this article was published in 1983) 'requires that the Data Analysis project ahead of the concrete work, the indispensable source of inspiration, a vision of science' (*Data Analysis* in uppercase indicating the particular sense of data analysis as—in Hayashi's terms and equally the spirit of Benzécri's work—quantification and classification). Benzécri as well as Hayashi developed data analysis as projecting a vision of science.

Benzécri (1983) continues: 'This vision is philosophical: it is not a matter of translating directly in mathematical terms the system of concepts of a particular discipline but of linking these concepts in the equations of a model. Nor is it a matter of accepting the data such as they are revealed, but instead of elaborating them in a deep-going synthesis which allows new entities to be discovered and simple relationships between these new entities'.

Finally, the overall domain of application of data analysis is characterized as follows: 'Through differential calculus, experimental situations that are admirably dissected into simple components were translated into so many fundamental laws. We believe that it is reserved for Data Analysis to express adequately the laws of that which, complex by nature (living being, social body, ecosystem), cannot be dissected without losing its very nature'.

While Hayashi and Benzécri shared a vision of science, they also shared greatly a view of methodology to be applied. Hayashi (1952, p. 70) referred to 'the problem of classification by quantification method' that is not direct and immediate clustering of data, but rather a careful combination of numerical encoding and representation of data as a basis for the clustering. Hayashi's aim was to discuss: '(1) the methods of quantification of qualitative statistical data obtained by our measurements and observations...; (2)...the patterns of behaviour must be represented by some numerical values; (3)...effective grouping is required'. Data analysis methods are therefore not applied in isolation. Benzécri (1983, p. 10) refers to correspondence analysis and hierarchical clustering, and indeed discriminant analysis ('so as not to be illusory, a discriminant procedure has to be applied using a first set of cases—the base set—and then trialled on other cases—the test set').

Benzécri (1983) refers, just a little, to the breakthrough results achieved in hierarchical clustering algorithms around this time, and described in the work of Juan (1982a, 1982b). These algorithmic results on hierarchical clustering are built on the work of de Rham (1980)—see the discussion in the section to follow.

8.2 Algorithms for Hierarchical Clustering: 1960s to 1980s and Their Legacy

8.2.1 Hierarchical Clustering

Given an observation set, X, we define dissimilarities as the mapping d from the Cartesian product of the observation set (i.e., all pairs of elements of X) into the set of positive real numbers. A dissimilarity is a positive, definite, symmetric measure (i.e., $d(x, y) > 0$; $d(x, y) = 0$ if $x = y$; $d(x, y) = d(y, x)$). If, in addition, the triangular inequality is satisfied (i.e., $d(x, y) \leq d(x, z) + d(z, y)$ for all x, y, z in X), then the dissimilarity is a true distance. A dissimilarity does not respect the triangular inequality. Lance and Williams (1967) use the term (i, j)-*measure* for a dissimilarity.

For d either a dissimilarity or a distance, it can be mapped onto an *ultrametric*, or tree distance, which defines a hierarchical clustering (and also an ultrametric topology, which goes beyond a metric geometry). An ultrametric differs from a distance in that the strong triangular inequality, commonly called the ultrametric inequality, is satisfied: $d(i, j) \leq \max\{d(i, k), d(k, j)\}$ (see Figure 8.1). In carrying out a hierarchical clustering, we say that a hierarchical clustering is induced on X.

A hierarchy, H, is defined as a binary, rooted, node-ranked tree, also termed a *dendrogram* (Benzécri, 1976; Johnson, 1967; Lerman, 1981; Murtagh, 1985). A hierarchy defines a set of embedded subsets of a given set of objects X, indexed by the set I. That is, object i in the object set X is denoted x_i, and $i \in I$. These subsets are totally ordered by an index function v, which is a stronger condition than the partial order required by the subset relation. The index function v is represented by the ordinate in Figure 8.1 (the height or level). There is a one-to-one correspondence between a hierarchy and an ultrametric space.

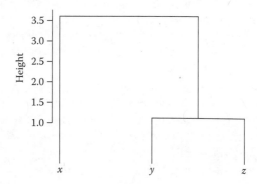

FIGURE 8.1
The strong triangular inequality defines an ultrametric. Every triplet of points satisfies the relationship $d(x, z) \leq \max\{d(x, y), d(y, z)\}$ for distance d. Reading off the hierarchy, this is verified for all x, y, z: $d(x, z) = 3.5$, $d(x, y) = 3.5$, $d(y, z) = 1.0$. In addition, the symmetry and positive definiteness conditions hold for any pair of points.

8.2.2 Ward's Agglomerative Hierarchical Clustering Method

For observations i in a cluster q, and a distance d (which can potentially be relaxed to a dissimilarity), we have the following definitions. We may want to consider a mass or weight associated with observation i, $p(i)$. By default we take $p(i) = 1/|q|$ when $i \in q$. With the context being clear, let q denote the cluster (a set) as well as the cluster's centre. We have this centre defined as $q = 1/|q| \sum_{i \in q} i$. Furthermore, and again the context makes this clear, i can be used for the observation label, or index, among all observations, and the observation vector. Some further definitions follow:

- Error sum of squares: $\sum_{i \in q} d^2(i, q)$
- Variance (or average centred sum of squares): $1/|q| \sum_{i \in q} d^2(i, q)$
- Inertia (or weighted average centred sum of squares): $\sum_{i \in q} p(i) d^2(i, q)$
  Inertia becomes variance if $p(i) = 1/|q|$, and becomes error sum of squares if $p(i) = 1$
- Euclidean distance squared using norm $||\ldots||$: if $i, i' \in R^{|J|}$, i.e., these observations have values on attributes $j \in \{1, 2, \ldots, |J|\}$, J is the attribute set, $|\ldots|$ denotes cardinality, then

$$d^2(i, i') = ||i - i'||^2 = \sum_j (i_j - i'_j)^2$$

Directly using error sum of squares or variance provides commonly used optimization criteria for data partitioning, or nonoverlapping, nonhierarchical clustering. For k-means, see Steinley (2006), Bock (2007), and Jain (2010). In Murtagh and Legendre (2014) it is described how the minimum variance compactness criterion is used in k-means (just as it is in competitive learning and the Kohonen self-organizing feature map—in all cases, providing for a suboptimal solution), and the Ward criterion similarly provides a solution based on the same compactness criterion (see Murtagh, 1985, for a discussion of the minimum variance agglomerative clustering criterion, in algorithmic terms).

Consider now a set of masses, or weights, m_i for observations i. Following Benzécri (1976, p. 185), the centred moment of order 2, $M^2(I)$ of the cloud (or set) of observations i, $i \in I$, is written: $M^2(I) = \sum_{i \in I} m_i ||i - g||^2$, where the centre of gravity of the system is $g = \sum_i m_i i / \sum_i m_i$. The variance, $V(I)$, is $V(I) = M^2(I)/m_I$, where m_I is the total mass of the cloud. Thanks to Huygen's theorem, the following can be shown (Benzécri, 1976, p. 186) for clusters q whose union makes up the partition Q:

$$M^2(Q) = \sum_{q \in Q} m_q ||q - g||^2$$

$$M^2(I) = M^2(Q) + \sum_{q \in Q} M^2(q)$$

$$V(Q) = \sum_{q \in Q} (m_q/m_I) \, ||q - g||^2$$

$$V(I) = V(Q) + \sum_{q \in Q} (m_q/m_I) \, V(q)$$

The $V(Q)$ and $V(I)$ definitions above are discussed in Jambu (1978, pp. 154–155). The last of the above can be seen to decompose (additively) total variance of the cloud I into (first term on the right-hand side) variance of the cloud of cluster centres ($q \in Q$), and (second term on the right-hand side) summed variances of the clusters. We can consider this last of the above relations as $T = B + W$, where B is the between-clusters variance, and W is the summed within-clusters variance.

A range of variants of the agglomerative clustering criterion and algorithm are discussed by Jambu (1978). These include minimum of the centred order 2 moment of the union of two clusters (p. 156), minimum variance of the union of two clusters (p. 156), maximum of the centred order 2 moment of a partition (p. 157), and maximum of the centred order 2 moment of a partition (p. 158). Jambu notes that these criteria for maximization of the centred order 2 moment, or variance, of a partition, were developed and used by numerous authors, with some of these authors introducing modifications (such as the use of particular metrics). Among authors referred to are Ward (1963), Orlóci (1967), and Wishart (1969). Joe H. Ward Jr., who developed this method, died on June 23, 2011, at age 84.

8.2.3 Lance–Williams Dissimilarity Update Formula

Lance and Williams (1967) (see also Anderberg, 1973) established a succinct form for the update of dissimilarities following an agglomeration. The parameters used in the update formula are dependent on the cluster criterion value. Consider clusters (including possibly singletons) i and j being agglomerated to form cluster $i \cup j$, and then consider the redefining of dissimilarity relative to an external cluster (including again possibly a singleton), k. The update formula is

$$d(i \cup j, k) = a(i) \, d(i, k) + a(j) \, d(j, k) + b \, d(i, j) + c \, |d(i, k) - d(j, k)|$$

where d is the dissimilarity used, which does not have to be a Euclidean distance to start with, insofar as the Lance and Williams formula can be used as a repeatedly executed recurrence, without reference to any other or separate criterion; coefficients $a(i)$, $a(j)$, b, and c are defined with reference to the clustering criterion used (see tables of these coefficients in Murtagh (1985, p. 68; Jambu, 1989, p. 366), and $|\ldots|$ here denotes absolute value.

The Lance–Williams recurrence formula considers dissimilarities and not dissimilarities squared. The original Lance and Williams paper (1967) does not consider the Ward criterion. It does, however, note that it allows one to 'generate an infinite set of new strategies' for agglomerative hierarchical clustering. Wishart (1969) brought the Ward criterion into the Lance–Williams algorithmic framework. Even starting the agglomerative process with a Euclidean distance will not avoid the fact that the intercluster (nonsingleton, i.e., with two or more members) dissimilarity does not respect the triangular inequality, and hence it does not respect this Euclidean metric property.

8.2.4 Generalizing Lance–Williams

The Lance and Williams recurrence formula has been generalized in various ways. See, for example, Batagelj (1988), who discusses what he terms generalized Ward clustering, which includes agglomerative criteria based on variance, inertia, and weighted increase in variance. Jambu (1989, pp. 356 et seq.) considers the following cluster criteria and associated Lance–Williams update formula in the generalized Ward framework: centred order 2 moment of a partition, variance of a partition, centred order 2 moment of the union of two classes, and variance of the union of two classes.

If using a Euclidean distance, the Murtagh (1985) and the Jambu (1989) Lance–Williams update formulas for variance and related criteria (as discussed by Jambu, 1989) are associated with an alternative agglomerative hierarchical clustering algorithm that defines cluster centres following each agglomeration, and thus does not require use of the Lance–Williams update formula. The same is true for hierarchical agglomerative clustering based on median and centroid criteria.

8.2.5 Hierarchical Clustering Algorithms

Average time implementations that come close to $O(n)$ (i.e., related linearly to number of objects) include the following. Rohlf (1973) discusses an $O(n \log \log n)$ expected time algorithm for the minimal spanning tree, which can subsequently be converted to a single-link hierarchic clustering in $O(n \log n)$ time (by sorting the edge distances and reading off the sorted list). Bentley et al. (1980) discuss an $O(n)$ expected time algorithm for the minimal spanning tree. In Murtagh (1983b) an $O(n)$ expected time algorithm is discussed for hierarchic clustering using the median method.

One could practically say that Sibson (1973) and Defays (1977) are part of the prehistory of clustering. At any rate, their $O(n^2)$ implementations of the single link and of a (non-unique) complete link method, respectively, were once widely cited.

Drawing on initial work (de Rham, 1980; Juan, 1982a) in the quarterly journal *Les Cahiers de l'Analyse des Données*, edited by J.-P. Benzécri, and running between the years 1976 and 1997 (for scanned copies of all 22 years of this journal, see the appendix to this chapter), Murtagh (1983a, 1983b, 1985) described implementations that required $O(n^2)$ time and either $O(n^2)$ or $O(n)$ space for the majority of the most widely used hierarchical clustering methods. These storage requirements refer, respectively, to whether dissimilarities or the initial data only need to be used.

These implementations are based on the quite powerful ideas of constructing nearest-neighbour chains and carrying out agglomerations whenever reciprocal nearest neighbours are encountered. The theoretical possibility of a hierarchical clustering criterion allowing such an agglomeration of reciprocal nearest neighbours to take place, without untoward local effects, is provided by the so-called *reducibility property*. This property of clustering criteria was first enunciated in Bruynooghe (1977) and is discussed in Murtagh (1985) and elsewhere. It asserts that a newly agglomerated pair of objects cannot be closer to any third-party object than the constituent objects had been. Whether or not this is always verified depends on the clustering criterion used.

In Day and Edelsbrunner (1984) and Murtagh (1983a, 1985), one finds discussions of $O(n^2)$ time and $O(n)$ space implementations of Ward's minimum variance (or error sum of squares) method (Murtagh and Legendre, 2013) and of the centroid and median methods. The latter two methods are termed the UPGMC and WPGMC methods by Sneath and Sokal (1973). Now, a problem with the cluster criteria used by these latter two methods is that the reducibility property is not satisfied by them. This means that the hierarchy constructed may not be unique as a result of inversions or reversals (nonmonotonic variation) in the clustering criterion value determined in the sequence of agglomerations.

Murtagh (1984) describes $O(n^2)$ time and $O(n^2)$ space implementations for the single-link method, the complete link method, and the weighted and unweighted group average methods (WPGMA and UPGMA). This approach is quite general vis-à-vis the dissimilarity used and can also be used for hierarchical clustering methods other than those mentioned.

Day and Edelsbrunner (1984) prove the exact $O(n^2)$ time complexity of the centroid and median methods using an argument related to the combinatorial problem of optimally packing hyperspheres into an m-dimensional volume. They also address the question of metrics: results are valid in a wide class of distances, including those associated with the Minkowski metrics.

The construction and maintenance of the nearest-neighbour chain, as well as the carrying out of agglomerations whenever reciprocal nearest neighbours meet, both offer possibilities for parallelization (see, for example, Griffiths et al., 1984).

8.3 Hierarchical Clustering as an Embedding of Observations in an Ultrametric Topology

8.3.1 Ultrametric Space for Representing a Hierarchy

Consider Figure 8.1 illustrating the ultrametric distance and its role in defining a hierarchy. An early, influential paper is Johnson (1967). The ultrametric topology was introduced by Marc Krasner (1944), after the ultrametric inequality (or strong triangular inequality) had been formulated by Hausdorff in 1934. Various terms are used interchangeably for analysis in and over fields such as p-adic, ultrametric, non-Archimedean, and isosceles. The natural geometric ordering of metric valuations is on the real line, whereas in the ultrametric case the natural ordering is a hierarchical tree.

8.3.2 Some Geometrical Properties of Ultrametric Spaces

We see from the following, based on Lerman (1981, Chapter 0, Part IV), that an ultrametric space is quite different from a metric one. In an ultrametric space everything 'lives' on a tree. Some further properties that are studied by Lerman (1981) are

1. Every point of a circle in an ultrametric space is a centre of the circle.
2. In an ultrametric topology, every ball is both open and closed (termed *clopen*).
3. An ultrametric space is 0-dimensional (see Chakraborty, 2005; Van Rooij, 1978).

It is clear that an ultrametric topology is very different from our intuitive, or Euclidean, notions.

8.3.3 Ultrametric Matrices and Their Properties

For an $n \times n$ symmetric matrix of positive real numbers to be a matrix of distances associated with an ultrametric distance, a necessary and sufficient condition is that a permutation of rows and columns satisfies the following form of the matrix:

- Above the diagonal term, equal to 0, the elements of the same row are nondecreasing.
- For every index k, if $d(k, k+1) = d(k, k+2) = \cdots = d(k, k+l+1)$, then $d(k+1, j) \leq d(k, j)$ for $k+1 < j \leq k+l+1$ and $d(k+1, j) = d(k, j)$ for $j > k+l+1$.

Under these circumstances, $l \geq 0$ is the length of the section beginning, beyond the principal diagonal, the interval of columns of equal terms in row k.

To illustrate the ultrametric matrix format, consider the small data set shown in Table 8.1. A dendrogram produced from this is in Figure 8.2. The ultrametric matrix that can be read off this dendrogram is shown in Table 8.2. Finally, a visualization of this matrix, illustrating the ultrametric matrix properties discussed above, is in Figure 8.3.

TABLE 8.1

Input Data: Eight Iris Flowers Characterized by Sepal and Petal Widths and Lengths

	Sepal Length	Sepal Width	Petal Length	Petal Width
Iris1	5.1	3.5	1.4	0.2
Iris2	4.9	3.0	1.4	0.2
Iris3	4.7	3.2	1.3	0.2
Iris4	4.6	3.1	1.5	0.2
Iris5	5.0	3.6	1.4	0.2
Iris6	5.4	3.9	1.7	0.4
Iris7	4.6	3.4	1.4	0.3

Source: Iris data in Fisher, R. A., *Annals of Eugenics*, 6, 179–188, 1936.

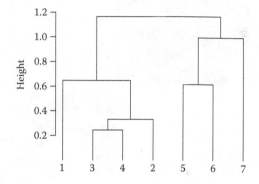

FIGURE 8.2
Hierarchical clustering of eight iris flowers using data from Table 8.1. No data normalization was applied and Euclidean distance was used. The agglomerative clustering criterion was the minimum variance or Ward one.

Clustering approaches have been developed that work by approximating an ultrametric structure of, say, a dissimilarity matrix. Implementation can be based on the principle of row and column permuting (leaving fully intact the row-column intersection value, and hence leading to a form of nondestructive or minimal intervention data analysis) such that small values are on or near the principal diagonal, allowing similar values to be near one another, and thereby facilitating visualization. See Deutsch and Martin (1971) and McCormick et al. (1972) for optimized ways to pursue this. Comprehensive surveys of clustering algorithms in this area, including objective functions, visualization schemes, optimization approaches, presence of constraints, and applications, can be found in Van Mechelen et al. (2004) and Madeira and Oliveira (2004). See too March (1983) for applications, and Murtagh (1985) for a short general review.

TABLE 8.2

Ultrametric Matrix Derived from the Dendrogram in Figure 8.2

	Iris1	Iris2	Iris3	Iris4	Iris5	Iris6	Iris7
Iris1	0	0.648	0.648	0.648	1.166	1.166	1.166
Iris2	0.648	0	0.332	0.332	1.166	1.166	1.166
Iris3	0.648	0.332	0	0.245	1.166	1.166	1.166
Iris4	0.648	0.332	0.245	0	1.166	1.166	1.166
Iris5	1.166	1.166	1.166	1.166	0	0.616	0.995
Iris6	1.166	1.166	1.166	1.166	0.616	0	0.995
Iris7	1.166	1.166	1.166	1.166	0.995	0.995	0

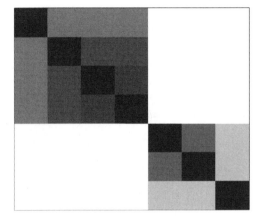

FIGURE 8.3
A visualization of the ultrametric matrix of Table 8.2, where bright or white = highest value and black = lowest value. Observables are ordered by row and column as the terminals in the dendrogram of Figure 8.2.

8.4 Then and Now: Continuing Importance of Clustering

8.4.1 Antecedents of Data Analytics

While antecedents of contemporary research and contemporary applications go back in time, I will conclude with the current major focus on 'big data' and data analytics that have come very much to the fore in recent times.

8.4.2 Benzécri's Data Analysis Project

Benzécri's early data analysis motivation sprung from text and document analysis. Research programs on correspondence analysis and 'aids to interpretation', as well as associated software programs, were developed and deployed on a broad scale in Benzécri's laboratory at the Université Pierre et Marie Curie, Paris 6, through the 1970s, 1980s, and 1990s. The hierarchical clustering programs distil the best of the reciprocal nearest-neighbours algorithm that was published in the early 1980s in Benzécri's journal, *Les Cahiers de l'Analyse des Données* (Journal of Data Analysis), and have not been bettered since then.

8.4.3 Pattern Recognition

With computers came pattern recognition and, at the start, neural networks. At a conference held in Honolulu in 1964 on methodologies of pattern recognition, which was attended by Benzécri, Rosenblatt's perceptron work was cited many times (albeit his work was cited, but not the perceptron as such). Frank Rosenblatt (1928–1971) was a pioneer of neural networks, including the perceptron and neuromimetic computing, which he developed in the 1950s. Early neural network research was simply what became known later as discriminant analysis. The problem of discriminant analysis, however, is insoluble if the characterization of observations and their measurements are not appropriate. This leads ineluctably to the importance of the data coding issue for any type of data analysis.

8.4.4 A Data Analysis Platform

The term *correspondence analysis* was first proposed in the fall of 1962. The first presentation under this title was made by Benzécri at the Collège de France in a course in the winter of 1963. By the late 1970s what correspondence analysis had become was not limited to the extraction of factors from any table of positive values. It also catered for data preparation, rules such as coding using complete disjunctive form, tools for critiquing the validity of results principally through calculations of contributions, provision of effective procedures for discrimination and regression, and harmonious linkage with cluster analysis. Thus, a unified approach was developed, for which the formalism remained quite simple, but for which deep integration of ideas was achieved with diverse problems. Many of the latter originally appeared from different sources, and some went back in time by many decades.

8.4.5 Role of Cluster Analysis in Science

The consequences of the data mining and related exploratory multivariate data analysis work overviewed in this article have been enormous. Nowhere have their effects been greater than in current search engine technologies. Also, wide swathes of database management, language engineering, and multimedia data and digital information handling are all directly related to the pioneering work described in this article.

I have looked at how early exploratory data analysis had come to play a central role in our computing infrastructure. An interesting and provocative view of how far clustering has come has been expressed by Peter Norvig, Google's research director (see Anderson, 2008), who finds that all of science has been usurped by exploratory, unsupervised clustering and data analysis, principally through Google's search facilities: 'Correlation supersedes causation, and science can advance even without coherent models, unified theories, or really any mechanistic explanation at all'. In Murtagh (2008), I discuss this thought-provoking view, which arises centrally out of clustering, or unsupervised classification, of data.

8.4.6 Conclusion

Cluster analysis has been at the origins of many areas of science (in the sense of natural, human, and social sciences). In Murtagh and Kurtz (2013) an analysis is carried out of about half a million published articles on cluster analysis and related data analysis. The very big disciplinary shift noted in that analysis of the clustering literature over the past two decades is from mathematical psychology, and biology and sociology, toward management, engineering, and also physics, economics, literature, and statistics.

The perennial importance and centrality of clustering comes out very clearly from the role played by cluster analysis in so many areas of scholarship and industry.

Appendix: Sources for Early Work

- Classification Literature Automated Search Service, a CD that was distributed with the *Journal of Classification*. See http://www.classification-society.org/clsoc for details of online availability.
- The following books have been scanned and are available in their entirety online to members of the Classification Society and the British Classification Society.
  1. *Algorithms for Clustering Data* (1988), AK Jain and RC Dubes
  2. *Automatische Klassifikation* (1974), HH Bock
  3. *Classification et Analyse Ordinale des Données* (1981), IC Lerman
  4. *Clustering Algorithms* (1975), JA Hartigan
  5. *Information Retrieval* (1979, 2nd ed.), CJ van Rijsbergen
  6. *Multidimensional Clustering Algorithms* (1985), F Murtagh
  7. *Principles of Numerical Taxonomy* (1963), RR Sokal and PHA Sneath

8. *Numerical Taxonomy: the Principles and Practice of Numerical Classification* (1973), PHA Sneath and RR Sokal

- *Les Cahiers de l'Analyse des Données* was the journal of Benzécri's lab from 1975 up to 1997, with four issues per year. Scanned versions of all issues are available at http://www.numdam.org.
- Some texts by Jean-Paul Benzécri and Françoise Benzécri-Leroy, published between 1954 and 1971, are available at http://www.numdam.org. Many other published works are available there by others also, including by Brigitte Escofier.

Section II

Contribution of Benzécri and the French School

9

Simple Correspondence Analysis

Pierre Cazes

CONTENTS

9.1 Factorial Analysis of a Cloud of Weighted Points 138
    9.1.1 Introduction and Notation 138
    9.1.2 Centred Matrix **Y** of the Cloud of Points 139
    9.1.3 Covariance Matrix **S** of the Cloud 139
    9.1.4 Total Inertia I_T of the Cloud 139
    9.1.5 Factorial Analysis of the Cloud: The Problem 140
    9.1.6 Solution of Factorial Analysis 140
    9.1.7 Aids to Interpretation 141
    9.1.8 Supplementary Elements 142
    9.1.9 Reconstitution Formula 142
    9.1.10 Torgerson Formula 143
9.2 Correspondence Analysis 143
    9.2.1 Introduction and Notation 143
    9.2.2 Centre of Gravity (Centroid) and Total Inertia of the Row Profiles 145
    9.2.3 Factorial Axial Vectors and Principal Coordinates 145
    9.2.4 Comparison of Factorial Analyses of Row and Column Profiles: Simultaneous Representation 146
    9.2.5 Distributional Equivalence Principle 146
    9.2.6 Aids to Interpretation 147
    9.2.7 Reconstitution Formula 147
    9.2.8 The Eigenvalues in CA Are Inferior or Equal to 1 147
    9.2.9 Symmetric Correspondence 148
    9.2.10 Burt Table 148

Correspondence analysis (CA) was introduced by Jean-Paul Benzécri (Benzécri et al., 1973) in the early 1960s to study contingency tables crossing the set of categories I of a qualitative variable with the set J of a second qualitative variable. Afterwards, this technique was extended to study homogeneous tables of positive numbers: tables of ranks, preferences, percentages, ratings, presences/absences (0/1), and so on. Lastly, methods of coding were developed to analyse any type of data, through the use of complete disjunctive coding (or dummy variable coding, also called 'crisp coding') and fuzzy coding (*codage flou* in French; see Chapter 15 by Greenacre).

CA can be viewed as a pair of factorial analyses or as a canonical analysis. Here I will present the former view, defining CA as the optimal representation of *I* and *J* points in a Euclidean space of low dimensionality, which was the way we introduced students to this technique in Benzécri's laboratory in Paris. The treatment is rather formal, but I have relaxed Benzécri's original notation of upper and lower indices (so-called Einstein notation, used mostly in physics) to use the simpler, but less precise, matrix-vector notation commonly used in multivariate analysis.

Section 9.1 summarizes the general theory of the factorial (or dimensional) analysis of a cloud of weighted points in a Euclidean space. In Section 9.2, this theory is applied to the case of CA.

9.1 Factorial Analysis of a Cloud of Weighted Points

9.1.1 Introduction and Notation

Consider the general problem of a cloud of n weighted points indexed by i ($1 \leq i \leq n$) in a real Euclidean space $E = \mathbb{R}^p$ of dimensionality p. One can think of the n points as representing sampling units and the dimensions as p variables. I will use the following notation:

- x_{ij} is the coordinate of the ith sampling unit (or case) on the jth dimension ($1 \leq j \leq p$) of $\mathbb{R}^p$, with the vector of p coordinates denoted by $\mathbf{x}_i$ ($p \times 1$).
- The n vectors $\mathbf{x}_i^\mathsf{T}$ are gathered as rows of the matrix $\mathbf{X}$ ($n \times p$).
- w_i is the (non-negative) weight associated with point i, and the weights sum to 1:

$$\sum_{i=1}^{n} w_i = \mathbf{1}^\mathsf{T}\mathbf{w} = 1,$$

where $\mathbf{1}$ is a vector of n 1s, and $\mathbf{w}$ is the vector of n weights.
- $\mathbf{D}_w$ ($n \times n$) is the diagonal matrix of the weights w_i.

Simple Correspondence Analysis

- **M** $(p \times p)$ is the symmetric positive-definite matrix that defines the inner product, and thus the norm and distance metric of $\mathbb{R}^p$: the scalar product between two vectors **u** and **v** of $\mathbb{R}^p$ is given by $<\mathbf{u}, \mathbf{v}>_M = \mathbf{u}^T\mathbf{M}\mathbf{v}$, the squared norm of a vector **u** is $\|\mathbf{u}\|^2_M = <\mathbf{u}, \mathbf{u}>_M = \mathbf{u}^T\mathbf{M}\mathbf{u}$, and the squared distance between **u** and **v** is $\|\mathbf{u} - \mathbf{v}\|^2_M = (\mathbf{u} - \mathbf{v})^T\mathbf{M}(\mathbf{u} - \mathbf{v})$.

9.1.2 Centred Matrix Y of the Cloud of Points

The centre of gravity (or centroid, or barycentre) of the cloud is given by

$$\mathbf{c} = \sum_{i=1}^{n} w_i \mathbf{x}_i$$

with coordinates c_j, where $\mathbf{c} = \mathbf{X}^T\mathbf{w}$ (remember that the point vectors are in the rows of **X**).

Each vector $\mathbf{x}_i$ is centred with respect to **c**: $\mathbf{y}_i = \mathbf{x}_i - \mathbf{c}$, and the centred vectors are gathered in the centred table $\mathbf{Y} = \mathbf{X} - \mathbf{1}\,\mathbf{c}^T = \mathbf{X} - \mathbf{1}\,\mathbf{w}^T\mathbf{X} = (\mathbf{I} - \mathbf{1}\,\mathbf{w}^T)\mathbf{X}$, where **1** denotes the vector of n 1s, and **I** the $n \times n$ identity matrix. The cloud of the rows of **Y** is called the *centred cloud*, and has centre of gravity at the origin of $\mathbb{R}^p$, the vector of p zeros. Working on the centred cloud is equivalent to working on the original one, and we continue with the centred one.

9.1.3 Covariance Matrix S of the Cloud

The (weighted) covariance matrix **S** $(p \times p)$ of the cloud of points is the symmetric matrix $\mathbf{S} = \mathbf{Y}^T\mathbf{D}_w\mathbf{Y} = \mathbf{X}^T\mathbf{D}_w\mathbf{X} - \mathbf{c}\mathbf{c}^T$. The diagonal values are weighted variances, and the off-diagonal values weighted covariances.

9.1.4 Total Inertia I_T of the Cloud

The total inertia I_T of the cloud is the weighted average squared distance of the n points to their centre of gravity:

$$I_T = \sum_{i=1}^{n} w_i \|\mathbf{x}_i - \mathbf{c}\|^2_M = \sum_{i=1}^{n} w_i \|\mathbf{y}_i\|^2_M.$$

The total inertia can be shown to be equal to

$$\text{trace}(\mathbf{S}\mathbf{M}) = \text{trace}(\mathbf{M}\mathbf{S}) = \sum_k \lambda_k,$$

where λ_k is the kth eigenvalue of **SM** (or **MS**).

9.1.5 Factorial Analysis of the Cloud: The Problem

The object of factorial analysis of the cloud is to find an optimal affine subspace E_k of $E = \mathbb{R}^p$ of small dimension k ($k = 2, 3$, etc.) to visualize the cloud's projection onto E_k. Optimality is defined as minimizing the weighted average squared distance of the points to the subspace. Suppose that $\hat{\mathbf{x}}_i$ denotes the projection of $\mathbf{x}_i$ on E_k. Since it can be easily proved that the subspace E_k contains the centre of gravity $\mathbf{c}$ (see, for example, Greenacre, 1984, Section 2.6.3), we can equivalently work with the centred points $\mathbf{y}_i$ and their projections $\hat{\mathbf{y}}_i$. E_k is then a vector subspace because the centre of gravity is at the origin. From Pythagoras's theorem, $\|\mathbf{y}_i\|_M^2 = \|\hat{\mathbf{y}}_i\|_M^2 + \|\mathbf{y}_i - \hat{\mathbf{y}}_i\|_M^2$; hence, the weighted average of the left-hand side (the total inertia) is decomposed into two parts, the inertia of the projected points in the subspace plus the weighted average squared distance of the points to the subspace. Hence, the objective of minimizing

$$\sum_{i=1}^{n} w_i \|\mathbf{y}_i - \hat{\mathbf{y}}_i\|_M^2$$

is equivalent to maximizing the inertia in the subspace,

$$I_k = \sum_{i=1}^{n} w_i \|\hat{\mathbf{y}}_i\|_M^2,$$

since the total inertia is fixed. It is this quantity to be maximized that we keep to characterize E_k.

9.1.6 Solution of Factorial Analysis

The subspace E_k can be defined by k orthonormal basis vectors $\mathbf{v}_1, \mathbf{v}_2, \ldots, \mathbf{v}_k$, which define the directions of k principal axes. These basis vectors can be shown to be the eigenvectors, in the metric $\mathbf{M}$, of the matrix $\mathbf{SM}$, associated with eigenvalues λ_k in decreasing order:

$$\mathbf{SM}\mathbf{v}_k = \lambda_k \mathbf{v}_k$$

$$\mathbf{v}_k^T \mathbf{M} \mathbf{v}_{k'} = 1 \text{ if } k = k', 0 \text{ otherwise}$$

$$\lambda_1 \geq \lambda_2 \geq \cdots \geq \lambda_k \geq 0$$

Notice that the matrix $\mathbf{SM}$ is not symmetric. The symmetric version of this problem can be obtained by premultiplying the above eigen-equation by $\mathbf{M}^{1/2}$:

$$(\mathbf{M}^{1/2}\mathbf{S}\mathbf{M}^{1/2})\mathbf{M}^{1/2}\mathbf{v}_k = \lambda_k \mathbf{M}^{1/2}\mathbf{v}_k$$

Simple Correspondence Analysis

so the eigenvectors of the symmetric matrix $\mathbf{M}^{1/2}\mathbf{S}\mathbf{M}^{1/2}$ are equal to $\mathbf{v}_k^* = \mathbf{M}^{1/2}\mathbf{v}_k$, from which $\mathbf{v}_k$ can be obtained as $\mathbf{v}_k = \mathbf{M}^{-1/2}\mathbf{v}_k^*$.

The coordinate of $\mathbf{y}_i$ on axis k is the scalar product in the metric $\mathbf{M}$ of the vector $\mathbf{y}_i$ with the basis vector $\mathbf{v}_k$: $f_{ik} = \mathbf{y}_i^T \mathbf{M} \mathbf{v}_k$. For the set of n points in the rows of $\mathbf{Y}$ these coordinates are gathered in $\mathbf{f}_k = \mathbf{Y}\mathbf{M}\mathbf{v}_k$, the so-called *principal coordinates* on axis k. The principal coordinates vector $\mathbf{f}_k$ can also be obtained as an eigenvector of $\mathbf{Q}\mathbf{D}_w$ (with $\mathbf{Q} = \mathbf{Y}\mathbf{M}\mathbf{Y}^T$), whose general term $q_{ii'} = \mathbf{y}_i^T \mathbf{M} \mathbf{y}_{i'}$ is the scalar product between $\mathbf{y}_i$ and $\mathbf{y}_{i'}$) relative to the eigenvalue λ_k ($\mathbf{Q}\mathbf{D}_w$ and $\mathbf{S}\mathbf{M}$ have the same nonnull eigenvalues).

The inertia of the points on axis k is equal to the eigenvalue λ_k:

$$\sum_{i=1}^{n} w_i f_{ik}^2 = \mathbf{f}_k^T \mathbf{D}_w \mathbf{f}_k = \mathbf{v}_k^T \mathbf{M} \mathbf{Y}^T \mathbf{D}_w \mathbf{Y} \mathbf{M} \mathbf{v}_k = \mathbf{v}_k^T \mathbf{M} \mathbf{S} \mathbf{M} \mathbf{v}_k = \lambda_k \mathbf{v}_k^T \mathbf{M} \mathbf{v}_k = \lambda_k$$

The inertia $I(E_k)$ of the points projected onto the subspace defined by the k first factorial axes is then equal to $\lambda_1 + \lambda_2 + \ldots + \lambda_k$. Since the principal coordinates are centred as a result of the $\mathbf{y}_i$ being centred, the covariance between coordinates $\mathbf{f}_k$ and $\mathbf{f}_{k'}$ on two different axes is equal to

$$\sum_{i=1}^{n} w_i f_{ik} f_{ik'} = \mathbf{f}_k^T \mathbf{D}_w \mathbf{f}_{k'} = \lambda_k \mathbf{v}_k^T \mathbf{M} \mathbf{v}_{k'} = 0,$$

because $\mathbf{v}_k^T \mathbf{M} \mathbf{v}_{k'} = 0$ if $k \neq k'$. Thus, the principal coordinates $\mathbf{f}_k$ are centred, uncorrelated, and have variance λ_k.

Knowing the factorial axes, the cloud can be visualized by projecting it onto the plane of the first two factorial axes, onto the plane on the factorial axes 1 and 3, 2 and 3, and so on.

Aids to interpretation, which are given below, are useful to help the statistician to analyse these visualizations.

9.1.7 Aids to Interpretation

The elements given below are defined for factorial axis k:

- Part of inertia of axis k: $\tau_k = \lambda_k/I_T = \lambda_k/\sum_m \lambda_m$, with $\tau_1 \geq \tau_2 \geq \ldots \geq \tau_k \geq \ldots \geq \tau_p$, summing to 1: $\sum_m \tau_m = 1$. This quantity, usually expressed as a percentage, gives the importance of axis k.
- Relative contribution of point i to the inertia of axis k: $\text{CTR}_k(i) = w_i f_{ik}^2 / \lambda_k$, summing to 1: $\sum_i \text{CTR}_k(i) = 1$.
- Relative contribution of axis k to the squared distance d_i^2 of $\mathbf{y}_i$ to the origin (which is the centre of gravity of the $\mathbf{y}_i$): $\text{COR}_k(i) = f_{ik}^2/d_i^2 = \cos^2(\theta_{ik})$, where $\cos(\theta_{ik})$ is the cosine of the angle between $\mathbf{y}_i$

and the factorial axis $\mathbf{v}_k$. Since $d_i^2 = \|\mathbf{y}_i\|^2_M = \Sigma_k f_{ik}^2$, the sum of these contributions over the axes is 1: $\Sigma_k \text{COR}_k(i) = 1$. If $\text{COR}_k(i) = 1$, the point i is on the axis k.

The results above can be easily generalized to the subspace E_k defined by the first k factorial axes, with the following contributions:

- Part of inertia of subspace E_k (or cumulated part of inertia):

$$\tau_{1,2,\ldots,k} = \sum_{m=1}^{k} \tau_m .$$

If this quantity is equal to 1, the cloud is perfectly represented in E_k.

- Quality of representation of y_i in the subspace E_k:

$$\text{QLT}_{1,2,\ldots,k}(i) = \sum_{m=1}^{k} \text{COR}_m(i).$$

If this quantity is equal to 1, the point i is perfectly represented in E_k. Finally, each point i makes a part contribution of $\text{INR}(i) = w_i d_i^2 / I_T$ to the total inertia, and these parts sum to 1: $\Sigma_i \text{INR}(i) = 1$.

9.1.8 Supplementary Elements

In the space $\mathbf{R}^p$ we may project any point vector $\mathbf{x}_s$ on the factorial axes, the coordinate f_{sk} of $\mathbf{x}_s$ on the kth factorial axis being equal to $\mathbf{x}_s^T \mathbf{M} \mathbf{v}_k$. An additional point projected on the axes in this way is called a *supplementary*, or *passive*, point, as opposed to the *active* points that have been used to define the factorial axes. It is equivalent to consider the supplementary point as part of the cloud, now of $n + 1$ points, with the added point $\mathbf{x}_s$ having a null weight. Supplementary elements are often used to aid the interpretation of factorial axes, for example, when the supplementary point is the centroid of a subset of individuals, the subset of men, or women, or individuals associated to an income class, etc. (see Chapter 13 by Blasius and Schmitz in this book).

9.1.9 Reconstitution Formula

Suppose that $\mathbf{Y}$ is of rank r ($r \leq p$). Since each row of $\mathbf{Y}$ can be written as a linear combination of the basis vectors $\mathbf{v}_k$, with coefficients equal to the principal coordinates, the elements of $\mathbf{Y}$ can be written as

$$y_{ij} = \sum_{k=1}^{r} f_{ik} v_{jk}$$

this is called the *reconstitution formula*. If we define the *standard coordinates* $\mathbf{u}_k = \mathbf{f}_k / \sqrt{\lambda_k}$ and if $\Lambda = \text{diag}(\lambda_1, \lambda_2, \cdots, \lambda_r)$ is the diagonal matrix of the r nonzero

Simple Correspondence Analysis

eigenvalues, then the reconstitution formula can be written in matrix form as $Y = FV^T = U\Lambda^{1/2}V^T$ with $F = (f_1 \cdots f_k \cdots f_r)$, $U = (u_1 \cdots u_k \cdots u_n)$, $V = (v_1 \cdots v_k \cdots v_n)$. This last matrix expression corresponds to the generalized singular value decomposition (SVD) of Y when we have the metrics M in the p-dimensional row space and D_w in the n-dimensional column space: $U^T D_w U = V^T M V = I_r$, where I_r is the $r \times r$ identity matrix. The original uncentred matrix X can then be reconstructed as a function of U, Λ, V, and the centroid c: $X = U\Lambda^{1/2}V^T + 1c^T$.

If we keep only the first r^* ($r^* < r$) factorial axes, we have an approximate reconstitution of Y (or X) noted Y^* (or X^*)—it suffices to replace r by r^* in the preceding formulae. The quality of the approximation is then given by the cumulated rate inertia $\tau_{1,2,\ldots,r^*}$ on the first r^* factorial axes.

9.1.10 Torgerson Formula

The representation of the cloud of points with respect to the basis of factorial axes depends only on the distances between the points and the weights of these points. This results in the formula by Torgerson (1958), which shows that the general term $q_{ii'}$ of the scalar matrix Q depends only on the interpoint distances and the point weights. This formula is given by the equation

$$q_{ii'} = -\tfrac{1}{2}(d_{ii'}^2 - d_{i\bullet}^2 - d_{\bullet i'}^2 + d_{\bullet\bullet}^2)$$

where $d_{ii'}^2$ is the square of the distance between y_i and $y_{i'}$, and the subindex $\bullet$ indicates weighted averaging, for example,

$$d_{i\bullet}^2 = \sum_{i'} w_{i'} d_{ii'}^2 .$$

Now, given a set of individuals $I = \{1, 2, \ldots, n\}$, each individual i with a weight w_i, and their interpoint distances $d_{ii'}$, the Torgerson formula, which is the basis of classic metric multidimensional scaling, also known as principal coordinate analysis, gives the matrix Q of scalar products between points. If Q is positive semi-definite, then the set I of individuals can be represented as points in a Euclidean space. There is an infinity of possible representations, but the representation in the system of factorial axes is unique, up to the signs of the axes.

9.2 Correspondence Analysis

9.2.1 Introduction and Notation

To introduce the concepts of correspondence analysis, we take the example of a contingency table N crossing two sets, for example, a set of I regions and a set of J political parties (nationalist, democrat, liberal, etc.), the general term n_{ij} of **N** being the number of voices obtained by the party j in region i.

We use the following notation:

- A subindex + denotes summation over the corresponding index, for example, $n_{i+} = \sum_j n_{ij}$, the total of row i of **N**, corresponding in the example to the number of voters in region i; similarly, n_{++} is the grand total of the table, which we shall simply denote by n.
- $\mathbf{P} = (1/n)\,\mathbf{N}$, the table of relative frequencies p_{ij}, summing to 1 over the whole table: $\sum_i \sum_j p_{ij} = 1$.
- p_{i+}, the ith row sum of **P**, equal to n_{i+}/n, which we shall denote simply by r_i, called the ith *row mass*. **r** is the vector of the r_i and $\mathbf{D}_r$ the diagonal matrix of the r_i. Similarly, p_{+j}, the jth column sum of **P**, equal to n_{+j}/n, denoted by c_j, is the jth *column mass*. **c** is the vector of the c_j and $\mathbf{D}_c$ the diagonal matrix of the c_j.
- $\mathbf{p}_i^r$, the profile of the ith row, which is the vector of components $n_{ij}/n_{i+.} = p_{ij}/r_i\,(j = 1, \ldots, J)$, with row profiles gathered in the rows of the matrix $\mathbf{D}_r^{-1}\mathbf{P}$. Similarly, $\mathbf{p}_j^c$ is the profile of the jth column, which is the vector of components $n_{ij}/n_{+j} = p_{ij}/c_j\,(i = 1, \ldots, I)$, with column profiles gathered in the columns of the matrix $\mathbf{P}\mathbf{D}_c^{-1}$.

If we want to differentiate two regions (rows) i and i' according their voting patterns, it is natural to compare the profiles of these two rows to suppress the size effect, for example, where region i is small (few voters) and i' is large (many voters). But it is logical to keep the importance of these regions for the final result by giving weights r_i and $r_{i'}$ to i and i', respectively. Similarly, to study the differences between two parties (columns) j and j', we compare their profiles and give weights c_j and $c_{j'}$ to j and j', respectively.

Correspondence analysis of **N** is the simultaneous study of the cloud of the I row profiles in $\mathbf{R}^J$ and the cloud of the J column profiles in $\mathbf{R}^I$ with the above-mentioned weights and appropriate metrics. This analysis is independent of the grand total n of **N**, so we also speak of the correspondence analysis of the matrix **P** of relative frequencies, which for the case of a two-way contingency table is a discrete bivariate distribution.

We first study the cloud of row profiles in the matrix $\mathbf{D}_r^{-1}\mathbf{P}$—the case of the column profiles will follow by symmetry. The row profile matrix $\mathbf{D}_r^{-1}\mathbf{P}$ corresponds to the matrix **X** of Section 9.1, while the row masses **r** correspond to the weights **w**, and the matrix $\mathbf{D}_c^{-1}$ to the metric **M**. This metric is called the *chi-square metric* of centre **c**. To perform the factorial analysis of the row profiles, we simply apply all the preceding results to $\mathbf{X} = \mathbf{D}_r^{-1}\mathbf{P}$, with weights $\mathbf{w} = \mathbf{r}\,(\mathbf{D}_w = \mathbf{D}_r)$ and metric $\mathbf{M} = \mathbf{D}_c^{-1}$.

9.2.2 Centre of Gravity (Centroid) and Total Inertia of the Row Profiles

The centre of gravity of the row profiles is equal to $(\mathbf{D}_r^{-1}\mathbf{P})^T\mathbf{r} = \mathbf{P}^T\mathbf{1} = \mathbf{c}$. The total inertia of the row profiles is equal to

$$\sum_i r_i \left\| \mathbf{p}_i^r - \mathbf{c} \right\|_{\mathbf{D}_c^{-1}}^2 = \sum_i r_i \sum_j \left((p_{ij}/r_i) - c_j \right)^2 / c_j = \sum_i \sum_j (p_{ij} - r_i c_j)^2 / (r_i c_j).$$

The last expression is symmetric in i and j, and is easily seen to be equal to χ^2/n, where χ^2 is the chi-square statistic used to test independence between the row and column variables.

9.2.3 Factorial Axial Vectors and Principal Coordinates

We apply the results of Section 9.1, replacing $\mathbf{Y}$ by the centred matrix $\mathbf{D}_r^{-1}\mathbf{P} - \mathbf{1}\mathbf{c}^T$, the metric $\mathbf{M}$ by $\mathbf{D}_c^{-1}$, and $\mathbf{D}_w$ by $\mathbf{D}_r$.

- Covariance matrix $\mathbf{S} = \mathbf{P}^T\mathbf{D}_r^{-1}\mathbf{P} - \mathbf{c}\mathbf{c}^T$.
- Matrix of scalar products $\mathbf{Q} = \mathbf{D}_r^{-1}\mathbf{P}\mathbf{D}_c^{-1}\mathbf{P}^T\mathbf{D}_r^{-1} - \mathbf{1}\mathbf{1}^T$.
- The factorial axis vectors $\mathbf{v}_k$ are eigenvectors of $\mathbf{S}\mathbf{D}_c^{-1} = (\mathbf{P}^T\mathbf{D}_r^{-1}\mathbf{P} - \mathbf{c}\mathbf{c}^T)\mathbf{D}_c^{-1} = (\mathbf{P}^T\mathbf{D}_r^{-1})(\mathbf{P}\mathbf{D}_c^{-1}) - \mathbf{c}\mathbf{1}^T$ corresponding to eigenvalues λ_k, where the matrices in parentheses are the row profile matrix (transposed) and column profile matrix, respectively. Notice that $\mathbf{c}$ is an eigenvector of $(\mathbf{P}^T\mathbf{D}_r^{-1})(\mathbf{P}\mathbf{D}_c^{-1})$—it can be verified that $(\mathbf{P}^T\mathbf{D}_r^{-1})(\mathbf{P}\mathbf{D}_c^{-1})\mathbf{c} = \mathbf{c}$—so it is equivalent to find the eigendecomposition of the uncentred $(\mathbf{P}^T\mathbf{D}_r^{-1})(\mathbf{P}\mathbf{D}_c^{-1})$, in which case the centroid $\mathbf{c}$ emerges as the so-called *trivial solution*, which can be denoted by $\mathbf{v}_0$. All the eigenvectors are normalized in the chi-square metric: $\mathbf{v}_k^T\mathbf{D}_c^{-1}\mathbf{v}_k = 1$.
- The principal coordinates of the row profiles on axis k are $\mathbf{f}_k = (\mathbf{D}_r^{-1}\mathbf{P})\mathbf{D}_c^{-1}\mathbf{v}_k$, and the standard coordinates are $\boldsymbol{\varphi}_k = \mathbf{f}_k/\sqrt{\lambda_k}$. Principal and standard coordinates have normalizations $\mathbf{f}_k^T\mathbf{D}_r\mathbf{f}_k = \lambda_k$ and $\boldsymbol{\varphi}_k^T\mathbf{D}_r\boldsymbol{\varphi}_k = 1$, respectively. The principal coordinates in $\mathbf{f}_k$ are eigenvectors of $\mathbf{Q}\mathbf{D}_r = \mathbf{D}_r^{-1}\mathbf{P}\mathbf{D}_c^{-1}\mathbf{P}^T - \mathbf{1}\mathbf{r}^T$ and of $\mathbf{D}_r^{-1}\mathbf{P}\mathbf{D}_c^{-1}\mathbf{P}^T$ relative to eigenvalue λ_k because $\mathbf{D}_r^{-1}\mathbf{P}\,\mathbf{D}_c^{-1}\mathbf{P}^T\,\mathbf{1} = \mathbf{1}$, in which case the trivial solution is given by $\mathbf{f}_0 = \mathbf{D}_r^{-1}\mathbf{P}\mathbf{D}_c^{-1}\mathbf{c} = \mathbf{D}_r^{-1}\mathbf{P}\,\mathbf{1} = \mathbf{D}_r^{-1}\,\mathbf{r} = \mathbf{1}$, because $\mathbf{v}_0 = \mathbf{c}$.

Similarly, all of the above can be repeated for the centred matrix of column profiles, transposed so that column profiles are in the rows of $\mathbf{D}_c^{-1}\mathbf{P}^T - \mathbf{1}\mathbf{r}^T$, with chi-square metric in $\mathbf{D}_r^{-1}$ and weights in $\mathbf{D}_c$. The solution follows in a symmetric way, leading to eigenvectors $\mathbf{u}_k$ of the matrix $(\mathbf{P}\mathbf{D}_c^{-1})(\mathbf{P}^T\mathbf{D}_r^{-1})$, including the trivial solution $\mathbf{u}_0 = \mathbf{r}$, with identical eigenvalues λ_k, principal coordinates $\mathbf{g}_k = (\mathbf{D}_c^{-1}\mathbf{P}^T)\mathbf{D}_r^{-1}\mathbf{u}_k$, and standard coordinates $\boldsymbol{\gamma}_k = \mathbf{g}_k/\sqrt{\lambda_k}$.

9.2.4 Comparison of Factorial Analyses of Row and Column Profiles: Simultaneous Representation

The rows and columns analyses, in $\mathbf{R}^J$ and $\mathbf{R}^I$, respectively, are intimately connected: both have the same total inertias, and the same decompositions of the total inertia according to the eigenvalues λ_k. From the eigenequations for the row profiles and the column profiles, the following *transition formulae* are easily deduced, linking the two solutions:

$$\mathbf{g}_k = (\mathbf{D}_c^{-1}\mathbf{P}^T)\,\boldsymbol{\varphi}_k$$

(i.e., the column profiles in principal coordinates are the weighted averages—or *barycentres*—of the row points in standard coordinates, with weights given by the profile elements);

$$\mathbf{f}_k = (\mathbf{D}_r^{-1}\mathbf{P})\,\boldsymbol{\gamma}_k$$

(i.e., the row profiles in principal coordinates are the weighted averages of the column points in standard coordinates, with weights given by the profile elements).

The transition formulae between principal coordinates is as follows:

$$\mathbf{g}_k = (\mathbf{D}_c^{-1}\mathbf{P}^T)\,\mathbf{f}_k/\sqrt{\lambda_k}$$

$$\mathbf{f}_k = (\mathbf{D}_r^{-1}\mathbf{P})\,\mathbf{g}_k/\sqrt{\lambda_k}$$

The classic simultaneous representation of the rows and columns superimposes the principal coordinates, for example, the optimal two-dimensional display shows the row points displayed using $[\mathbf{f}_1\ \mathbf{f}_2]$ and the column points using $[\mathbf{g}_1\ \mathbf{g}_2]$. The advantage of this display, called the *symmetric map*, is the same levels of dispersion of the row and column points on each axis, equal to λ_1 and λ_2 here. Other displays show one set of points in principal coordinates and the other in standard coordinates, with the barycentric relationship between them—these are called *asymmetric maps*, which are biplots.

9.2.5 Distributional Equivalence Principle

This principle is one of the foundations of CA and can be stated as follows: if two rows or two columns of table $\mathbf{N}$ are proportional, the CA of $\mathbf{N}$ is unchanged if these two rows or these two columns are amalgamated. Amalgamating two proportional rows, i.e., with the same profiles, will not change the chi-square distances between the columns, and amalgamating two proportional columns will not change the chi-square distances between two rows (see Chapter 10 by Bécue-Bertaut in this volume). CA is not the only method with this property, which can be found in the analysis of *Hellinger distances* between profiles as well as *weighted log-ratio distances* for strictly positive data—see Escofier (1978) and Greenacre and Lewi (2009).

9.2.6 Aids to Interpretation

The aids to interpretation are the same as those described in Section 9.1, and are routinely used to support the interpretation of a CA map. Again, the respective quantities just need to be plugged in for the general case described previously. For example, the relative contribution of column j to the inertia λ_k of the kth factorial axis is $\text{CTR}_k(j) = c_j g_{jk}^2 / \lambda_k$. The relative contribution of axis k to the inertia of the jth column is

$$g_{jk}^2 / \sum_k g_{jk}^2,$$

where the denominator is the square of the chi-square distance of the column profile $\mathbf{p}_j^c$ to the centroid in the full space. We may also use supplementary elements to facilitate the interpretation. If, for instance, we add a column s to table $\mathbf{N}$, we may project on the factorial axes the profile of this column. The coordinate g_{sk} of this profile $\mathbf{p}_s^c$ on the kth factorial axis is equal to $(\mathbf{p}_s^c)^T \boldsymbol{\varphi}_k = (\mathbf{p}_s^c)^T \mathbf{f}_k / \sqrt{\lambda_k}$. Similarly, we can project onto the factorial axes the profiles of supplementary rows.

9.2.7 Reconstitution Formula

The reconstitution formula for the elements of matrix $\mathbf{P}$, in terms of the principal coordinates, is $p_{ij} = r_i c_j (1 + \sum_k f_{ik} g_{jk} / \sqrt{\lambda_k})$, where the summation is over all $r = \min\{I-1, J-1\}$ factorial axes. In matrix notation, where $\boldsymbol{\Lambda}$ is the diagonal matrix of the eigenvalues, $\mathbf{F} = [\mathbf{f}_1 \ldots \mathbf{f}_r]$, and $\mathbf{G} = [\mathbf{g}_1 \ldots \mathbf{g}_r]$, this formula is $\mathbf{P} = \mathbf{D}_r (\mathbf{1}\mathbf{1}^T + \mathbf{F} \boldsymbol{\Lambda}^{-\frac{1}{2}} \mathbf{G}^T) \mathbf{D}_c$. If we let $\mathbf{U} = \mathbf{D}_r^{\frac{1}{2}} \mathbf{F} \boldsymbol{\Lambda}^{-\frac{1}{2}}$ and $\mathbf{V} = \mathbf{D}_c^{\frac{1}{2}} \mathbf{G} \boldsymbol{\Lambda}^{-\frac{1}{2}}$, then the reconstitution formula can be reexpressed as $\mathbf{D}_r^{-\frac{1}{2}} (\mathbf{P} - \mathbf{r}\,\mathbf{c}^T) \mathbf{D}_c^{-\frac{1}{2}} = \mathbf{U} \boldsymbol{\Lambda}^{\frac{1}{2}} \mathbf{V}^T$, which is the classic SVD form of the standardized residuals, often used for computation of the CA solution (for example, in the ca package in R by Nenadić and Greenacre (2007)).

9.2.8 The Eigenvalues in CA Are Inferior or Equal to 1

The singular values of the CA solution can be shown to be correlation coefficients between the row and column variables, and hence the eigenvalues are all less than or equal to 1. If there is a nontrivial eigenvalue equal to 1, the table $\mathbf{N}$ is block diagonal and there is a partition of the rows $I = I_1 \cup I_2$ and of the columns $J = J_1 \cup J_2$ in two sets such that the cells of the matrix in the block crossing I_1 with J_2 are zero, just as the block crossing I_2 with J_1. The simultaneous representation of I and J on the factorial axis relative to the eigenvalue 1 is very particular and shows the block partition: all the elements of I_1 and J_1 have the same coordinate on this axis, and likewise, all the elements of I_2 and J_2 have the same coordinate on this axis, but with opposite sign.

9.2.9 Symmetric Correspondence

If the rows and columns refer to the same entities so that $I = J$ and $\mathbf{N}$ is symmetric ($\mathbf{N} = \mathbf{N}^T$ and also $\mathbf{P} = \mathbf{P}^T$), the CA of $\mathbf{N}$ has particular properties because $\mathbf{r} = \mathbf{c}$, $\mathbf{D}_r = \mathbf{D}_c$, and both sets of profiles are the same. It suffices then to diagonalize one profile matrix, not the product of the row and column profile matrices as for the rectangular case. The eigenvalues are then the square roots of the eigenvalues of the CA; that is, they seem to be the singular values except for the fact that they can be negative or positive. This leads to the concept of *direct* and *inverse* factors for a symmetric matrix. Direct factors correspond to axes where the row and column solution is identical (positive eigenvalue), while for inverse factors the sign of the eigenvectors is reversed for the row and column solution (negative eigenvalue, with the negative sign passing to the eigenvector). The transition formula for this case is

$$\varepsilon_k \sqrt{\lambda_k}\, \mathbf{g}_k = \sqrt{\lambda_k}\, \mathbf{f}_k = (\mathbf{D}_r^{-1}\mathbf{P})\mathbf{g}_k$$

If $\varepsilon_k = 1$, $\mathbf{f}_k = \mathbf{g}_k$ and the kth factor is direct; otherwise, if $\varepsilon_k = -1$, $\mathbf{f}_k = -\mathbf{g}_k$ and the kth factor is inverse. Direct factors correspond to a positive association, while inverse factors correspond to negative association.

9.2.10 Burt Table

A Burt table associated with table $\mathbf{P}$ is the symmetric table $\mathbf{B} = \mathbf{P}^T\mathbf{D}_r^{-1}\mathbf{P}$, with general term $b_{jj'} = \sum_i p_{ij}\, p_{ij'}/r_i$. The CA of $\mathbf{B}$ is equivalent to the CA of $\mathbf{P}$, as we now show, along with the properties of the Burt table. The total of each column (or row) j of $\mathbf{B}$ is equal to c_j because $\mathbf{B1} = \mathbf{P}^T\mathbf{D}_r^{-1}\mathbf{P1} = \mathbf{P}^T\mathbf{D}_r^{-1}\mathbf{r} = \mathbf{P}^T\mathbf{1} = \mathbf{c}$. Thus, the grand total of $\mathbf{B}$ is 1, and in J-dimensional space the chi-square metric in CA of $\mathbf{B}$ is the same as that of the CA of $\mathbf{P}$, namely, $\mathbf{D}_c^{-1}$. The matrix of column profiles of $\mathbf{B}$ (equal to the row profiles matrix because $\mathbf{B}$ is symmetric) is equal to $\mathbf{BD}_c^{-1} = \mathbf{P}^T\mathbf{D}_r^{-1}\mathbf{PD}_c^{-1}$, which is exactly the matrix that we diagonalized to obtain the CA solution of $\mathbf{P}$, except that the CA of $\mathbf{B}$ will yield eigenvalues the squares of those of the CA of $\mathbf{P}$. The eigenvectors will be the same as before, and since all the eigenvalues of $\mathbf{BD}_c^{-1}$ are positive or null, all the factors of $\mathbf{B}$ are direct.

The same Burt table associated with the original table $\mathbf{N}$ is $\mathbf{N}^T\mathbf{D}_n^{-1}\mathbf{N} = n\mathbf{B}$, where $\mathbf{D}_n$ is the diagonal matrix of the row totals of $\mathbf{N}$. Notice that this Burt table is the same as the classic Burt matrix, the block matrix of all two-way tables, when $\mathbf{N}$ is a disjunctive complete table, or indicator matrix, of 0s and 1s, coding a multivariate data set of categorical data; see Chapter 3 by Lebart and Saporta.

10

Distributional Equivalence and Linguistics

Mónica Bécue-Bertaut

CONTENTS

10.1 Encoding the Corpus through Some Examples 150
10.2 Spaces of Row and Column Profiles 153
10.3 Distributional Equivalence and Chi-Square Distance 154
10.4 Correspondence Analysis and Transition Relationships 156
10.5 Selecting Words and Sequences 157
10.6 Application to a Forensic Closing Speech 159
10.7 Concluding Remarks 163

Correspondence analysis (CA) was conceived by Benzécri (1977b; Benzécri et al., 1973, 1981) as an inductive method able to infer from their own texts of interest the linguistic norms that rule them. He wanted to offer a powerful tool to tackle all kinds of issues related to the form, meaning, and style of texts.

To begin with, textual data were considered as sequences of words or other linguistic units. Among all the sequences that may be built, only some are grammatically and semantically licit, and thus only some associations between linguistic units are allowed. Thereby, the units are characterized by all the contexts in which they can be inserted to produce a correct sequence. Considering a large enough corpus—that is, a large collection of written or oral documents—would precisely inform us about the allowed contexts. These considerations led to

- Encoding a large range of corpora into frequency tables, whose row and column elements are chosen in accordance to the selected point of view. The tables are mainly of the form sequences × linguistic units or units × units. *Sequence* can mean document, theatrical play, book chapter, paragraph, or sentence, but also oral fragment or musical piece. *Linguistic unit* can mean letter, word, lemma, phoneme, and so on.

- Transforming the frequencies into proportions and computing the profiles of both sets of rows and columns.
- Defining an appropriate distance between profiles in the form of the chi-square distance.
- Conceiving CA as a method to account for and visualize the similarities between row profiles, on the one hand, and column profiles, on the other, as well as their mutual associations.

In the following, we first develop the points of this rationale with the support of a short fictitious example proposed by Benzécri et al. (1981). Then, we show how distributional equivalence allows for addressing either the selection of the words or the sequencing of the corpus through commenting on different options, giving several important literature sources in the field of textual analysis.

10.1 Encoding the Corpus through Some Examples

Written and oral texts are basic objects for linguistics. CA, conceived as a contribution to linguistic concerns, is in accordance with the distributional linguistics introduced by Harris (1954) as a tool to infer the rules of any language from corpora. The process relies on uncovering the distribution of each linguistic unit, defined by Harris as 'the sum of all its environments', or, equivalently, the set of its co-occurrences. According to Harris, interword dependencies suffice to determine correspondences in the set of textual sequences without regard to meaning. This process is what CA implements, providing that the corpus is encoded into a table crossing I rows and J columns.

A wide array of choices is offered leading to different sets I and J. Just a few of the earliest studies, mentioned and commented on in Benzécri et al. (1981), are cited hereafter. The first is used to introduce the basic notation.

- Table 10.1 shows the count of the co-occurrences of nouns as subjects and verbs as predicates in a short fictitious corpus in which all the sentences only consist of a noun (subject) and a verb (predicate). Thus, in every sentence, the context (or environment) of the noun is only the verb of which it is the subject, and the context of the verb is only the noun which is its subject. This example was proposed by Benzécri et al. (1981). The corpus is encoded into a contingency table, first called a correspondence table by Benzécri et al. (1981, pp. 74, 559–560), crossing the set I of nouns and the set J of verbs. Row i refers to noun i ($I = 1, \ldots, I$) and column j to verb j ($j = 1, \ldots, J$), with $I = 5$ and $J = 3$—or the other way round, given that rows and

TABLE 10.1

Encoding the Fictitious Corpus into a Contingency Table

	Aboyer	Dormir	Ronfler	Sum
Chacal	1	1	1	3
Chat	0	5	1	6
Chien	3	3	3	9
Coyotte	2	1	1	4
Moteur	1	0	8	9
Sum	7	10	14	31

Note: The row names are *jackal, cat, dog, coyote,* and *engine,* and the column names are *bark, sleep,* and *snore.*

columns play a symmetric role. At the intersection of row i and column j, the cell (i, j) contains the frequency n_{ij}, with which the noun i is associated with verb j;

$$\sum_{i=1}^{I} \sum_{j=1}^{J} n_{ij} = n = 31$$

is the grand total of co-occurrences of nouns and verbs (Table 10.1). The margin on the columns contains the total counts of the verbs

$$n_{+j} = \sum_{i=1}^{I} n_{ij}; j = 1, \ldots, J.$$

The margin on the rows contains the total counts of the nouns

$$n_{i+} = \sum_{j=1}^{J} n_{ij}; i = 1, \ldots, I.$$

The original French words are given in Table 10.1.

- Corpus of sentences extracted from 'news in brief' published in a Spanish newspaper. Every sentence is segmented into *functional blocks*; these blocks are grouped into $I = 40$ categories depending on their function (subject, circumstantial object, verb, and so on). Set J schematically indicates the location in the sentence, such as 'immediately before a verb', 'immediately before a subject', etc. n_{ij} is the number of times that category i appears in location j (Huynh-Armanet, 1976).

- Poetic texts aiming at studying the distribution of the phonemes among the verses; I is the set of phonemes, J is the set of verses, and n_{ij} is the number of times that phoneme i appears in verse j. Presence in a context means presence in a verse (Hathout and Reinert, 1981).

- Fifty-two theatrical plays by Molière (30 plays), Corneille (11), Racine (9), and another 2 plays, each of which is written by another contemporary author whose name is not given. I is the set of the 62 nouns and qualifying adjectives whose frequency is over 125 in this corpus. J is the set of the 52 plays. n_{ij} is the number of times that word i appears in play j. Here, presence in a context means presence in a play (Sainte-Marie, 1973).

The different words are identified from their graphical form—the only rigorous definition of *word* that can be provided—as a sequence of letters, bounded on the left and right by a space or punctuation mark. The words can be grouped into *lemmas* (or dictionary entries). Some users take into account compound words, phrasal verbs, and stereotyped expressions as if they were single words. Muller (1977) discusses the difficulties involved in setting standards for this type of grouping and notes that, in order not to lose its effectiveness, statistical analysis requires simplifying and reducing the distinctions. In the following, we refer to the chosen linguistic unit, whatever it is, by *word*.

Lebart et al. (1998, p. 114) proposes a composed unit, the *repeated segment*, that is, repeated succession of exactly the same words in the corpus. The repeated segments enlighten different usages of a same word, such as *peace* in either 'peace of mind' or 'peace in the world'. The repeated segments integrate the context of the words, which gives them a different status than words. So, they are only used as supplementary units in CA and not to distort the structure highlighted from the word distributions. The repeated segments inform about the context of the words and ease the reading of CA maps.

Either only the function words (articles, prepositions, and so on) or only the content words (mainly nouns, verbs, and adjectives) can be selected for the analysis, depending on the aims of the study and other considerations of the data (Lebart et al., 1998, p. 12; Murtagh, 2005, p. 162). No clear distinction exists between both types of words. For example, adverbs belong to one or the other category. Minimum thresholds are usually imposed on the word frequencies and on the minimum number of documents in which the words occur to make comparisons between documents meaningful from a statistical point of view (Lebart et al., 1998, p. 104; Murtagh, 2005, chap. 5).

Beyond these considerations, CA adapts to the selected unit, whatever that might be. Nevertheless, the user must keep in mind the choices that were made, as well as the reasons that support them, during the whole analytic process. We must emphasize that, in all the cases, we have the same data structure, a crossed table containing nonnegative values such that both row sums and column sums make sense as weight measures for the analysis.

10.2 Spaces of Row and Column Profiles

We go back to the example of Table 10.1. Both sets of nouns and verbs are taken into account through their profiles, i.e., their sets of relative frequencies. *Chacal* has three occurrences and *Chien* nine. Of the occurrences of *Chacal*, 1/3 correspond to verb *Aboyer*, 1/3 to verb *Dormir*, and 1/3 to verb *Ronfler*. Concerning *Chien*, we observe the same distribution of its occurrences among the three verbs. Both *Chacal* and *Chien* have the same profile (1/3, 1/3, 1/3) and are said to be *distributional synonyms*. The other nouns present different profiles (Table 10.2). The profile of row i is computed by dividing the row i by its total

$$\left(\frac{n_{ij}}{n_{i+}}, j = 1, \ldots, J\right).$$

In a symmetric way, the profile of verb j is computed by dividing the column j by its total

$$\left(\frac{n_{ij}}{n_{+j}}, i = 1, \ldots, I\right).$$

The three verbs show different profiles. Equivalently, the profiles could be computed from the table **P** of proportions with general term

$$p_{ij} = \frac{n_{ij}}{n},$$

issued from the contingency table divided by its grand total n, so that

$$\sum_{i=1}^{I} \sum_{j=1}^{J} p_{ij} = 1.$$

Row and column profiles are expressed as

$$\frac{p_{ij}}{p_{i+}}, i = 1, \ldots, I, \text{ and } \frac{p_{ij}}{p_{+j}}, j = 1, \ldots, J.$$

Table 10.2 shows the rows and columns profiles as well as the profiles of the margins in the case of the fictitious example.

CA looks for a geometrical representation of the similarities and differences between profiles. The cloud of nouns $N(I)$ is a set of five weighted points in a space of three dimensions or, more precisely, in the simplex of profiles included in it whose dimensions are reduced to two. Every row profile i is

TABLE 10.2

Profiles of Row Nouns (a) and Column Verbs (b)

a.	Aboyer	Dormir	Ronfler	b.	Aboyer	Dormir	Ronfler	Margin
Chacal	1/3	1/3	1/3	Chacal	1/7	1/10	1/14	3/31
Chat	0	5/6	1/6	Chat	0	5/10	1/14	6/31
Chien	1/3	1/3	1/3	Chien	3/7	3/10	3/14	9/31
Coyote	2/4	1/4	1/4	Coyote	2/7	1/10	1/14	4/31
Moteur	1/9	0	8/9	Moteur	1/7	0	8/14	9/31
Margin	7/31	10/31	14/31					

provided with a weight corresponding to the proportion of its occurrences on the grand total of occurrences, that is,

$$p_{i+} = 1/N \sum_{j=1}^{J} n_{ij}.$$

The margin profile of the rows is the centroid of the cloud.

Symmetrically, the cloud of verbs $N(J)$ is composed of three weighted points. Column profile j is weighted with

$$p_{+j} = \sum_{i=1}^{I} p_{ij}.$$

The margin profile of the columns is the centroid of this cloud. $N(J)$ is in a space of five dimensions, but as there are only three points, they are in a subspace with two dimensions. Both clouds, of rows and columns profiles, belong to subspaces with dimensions equal or less than $\min(I-1, J-1)$.

We have two clouds of profiles placed in two different spaces. Their centroids are given by the respective margins. These spaces have to be endowed with distances that induce the proximities between rows or between columns.

10.3 Distributional Equivalence and Chi-Square Distance

A distance measure on I (respectively, J) is defined on the profiles such that the distance between two rows (respectively, two columns), for example, i and i', is zero if the associated profiles are equal and small when they are similar. Distributional synonyms are rows and columns with identical profiles, as illustrated by the nouns *Chacal* and *Chien* in Table 10.1. Distances should also fulfil the *distributional equivalence principle*. This principle imposes that

Distributional Equivalence and Linguistics

the distance between two nouns i and i' does not change if we merge two verbs j and j' that are distributional synonyms. Similarly, on J, the distance between two verbs j and j' does not change if two nouns i and i' that are distributional synonyms are merged. The distributional equivalence principle and the requirement for a quadratic form lead Benzécri (1977b; Benzécri et al., 1973, 1981) to choose the following distances between rows (10.1a) and between columns (10.1b):

$$d^2(i,i') = \sum_{j=1}^{J} \frac{1}{p_{\cdot j}} \left(\frac{p_{ij}}{p_{i+}} - \frac{p_{i'j}}{p_{i'+}} \right)^2 \qquad (10.1a)$$

$$d^2(j,j') = \sum_{i=1}^{I} \frac{1}{p_{i\cdot}} \left(\frac{p_{ij}}{p_{+j}} - \frac{p_{ij'}}{p_{+j'}} \right)^2 \qquad (10.1b)$$

A quadratic formulation is required to take advantage of the multidimensional Euclidean geometry, without which the computing process would be tedious and the results lose properties. The distance between rows (between columns), first called distributional distance by Bénzécri (1977b), is known as *chi-square distance*. Other distances between profiles that obey the distributional equivalence principle are discussed in Escofier (2003, chap. 3) and Greenacre and Lewi (2009). Table 10.3 shows the frequencies after collapsing *Chacal* and *Chien*.

The chi-square distance between any pair of verbs j and j' is the same before and after merging two distributional synonyms i and i': the terms corresponding to the other nouns do not change in (10.1b); the sum of both terms issued from both nouns i and i' is equal to the term issued from merging them in (10.1b) (cf. Benzécri et al., 1981, pp. 94–95). Similarly, collapsing columns with equal profiles would not modify the distances between rows. As *Chacal* and *Chien* are distributional synonyms, the corresponding rows of Table 10.1 are collapsed into one row in Table 10.3.

TABLE 10.3

Collapsing *Chacal* and *Chien* into *Chacal/Chien*

	Aboyer	Dormir	Ronfler	Sum
Chacal/Chien	4	4	4	12
Chat	0	5	1	6
Coyotte	2	1	1	4
Moteur	1	0	8	9
Sum	7	10	14	31

10.4 Correspondence Analysis and Transition Relationships

CA computes the dispersion axes of both the clouds of rows and columns, as discussed in Husson and Josse (in Chapter 11 of this book), and embeds the spaces of rows and columns endowed with chi-square distances into classical Euclidean spaces. Thus, the distances between the points of each space can be visually evaluated. Starting from either Table 10.1 or Table 10.3 leads to the same principal plane, reproduced on page 161, which displays the exact distances because, in this particular case, rows and columns are in a two-dimensional space.

The simultaneous representation of rows and columns relies on the transition relationships ((10.2a) and (10.2b)) linking the coordinates f_{ik} of row points i ($i = 1, \ldots, I$) and the coordinates g_{jk} of column points j ($j = 1, \ldots, J$) on the dispersion axes k ($k = 1, \ldots, min(I-1, J-1)$).

$$F_s(i) \propto \sum_{j=1}^{J} \frac{p_{ij}}{p_{i+}} g_{jk} \tag{10.2a}$$

$$G_s(j) \propto \sum_{i=1}^{I} \frac{p_{ij}}{p_{+j}} f_{ik} \tag{10.2b}$$

where $\propto$ stands for 'is proportional to'. On each axis s, each row point i is, except for a constant, at the weighted average of the columns j and vice versa. The weights are nothing but the components of the row profile of element i (the column profile of element j). The transition relationships allow for passing from one set to the other.

Figure 10.1 shows that, in the column space, the first axis opposes the verbs *Ronfler* (at the top left) to *Dormir* (at the top right), indicating that they are associated with different nouns. In fact, their profiles are the furthest apart and both verbs have high weights. Then, this opposition is the highest inertial opposition. The second axis opposes *Ronfler* and *Dormir* (on the positive part of dimension 2) to *Aboyer* (on the negative part of dimension 2).

In the row space, the first axis opposes *Moteur* (on the left) and *Chat* (on the right), which are the nouns whose profiles are the farthest apart. *Coyotte* stands out on the second axis. *Chacal* and *Chien* lie exactly at the same position, as expected for being distributional synonyms, at the centroid of the three verbs to which they are associated in an equal proportion (1/3, 1/3, 1/3).

The superimposed representation of verbs and nouns reveals the associations between elements of both spaces that explain the similarities and oppositions. For example, *Moteur* and *Ronfler* frequently co-occur: 8 of 14 occurrences of *Ronfler* are linked to *Moteur*, while 8 of 9 occurrences of *Moteur* are subjects of *Ronfler* (Table 10.1).

Distributional Equivalence and Linguistics

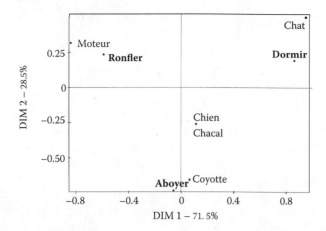

FIGURE 10.1
Simultaneous representation of nouns and verbs issued from CA applied to either Table 10.1 or Table 10.3.

The type of encoding of the fictitious example is seldom used. As in the studies briefly commented on in Section 10.1, the most usual encoding structure is a table crossing textual sequences and linguistic units. CA analyses this table and thus compares the word profiles and sequence profiles, providing a simultaneous representation of both sets of words and sequences. In this case, the transition equations ((10.2a) and (10.2b)) link the cloud of sequences and the cloud of words. So, on the graphical displays, two sequences (rows) are all the closer that they use the same words with the same proportion; two words (columns) are all the closer that they are used in the same sequences with similar frequency. The transition equations also suggest an interpretation for an observed proximity between two words never used in the same sequences, which should move them away. In fact, this happens when these two words co-occur with the same other words. The transition equation (10.2a) shows that the sequences containing them will get closer because they have similar profiles, except for these two words. From transition equation (10.2b), these two words get closer because they lie at the centroid of close sentences, except for the same constant. The latter assertion means that CA is able to retrieve synonymy relationships when the synonyms co-occur with the same other words in the corpus under study.

10.5 Selecting Words and Sequences

If CA is applied to textual data, preprocessing and encoding are crucial steps. Sequencing can be chosen differently, at more or less fine levels, such as functional blocks (see Section 10.1), phrases, sentences, paragraphs, chapters, and so on. The words can be either unified or not, depending on their lemma or meaning. Both decisions on sequencing level and word unification, essential to display clear patterns on the first planes, are mainly supported by the distributional equivalence principle. Ensuring results that are robust when collapsing sequence rows (or word columns) with close profiles, this principle allows for refining first partitions of rows or columns.

Benzécri et al. (1973) proposed a strategy to compare different partitions on the sets of rows or columns. This can be adapted to textual domain as follows. CA is performed on an $I \times (J \cup J')$ crossed table, where J is the set of active column words and J' the set of supplementary column lemmas (or vice versa). That allows for either comparing J and J' or deciding which words have to be merged into their lemmas. The latter would depend on the similarities or dissimilarities between the words issued from a common lemma, and the lemma itself on the principal planes. A similar rationale could also be used to merge words with a close meaning. In a symmetrical way, CA performed on $(I \cup I') \times J$, where I and I' correspond to sequencing the corpus at two different levels, for example, into either sentences or paragraphs, allows for selecting which sequencing is suitable. Other strategies can be adopted to address the corpus sequencing. Some of them are detailed hereafter.

Open-ended questions in surveys, i.e., questions without given response categories, allow for selecting more or less fine sequencings through resorting to the characteristics of the individuals known from closed categorical questions (Lebart et al., 1998). CA can be performed at a fine level directly on the lexical, or responses × words, table (CA-LT). Then the categories are placed on the principal planes as supplementary rows at the centroid of the individuals that belong to them, providing a category pattern. A coarse level can be chosen by applying CA to an aggregated lexical table crossing categories (rows) and words (columns) (CA-ALT). This table is built by merging the individuals' responses belonging to a same category into a single sequence. Categories may be issued from either a unique categorical variable or several variables. In the latter case, the individuals are clustered from their coordinates on the dispersion axes computed by multiple correspondence analysis (MCA), taking the selected variables as active. MCA is here an intermediate step (Lebart et al., 1998, pp. 118–121). The clusters form the categories used to build an aggregated lexical table (Lebart et al., 1998, pp. 110–128). CA-ALT offers another category pattern. Both category patterns, as issued from CA-LT or CA-ALT, are roughly similar provided that the individuals belonging to a same category are fairly homogeneous with respect to vocabulary. However, the pattern is clearer when issued from CA-ALT, as the information is spread over a smaller number of axes.

The chronological analysis of a single text is a challenging task. To track the evolution of the word usage in Aristotle's *Categories*, Murtagh (2005, pp. 207–218) proposes to sequence the text at four nested levels: sections (3), parts (15), paragraphs (141), and finally, sentences (701); 166 words are retained. CA is performed on the table juxtaposing the four sets of sequences (rows) and the set of words (columns). Words and the four levels of sequences (sections, parts, paragraphs, and sentences) can be represented on a same graphic. Passing through the four sequencing levels allows us to follow how the associations between words evolve.

Confidence areas, as computed by bootstrap methods (Lebart, 2003), offer another rationale. Sequences having no significantly different profiles, as assessed by bootstrap replicates that enclose them into overlapping confidence area, will be profitably merged into a single one. A similar strategy can be used for possible merging of words.

Distributional equivalence is at the basis of an automated method for sequencing a long, undivided single text recently proposed by Bécue-Bertaut et al. (in press). This process is summarized here. A first and too fine division into arbitrary equal-length passages (50–70 words long) is performed. Then, these passages are grouped through chronological clustering (CC) (Legendre and Legendre, 1998, pp. 696–699), that is, a hierarchical algorithm with a temporal contiguity restriction (complete linkage algorithm) but adapted to textual data by selecting chi-square distance between passages. Every aggregation of nodes has to be validated by a permutation test to ensure that only lexically homogeneous nodes, that is, approximate equivalent distributional nodes, are grouped. The algorithm stops when no more aggregations can be performed, providing a final sequencing into homogeneous blocks. Short or singular blocks can be cut out in the text, depending on the threshold on the p-value associated with the permutation test. The shape of the text, which is the trace of the discourse organization, is shown through the trajectory of the blocks. Combining CA and CC is in accordance with the hierarchical model in discourse analysis (Kuyumcuyan, 1999). The sequencing is deduced from the own corpus, without any reference to content, through only the words distribution.

10.6 Application to a Forensic Closing Speech

As to enhance the relevance of the sequencing step, we present the analysis of a closing speech for the prosecution, videotaped and copied out for the purposes of a research project about Spanish judicature (Casanovas et al., 1995). This closing speech, 4,010 occurrences long, was addressed in a trial for murder at the Court of Justice in Barcelona. The prosecutor charges a defendant, a young emigrant woman, LV, with a double offence of breaking and entering successively committed against two elderly men, EN and RV, living on their own. In both cases, this woman used the same strategy. She simulated a sentimental relationship with the victims and later on went to live with each one of them at their own expense. When, after a time, she was asked to leave, not only did she refuse, but she first blackmailed them with committing suicide, then threatened to slash them. This woman was neither working nor contributing to pay the rent. After being forced to move out of the first house by the police, she immediately repeated the same demeanor with her second victim. DS, a common relationship of these two men, followed all the vicissitudes that led her to testify in Court of Justice as a witness for the prosecution.

The prosecutor delivers an unprepared speech, built while in progress. He has to develop a demonstrative rationale connecting all arguments effectively. Its main assets rely on much information in favour of the implausibility of the sentimental relationship between the defendant and her victims issued from the statement of the witness and from the facts.

The speech was initially divided into 80 short passages by the researchers through classical reading, looking for a balanced length of about 50 occurrences. Words are only differentiated from one another through their spelling. Neither lemmatization nor stemming is performed. Only the articles are considered as stop words. The frequency threshold is equal to 4. Thus, 160 different words are kept. The most frequent content words concern the description of the case. The persons involved in it are repeatedly cited: RV (victim), EN (victim), LV (defendant), and DS (witness for the prosecution). With significance level α equal to 0.1, CC clusters the passages into 10 lexically homogeneous blocks noted BLj, where $j = 1, \ldots, 10$.

CA is performed on the lexical table blocks × words (10 blocks, 169 words). The trajectory of the parts on the first CA plane, which accounts for 30% of the total inertia, constitutes the narrative drive scheme (Figure 10.2).

This scheme has to be compared with the pattern expected in the case of a well-controlled speech presenting a smooth and gradual renewal of the vocabulary. Adjacent blocks would present profiles more similar to one another than blocks separated by a long time interval, and the trajectory would follow an *arch pattern* on the first CA plane. This pattern is observed when a contingency table is characterized by a smooth gradation on rows and columns (Lebart et al., 1998, p. 150).

Departures from this basic arch are what we look for and interpret as characterizing the argumentative drive. A first reading of this plane with no reference to the vocabulary allows for a better understanding of the rhythm and progression of the speech.

Figure 10.2 shows a high regularity in the progression of the first four blocks (BL1–BL4). BL5 and BL6 uses the words introduced by BL4 without significant influx of new words. BL7 regains the thread of the speech, while BL8 and BL9 recover much of the words used in the initial blocks. BL10 completely breaks the trend, showing that the speech closure is little linked to the rest of the speech.

Thus, we have an orderly first part of the speech, followed by two loops, the first (BL4–BL6) being very constricted while the second (BL7–BL9) is broader. These loops correspond to two moments of the speech where the prosecutor revolves around a same topic with a certain difficulty for an ordered

Distributional Equivalence and Linguistics

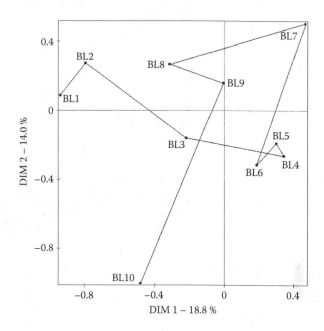

FIGURE 10.2
Trajectory of the blocks on the first principal plane.

presentation of the arguments. Substituting these loops by their centroid, an arch effect would be identified from BL2 to BL9, followed by a final rupture without much continuity with the argument's progression in the speech. The arch or Guttman effect no longer appears on the configuration provided by the third axis that mainly underlines and details some specific aspects of the blocks. The observed patterns are quite stable to small changes in the frequency threshold or in the location of the break points.

Now, we place the words, and some few repeated segments as supplementary elements, on the graphic in order to capture the meaning of the speech of which we only had the shape (Figure 10.3). So as not to produce an unreadable graphic, we only represent the 20% most contributory words and some repeated segments. The use of this new information allows us to follow the rationale of the prosecutor.

In BL1–BL3, the prosecutor communicates to the jurors the role of each intervener in the Court of Justice, explaining why the defendant is charged with breaking and entering. He outlines the legal framework in which the accusation is placed. BL4–BL6 detail the facts that occurred with the first victim, including threatening and final arrest of the defendant by the police. All the facts are supported by evidence.

BL7 and BL8 describe the events related to the second victim. The prosecutor has to insist on the similarities between both sequences of facts to convince the jurors that they are dealing with a premeditated strategy set up by

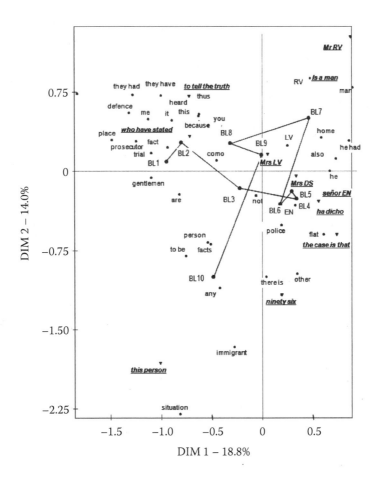

FIGURE 10.3
Words (active), segments (supplementary, in italics, bold, and underlined letters), and blocks (active) on the first principal plane.

an unscrupulous but not a helpless and desperate woman. He insists on this argument along BL9, showing that her situation does not justify her behaviour.

Finally, BL10 provides noncoordinated information about the defendant, the victims, and the difficult situation of many immigrants who do not commit this type of offence. The results put to the fore a quite controlled speech, despite some departures from the arch effect, that nevertheless presents a closure badly linked to the former blocks.

In this type of study, CA proves to be a powerful method to effectively capture the pattern of argumentation, which is the main concern. The same strategy can be applied whenever the argumentative or narrative structure of a text is the objective.

10.7 Concluding Remarks

The short fictitious example has allowed for exposing the approach to texts (or other linguistic data) offered by CA. This is in accordance with linguistic principles considering that the meaning of a word derives from its context. Sequencing into homogeneous blocks is deduced from the differences between one block and the neighbouring ones. Distributional equivalence and transition relationships combine to make CA a profitable tool in textual analysis.

Data coding is crucial in CA (Murtagh, 2005, chap. 3). This is particularly complex in textual domains where very different choices can be made to sequence the corpus and choose the linguistic units. This task is largely supported by distributional equivalence.

The journal *Les Cahiers de l'Analyse des Données* still presently constitutes the major source of works related to textual analysis relying on CA. The proceedings of JADT conferences (*Journées Internationales d'Analyse de Données Textuelles*, http://lexicometrica.univ-paris3.fr/jadt/, available since 1990) offer a further collection of recent works in textual statistics.

11

Multiple Correspondence Analysis

François Husson and Julie Josse

CONTENTS

11.1 Studying Individuals 166
    11.1.1 Cloud of Individuals and Distance between Individuals 166
    11.1.2 Fitting the Cloud of Individuals 168
    11.1.3 Interpreting the Cloud of Individuals Using the Variables 169
    11.1.4 Interpreting the Cloud of Individuals from the Categories 170
11.2 Studying Categories 173
    11.2.1 The Cloud of Categories and Distance between Categories ... 173
    11.2.2 Fitting the Cloud 174
11.3 Link between the Study of the Two Clouds—Duality 176
11.4 CA on the Burt Table 177
11.5 Example 178
11.6 Discussion 183

Multiple correspondence analysis (MCA) is a method of *analyse des données* used to describe, explore, summarize, and visualize information contained within a data table of N individuals described by Q categorical variables. This method is often used to analyse questionnaire data. It can be seen as an analogue of principal components analysis (PCA) for categorical variables (rather than quantitative variables) or even as an extension of correspondence analysis (CA) to the case of more than two categorical variables.

The main objectives of MCA can be defined as follows: (1) to provide a typology of the individuals, that is, to study the similarities between the individuals from a multidimensional perspective; (2) to assess the relationships between the variables and study the associations between the categories; and (3) to link together the study of individuals and that of variables in order to characterize the individuals using the variables.

FIGURE 11.1

	x_a	x_b	x_c	x_d		Label	a	A	b	B	c	C	γ	d	D	δ	Δ	
1	a	b	c	d		abcd1	1	0	1	0	1	0	0	1	0	0	0	Q = 4
2	a	b	c	d		abcd2	1	0	1	0	1	0	0	1	0	0	0	Q = 4
3	A	b	c	d		Abcd	0	1	1	0	1	0	0	1	0	0	0	Q = 4
4	a	B	c	D	⇒	aBcD	1	0	0	1	1	0	0	0	1	0	0	Q = 4
5	A	B	c	D		ABcD	0	1	0	1	1	0	0	0	1	0	0	Q = 4
6	a	B	C	δ		aBCδ	1	0	0	1	0	1	0	0	0	1	0	Q = 4
7	A	B	C	δ		ABCδ	0	1	0	1	0	1	0	0	0	1	0	Q = 4
8	a	B	γ	Δ		aByΔ	1	0	0	1	0	0	1	0	0	0	1	Q = 4
							N_1	N_2	N_3	N_4	N_5	N_6	N_7	N_8	N_9	N_{10}	N_{11}	
							=	=	=	=	=	=	=	=	=	=	=	
							5	3	3	5	5	2	1	3	2	2	1	

Example of converting raw data into an indicator matrix. The response patterns are used as respondent labels of the indicator matrix.

To illustrate the method let us consider a toy data set containing $N = 8$ individuals described by $Q = 4$ categorical variables (see left table in Figure 11.1) with $J_1 = 2$, $J_2 = 2$, $J_3 = 3$, and $J_4 = 4$ categories, respectively. For example, the first two individuals give responses a for the first variable, b for the second, c for the third, and d for the fourth.

The first step in MCA is to recode the data. Usually, data are coded using the indicator matrix of dummy variables, denoted **Z** of size $N \times J$, with

$$J = \sum_{q=1}^{Q} J_q,$$

as illustrated in Figure 11.1. The row margins of **Z** are equal to Q. The column margins are equal to (N_j), $j = 1, \ldots, J$, where N_j is the number of individuals taking the category j (i.e., giving that categorical response in a questionnaire, for example). The grand total is equal to NQ.

Historically (see Chapter 3 by Lebart and Saporta in this book), Lebart had the idea to apply correspondence analysis to this indicator matrix. This strategy yields very interesting results with new properties: that is how MCA was born, and this remains its most common definition.

11.1 Studying Individuals

11.1.1 Cloud of Individuals and Distance between Individuals

Studying the individuals involves looking at the data table row by row. Each row of the indicator matrix has J coordinates and defines a vector in $\mathbb{R}^J$. It may thus be represented as a point in $\mathbb{R}^J$, and the set of rows is the cloud of individuals. From a geometric point of view, studying the similarities

Multiple Correspondence Analysis

between individuals can be seen as studying the shape of this cloud of points. Are the individuals divided into groups? Is there a main direction in the cloud that opposes some specific individuals with others? The distance between points representing the individuals is thus crucial. The chi-square distance is used since MCA is presented as a CA of the indicator matrix. The square of the chi-square distance (see also Chapter 9 by Cazes in this book) between two rows is

$$d^2_{\chi^2}(\text{row profile } i, \text{row profile } i') = \sum_{j=1}^{J} \frac{1}{c_j}\left(\frac{p_{ij}}{r_i} - \frac{p_{i'j}}{r_{i'}}\right)^2$$

In much the same way as the MCA's contingency table is the indicator matrix $\mathbf{Z}$, the correspondence matrix $\mathbf{P}$ is equal to $\mathbf{Z}/(NQ)$, $\mathbf{D}_r$ is the diagonal matrix of row masses

$$r_i = \frac{Q}{NQ} = \frac{1}{N}, i=1,\ldots,N,$$

and $\mathbf{D}_c$ is the diagonal matrix of column masses

$$c_j = \frac{N_j}{NQ}, j=1,..,J.$$

Consequently, the squared chi-square distance between individual i and i' is

$$d^2_{\chi^2}(i,i') = \sum_{j=1}^{J} \frac{1}{N_j/(NQ)}\left(\frac{z_{ij}/(NQ)}{1/N} - \frac{z_{i'j}/(NQ)}{1/N}\right)^2$$

$$= \frac{N}{Q}\sum_{j=1}^{J} \frac{1}{N_j}(z_{ij}-z_{i'j})^2$$

Two individuals are close to each other if they answer a relatively large number of questions in the same way, and are farther apart if they have very different response profiles. The categories with low frequencies make higher contributions to the distances between those (few) individuals that give that response and the large majority that do not.

The cloud of points defined by these distances has a centre of gravity G_N (with coordinates $c_j, j = 1,\ldots,J$) and a total inertia that is equal to

$$\text{Total inertia} = \frac{1}{N}\sum_{i=1}^{N} d^2(i,G_N) = \frac{J-Q}{Q} = \frac{J}{Q}-1$$

The total inertia, a multidimensional extension of the concept of variance, is thus defined as the sum of the squared distances from individuals to the centre of gravity weighted by 1/N (the weight of the individuals). In MCA it is equal to the mean number of categories per variable minus 1, and thus does not depend on the relationships between the variables (which is unlike the total inertia in simple CA).

11.1.2 Fitting the Cloud of Individuals

The cloud of individuals cannot be visualized directly, as it lies in a high-dimensional space. As a consequence, like in PCA and in CA, it is projected onto a subspace of lower dimensionality by maximizing the variability of the projected points (projected inertia). The representation thus obtained provides the most accurate possible image of the cloud, that is, an image that deforms the initial distances between individuals as little as possible. Within this new space, the principal axis k ($\mathbf{v}_k$) with the eigenvalue λ_k is obtained by performing the eigenvalue decomposition of $\mathbf{P}^T\mathbf{D}_r^{-1}\mathbf{P}\mathbf{D}_c^{-1}$ (see Chapter 9 by Cazes), thus:

$$\frac{1}{Q}\mathbf{Z}^T\mathbf{Z}\mathbf{D}^{-1}\mathbf{v}_k = \lambda_k \mathbf{v}_k$$

where $\mathbf{D}$ is the diagonal matrix of the column sums of $\mathbf{Z}$, $N_j, j = 1, \ldots, J$. This is the eigenequation of a nonsymmetric matrix, and the normalization of the eigenvectors $\mathbf{v}_k$, gathered as columns of the matrix $\mathbf{V}$, is $\mathbf{V}^T\mathbf{D}\mathbf{V} = \mathbf{I}$. The eigenvectors in $\mathbf{V}$ define the principal axes, and the coordinates of the individuals are then the projections of the row profiles $(1/Q)\mathbf{Z}$ onto these principal axes.

Figure 11.2 provides the best representation of the cloud of individuals in two dimensions. We can see that points *abcd*1 and *abcd*2 are superimposed (as they have the same response categories) and are close to point *Abcd* (as they have three out of four categories in common). On the other hand, point *aBγΔ* (which has categories with low frequencies in the sample) is relatively far from both the centre of gravity and the other individuals.

The quality of the cloud's fit on the two-dimensional space is measured by the percentage of inertia, i.e., the ratio between the variability explained by the map and the total variability. The variability explained by an axis is equal to the variance of the coordinates of the individuals on this axis, i.e., to the eigenvalue associated with that axis. The percentage of inertia associated with the two-dimensional map is therefore

$$(\lambda_1 + \lambda_2) / \sum_{k=1}^{J-Q} \lambda_k = 68.0\%.$$

In this case, the percentage is high, as it is a small data set with strong links between variables. Often in MCA, however, the percentages of inertia associated with each axis are weak (Benzécri, 1977c).

Multiple Correspondence Analysis

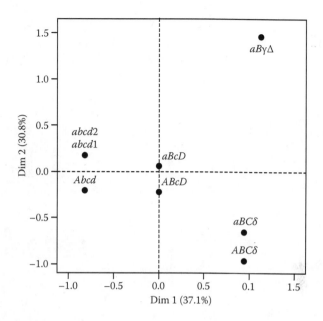

FIGURE 11.2
Representation of the individuals on the first two axes.

Notice that the maximum number of nonnull eigenvalues is $J - Q$, assuming that N is larger than this number; hence, the individuals lie in a $(J - Q)$-dimensional space (Lebart et al., 1997). This is because there are J columns of the matrix but there are Q linear restrictions on the columns since each group of columns corresponding to a variable has row sums equal to a constant 1.

11.1.3 Interpreting the Cloud of Individuals Using the Variables

In order to interpret the representation of the cloud of individuals, we can use the graph of variables (Figure 11.3). In the latter, each variable takes as coordinates the square correlation ratios (denoted as η^2) between the variable and the coordinates of the individuals on the first and second axes, respectively. More precisely, the vector of the coordinates of the N individuals on the axis k ($\mathbf{v}_k$) is denoted $\mathbf{f}_k$. Then each variable $\mathbf{x}_q$ has the coordinate $\eta^2(\mathbf{f}_k, \mathbf{x}_q)$ on dimension k. This latter quantity is usually used in analysis of variance to assess the relationship between the categorical variable $\mathbf{x}_q$ and the continuous variable $\mathbf{f}_k$. Thus, it represents the proportion of variance of the coordinates of the individuals on a dimension explained by the categorical variable, i.e., the ratio of the between-categories variability over the total variability. Here, for example, variable $\mathbf{x}_b$ has the coordinates (0.8, 0), which means that on the first axis the individuals are separated according to their responses to variable $\mathbf{x}_b$ (in other words, the first dimension of variability

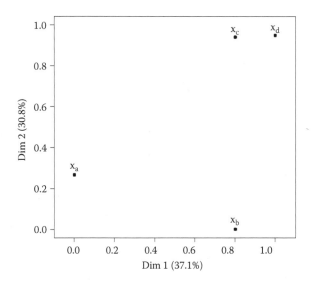

FIGURE 11.3
Square correlation ratios of the variables on the two dimensions.

opposes individuals carrying different categories for variable x_b). Variables x_c and x_d, on the other hand, can be used to explain oppositions between individuals on the first and second axes.

Note: The vector f_k of the coordinates of the N individuals on an axis k is a vector of $\mathbb{R}^N$, which can be considered a quantitative variable. The set of these new variables, known as principal components, verifies the following property:

$$\hat{f}_k = \underset{f_k \in \mathbb{R}^N}{\arg\max} \sum_{q=1}^{Q} \eta^2(f_k, x_q)$$

under the constraint that f_k is orthogonal to $f_{k'}$ for all $k' < k$. Thus, as with the other methods of *analyse des données*, the first principal components of the MCA are orthogonal variables that are the most closely linked to the set of variables, with the relationship measured by the squared correlation ratio.

11.1.4 Interpreting the Cloud of Individuals from the Categories

The information provided by the variables is general and thus insufficient for interpreting the similarities and differences between individuals. It is therefore natural to make a more detailed analysis using the categories of these variables. One instinctive way of representing a category on the graph of individuals is to position the category at the barycentre (i.e., average) of the individuals who have it in their response (left graph, Figure 11.4 for the

Multiple Correspondence Analysis

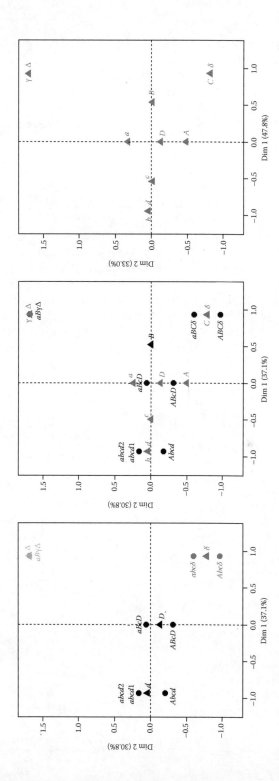

FIGURE 11.4
Representation of categories in Figure 11.2. Left: Representation of the categories of variable x_d; individuals are coloured (using grey levels) according to their category. Middle: Superimposed representation of the individuals and categories of all the variables. Right: Representation of the categories without the individuals. The different percentages of inertia will be explained Section 11.4.

categories of variable x_d). Generally the categories of all the variables are represented on a single graph (middle graph, Figure 11.4).

Two categories are close if, overall, the individuals carrying them are close, which means that their responses are similar over all the variables. Proximity between two categories is interpreted as proximity between two groups of individuals. Here, for example, the categories C and δ are superimposed as they are chosen by the same individuals.

Another way of constructing this barycentric representation is, for each category, to add up the rows of the indicator matrix relating to the individuals carrying that category and then to project these new vectors of row sums as supplementary elements on the graph of individuals (using the projection of supplementary rows from CA; see Chapter 9 by Cazes in this book). The table of these row sums, represented in the lower part of Table 11.1, is the *Burt table*. Presenting categories as new rows in this way justifies superimposing representations on the same graph of individuals and categories, which thus lie in the same space.

Often in MCA, we examine the representation of the cloud of individuals to have an idea of how the individuals are distributed, and we then focus on interpreting the representation of the cloud of barycentres alone (right graph, Figure 11.4). Indeed, we are often not interested in the details of the

TABLE 11.1

Burt Table as Supplementary Rows of the Indicator Matrix

	a	A	b	B	c	C	γ	d	D	δ	Δ
abcd1	1	0	1	0	1	0	0	1	0	0	0
abcd2	1	0	1	0	1	0	0	1	0	0	0
Abcd	0	1	1	0	1	0	0	1	0	0	0
aBcD	1	0	0	1	1	0	0	0	1	0	0
ABcD	0	1	0	1	1	0	0	0	1	0	0
aBCδ	1	0	0	1	0	1	0	0	0	1	0
ABCδ	0	1	0	1	0	1	0	0	0	1	0
aB$\gamma\Delta$	1	0	0	1	0	0	1	0	0	0	1
a	5	0	2	3	3	1	1	2	1	1	1
A	0	3	1	2	2	1	0	1	1	1	0
b	2	1	3	0	3	0	0	3	0	0	0
B	3	2	0	5	2	2	1	0	2	2	1
c	3	2	3	2	5	0	0	3	2	0	0
C	1	1	0	2	0	2	0	0	0	2	0
γ	1	0	0	1	0	0	1	0	0	0	1
d	2	1	3	0	3	0	0	3	0	0	0
D	1	1	0	2	2	0	0	0	2	0	0
δ	1	1	0	2	0	2	0	0	0	2	0
Δ	1	0	0	1	0	0	1	0	0	0	1

individual information itself, and therefore prefer to base the interpretation on the groups of individuals.

In this example, all the variables are active. Often in MCA, however, there are also supplementary variables, that is, variables that are not used to calculate the distances between individuals but rather to assist in interpreting the data. It is possible to represent the categories of these variables in the same way as the active categories of the active variables, i.e., at the barycentre of the individuals who have chosen it.

11.2 Studying Categories

11.2.1 The Cloud of Categories and Distance between Categories

The representation of the categories obtained previously was constructed from the optimal representation of individuals. It is also possible to construct a representation of the categories by analysing the cloud of the columns from the indicator matrix. From a geometric point of view, the category study can be seen as studying the shape of a cloud of points in $\mathbb{R}^N$. The distance between two columns is the chi-square distance, as defined above:

$$d^2_{\chi^2}(\text{column profile } j, \text{column profile } j') = \sum_{i=1}^{N} \frac{1}{r_i}\left(\frac{p_{ij}}{c_j} - \frac{p_{ij'}}{c_{j'}}\right)^2$$

The squared distance between two categories j and j' is thus

$$d^2_{\chi^2}(j, j') = \sum_{i=1}^{N} \frac{1}{1/N}\left(\frac{z_{ij}/(NQ)}{N_j/(NQ)} - \frac{z_{ij'}/(NQ)}{N_{j'}/(NQ)}\right)^2$$

$$= N\sum_{i=1}^{N}\left(\frac{z_{ij}}{N_j} - \frac{z_{ij'}}{N_{j'}}\right)^2$$

The distance between two categories decreases when most of the individuals carrying one category also carry the other. The distance between two categories depends on their frequencies. It must be noted that the distance between two categories does not depend on the other variables, unlike the distance between two categories when they are considered as groups of individuals, as described above. The chi-square distance between two categories in the indicator matrix is thus more difficult to justify (Greenacre, 2006, p. 60).

The squared distance of a category j from the centre of gravity G_I of the cloud of categories (of coordinates $r_i = 1/N$ for all the i) is

$$d^2(j, G_J) = N \sum_{i=1}^{N} \left(\frac{z_{ij}}{N_j} - \frac{1}{N} \right)^2 = \frac{N}{N_j} - 1$$

The smaller the frequency of a given category, the farther it is from the centre of gravity. The squared distance weighted by N_j/NQ (the weight of category j) is the inertia of category j:

$$\text{Inertia}(j) = \frac{N_j}{NQ} \times d^2(j, G_J) = \frac{1}{Q}\left(1 - \frac{N_j}{N}\right)$$

The less frequently a category occurs, the higher its inertia. It is often recommended to avoid categories with low frequencies that might too heavily influence the overall analysis of the results. However, it must be noted that the inertias between a low-frequency category, for example, carried by 1% of individuals, and one carried by 10% are not that different (0.99 compared with 0.9 divided by the number of variables). Generally speaking, in MCA, the impact of categories with low frequencies is often overestimated. This tends to stem from their great distance from the centre of gravity (and thus their extreme position on graphs). However, in calculating a category's inertia, its weight counterbalances the squared distance.

A variable's inertia is defined by the sum of inertias of its categories:

$$\text{Inertia}(q) = \sum_{j=1}^{J_q} \frac{1}{Q}\left(1 - \frac{N_j}{N}\right) = \frac{J_q - 1}{Q} \tag{11.1}$$

The higher the number of categories a variable has, the greater its inertia. By adding together the inertias of all the variables, we obtain the total inertia of the cloud of categories:

$$\text{Inertia} = \sum_{q=1}^{Q} \frac{J_q - 1}{Q} = \frac{J}{Q} - 1$$

As expected, the total inertia of the cloud of categories is the same as the one of the cloud of individuals. This is explained by the duality between the two analyses of rows and columns. Duality is an intrinsic property of methods of *analyse des données*.

11.2.2 Fitting the Cloud

As for the individuals, the cloud of categories lies into a high-dimensional space, and we want to find a subspace to represent the cloud of points as best as possible. Within this space, the eigenvectors $\mathbf{u}_k$, associated with the eigenvalues λ_k, are obtained from the eigenvalue decomposition of $\mathbf{PD}_c^{-1}\mathbf{P}^T\mathbf{D}_r^{-1}$, thus

$$\frac{1}{Q}\mathbf{ZD}^{-1}\mathbf{Z}^T\mathbf{u}_k = \lambda_k\mathbf{u}_k$$

Figure 11.5 provides the best representation of the indicator matrix cloud of categories in two dimensions. The variance of the coordinates of the category points on dimension k is equal to the eigenvalue λ_k. In this figure, the categories γ and Δ are superimposed and positioned on the edge of the graph: they are rare and chosen by the same individual.

It must be noted that the representation of categories (Figure 11.5) is not exactly the same as the representation of the barycentres (right graph, Figure 11.4), due to a dilation factor on each dimension. The relationship between these two representations is explained in Section 11.4.

A number of indicators (such as contribution and quality of representation; see Le Roux and Rouanet, 2010; Husson et al., 2011) can be used to enhance the analysis of the fitted cloud. The contribution of a given variable is interesting, as it is linked to the square correlation ratio between the variable and the principal component by the following relationship: $\eta^2(\mathbf{f}_k, \mathbf{x}_q) = Q \times \text{CTR}_k(q) \times \lambda_k$, where $\text{CTR}_k(q)$ is the sum of the contributions $\text{CTR}_k(j)$ for all categories j of variable q; see Chapter 9 by Cazes in this book for the definition of the contributions. Thus, Figure 11.3 can be used

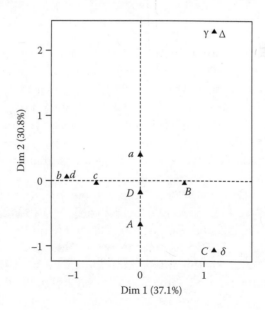

FIGURE 11.5
Representation of the categories produced by the CA of the indicator matrix.

to read the contributions of the variables directly. In addition, this relationship means that the projected inertia of a variable ($\text{CTR}_k(q) \times \lambda_k$) on a dimension is bounded by $1/Q$. This means that even if a variable has a high inertia (i.e., many categories, see (11.1)), the MCA solution will not be overwhelmed by the variable. It is therefore not essential to have variables with equal numbers of categories.

11.3 Link between the Study of the Two Clouds—Duality

Both studies, that of the individuals (Section 11.1) and that of the categories (Section 11.2), are linked by the usual transition formulae, obtained by applying the CA relationships to the indicator matrix (see Chapter 9, by Pierre Cazes in this book, on simple CA):

$$f_{ik} = \frac{1}{\sqrt{\lambda_k}} \sum_{q=1}^{Q} \sum_{j=1}^{J_q} \frac{z_{ij}}{Q} g_{jk}$$

$$f_{ik} = \frac{1}{\sqrt{\lambda_k}} \sum_{i=1}^{N} \frac{z_{ij}}{N_j} g_{jk}$$

with f_{ik} and g_{jk} the coordinates of the individual i and category j on dimension k. An individual is, up to the multiplicative factor $1/\sqrt{\lambda_k}$, at the barycentre of the categories that it carries. And a category is, up to the multiplicative factor $1/\sqrt{\lambda_k}$, at the barycentre of the individuals that carry it (Figure 11.5). These relationships are often used to construct a pseudobarycentric graphical representation, also called the symmetric map. The $1/\sqrt{\lambda_k}$ coefficient ensures that both barycentric properties are represented simultaneously on the same graph (Figure 11.6).

However, this pseudobarycentric representation is difficult to read, as the distances between an individual and a category cannot be interpreted. Indeed, the individuals and the categories do not lie in the same space ($\mathbb{R}^J$ for the individuals and $\mathbb{R}^N$ for the categories). Note, for example, that the individual $a\text{B}\gamma\Delta$, which is the unique individual who has chosen categories γ and δ, is not superimposed with these two categories. We therefore recommend not to construct this representation, but rather prefer a barycentric representation, such as Figure 11.4, where a category is at the exact barycentre of the individuals who have chosen it.

Multiple Correspondence Analysis

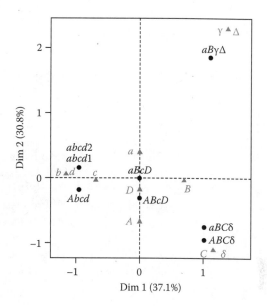

FIGURE 11.6
Pseudobarycentric representation.

11.4 CA on the Burt Table

MCA is also defined as a correspondence analysis of the Burt table $\mathbf{B} = \mathbf{Z}^T\mathbf{Z}$, the matrix of all pairwise associations between the variables, including on the diagonal associations between each variable and itself. In this table, the information about the individuals is not available, and only the information about the relationships between categories is present. The representation of the cloud of categories for this analysis coincides exactly with the representation of the cloud of barycentres (right graph, Figure 11.4). It is nonetheless possible to represent the individuals on the graph of categories and thus obtain the representation of the middle graph in Figure 11.4 by projecting the indicator matrix as supplementary elements. In this representation, it is possible to interpret the distances between an individual and a category, with a category considered a group of individuals. In addition, this representation is the optimum representation of the individuals in terms of the fit of the cloud of individuals, and the individuals' barycentre in terms of the fit of the Burt table.

There is a link between the CA of the Burt table and the CA of the indicator matrix (Lebart et al., 1997, p. 126). The associated eigenvalues are linked by the following relationship:

$$\lambda_k^{\text{Burt}} = \left(\lambda_k^{\text{Indicator matrix}}\right)^2$$

Up to the multiplicative factor $1/\sqrt{\lambda_k}$, the representation of the categories in the analysis of the Burt table (Figure 11.4) corresponds to that of the categories in the analysis of the indicator matrix (Figure 11.5). The percentages of inertia are different, as one corresponds to the fit of the cloud of individuals and the other to that of their barycentres. The percentages associated with the first dimensions of the Burt table are greater (see right graph, Figure 11.4), and the decrease in percentages of inertia is more noticeable (Benzécri, 1977c). One may consider that the inertias obtained from the Burt table are not satisfactory. Indeed, when analysing the Burt table, all the elements are fitted. However, one may not want to take into account the diagonal matrices on the diagonal part of the Burt matrix since they do not represent associations between variables. That is why adjusted percentages of inertia have been proposed (Greenacre, 2007, p. 150) to measure the inertia explained in the intervariable associations only.

11.5 Example

The example studied here is inspired by the work of the sociologist Bourdieu, who initially made MCA popular in the humanities. Here we examine the results of a survey on the construction of identity called 'Histoire de vie' ('Life Story'). This survey, in which 8,403 people aged 15 or over were questioned, was conducted in 2003 by the French National Institute of Statistics (Insee: http://www.insee.fr/). Identity describes the way in which individuals construct a place for themselves within a society that will both enable integration and influence the affirmation of one's own identity. The aim of the survey is to describe the different types of social relationships that enable individuals to fit in to French society at the start of the 21st century. Here we will examine an extract of this database relating to the hobbies of the French, available in the R package FactoMineR (Husson et al., 2013) under the name 'hobbies'. The data set contains 8,403 individuals and 24 variables divided into three groups. The first 18 variables relate to different hobbies. The question format was as follows: 'Have you done or been involved in the following hobby in the past 12 months, without ever having been obliged to do it?' with possible answers yes and no. The questions were related to going to the cinema; reading; listening to music; going to a show (theatre, concert, dance, circus, etc.); visiting an exhibition, museum, or historical monument; using a computer or game console; doing sport or physical activity; walking or hiking; travelling or visiting; playing music, painting, or other artistic activity (such as dance, drama, writing, photography); collecting; volunteering; doing manual work (mechanics, DIY, or decorating); gardening; knitting, embroidery, or sewing; cooking (as a hobby); fishing or hunting; and the number of hours spent watching television on average every day (no TV at all, 0–1h, 1–2h, 2–3h, more than 3h, where left boundaries of intervals are not included; labelled TV_0, TV_1, ..., TV_4).

The following four variables were used to label the individuals: sex (male, female), age (16–25, 26–35, 36–45, 46–55, 56–65, 66–75, 76–85, 86 or more), marital status (single, married, widowed, divorced, remarried), and profession (manual labourer, unskilled worker, technician, foreman, senior management, employee, other). And finally, a quantitative variable indicates the number of hobbies practiced out of the 18 possible choices.

The 18 categorical variables corresponding to the hobbies are considered active variables (the distances between individuals are calculated from these variables), and the sociodemographic variables are considered supplementary (or illustrative). These will be used to assist in interpreting the dimensions of the MCA. In this presentation, we will limit ourselves to interpreting the first two dimensions.

The representation of individuals (not presented here) is homogeneous, without any distinct groups of individuals. The square correlation ratios of variables on the two dimensions (Figure 11.7) indicate that the variables that contribute the most to the construction of the first dimension are exhibition, show, and cinema, and that which contributes the most to the construction of

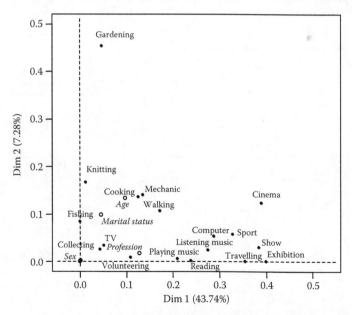

FIGURE 11.7
Map of categorical variables. Illustrative variables in italics.

the second dimension is gardening. The supplementary variables, in italics on the graph, are represented in the same way as the active variables. They are all somewhat linked to the first two dimensions (except the variable 'sex').

The representation of active categories (Figure 11.8) obtained by the CA of the Burt table shows an opposition on the first dimension between all the yes categories and the no categories. This dimension thus opposes those individuals who tend to have more activities with those who have none, or only very few. The second dimension separates different activities, with those hobbies that could be described as 'young' (cinema, computing, sport, shows, playing music) at the bottom, and the 'calm' activities (gardening, knitting, fishing) at the top of the graph. The proximity between the categories 'fishing', 'gardening', and 'knitting' means that individuals taking these categories have the same profile of responses. More precisely, it covers two situations: cases where the individuals are the same and cases where the individuals are different but have the same pattern of responses for the other categories. The former situation may correspond to the proximity between the barycentres of the individuals who fish and those who do gardening, indicating that most of those that do fishing also do gardening. The latter case may correspond to the proximity between 'knitting' and the two other categories. Indeed, we can suppose that it is not necessarily the same individuals who do both knitting and fishing. However, these individuals may have very similar profiles: they are not involved in any 'young' activities. Using the graphs, it is possible to appreciate the main tendencies of the analysis. The interpretation can be refined by using different indicators (contribution, quality of representation) or even by returning to the raw data (by conducting cross-tabulations, for

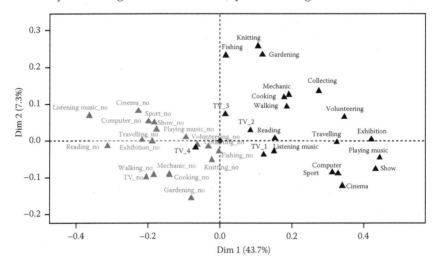

FIGURE 11.8
Map of active categories obtained from the Burt table. The graph of individuals is represented in the top left-hand corner.

example). Here we can check that 55.9% of people who do fishing also do gardening. We should also remark that the contribution of the category 'gardening' to the construction of the second dimension is greater than the contribution of 'fishing' (18.7 versus 9.5). This is also suggested by the graph of variables rather than the coordinate of the categories on the second dimension. This is due to the fact that the category 'fishing' is less common than the category 'gardening' (only 945 respondents compared with 3,356).

To make the results easier to interpret, we can use the supplementary elements. The graph of supplementary categories (Figure 11.9) shows a Guttman effect (or arch effect) for the categories of the variable 'age', which are perfectly sequenced. Young people, who typically have a lot of activities, are opposed on the first dimension with older people with fewer or no activities. The second dimension opposes the classes of extreme age with the classes of middle age. More specifically, the age categories of between 45 and 75 years tend to do more gardening, knitting, and fishing than the other age categories. In the same way, young people aged 16–35 tend to prefer activities relating to sport, cinema, computing, and shows, when compared with other age categories. These results can be substantiated by constructing tables with row or column percentages, for example, confronting the variables 'computer' and 'age' or 'gardening' and 'age' (Table 11.2).

On the graph, the different socioprofessional categories are confronted on the first dimension with, from left to right, manual labourers, unskilled workers, employees, technicians, foremen, and senior management. Overall, managers have more hobbies (by returning to the raw data, we can see that on average a manager has 8.94 hobbies) than manual labourers (5.06) or unskilled workers (5.79).

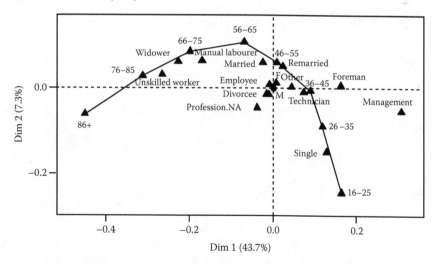

FIGURE 11.9
Map of supplementary categories.

TABLE 11.2

Column Percentages for the Cross-Tabulations Computer by Age and Gardening by Age

	16–25	26–35	36–45	46–55	56–65	66–75	76–85	86+
Computer no	0.25	0.43	0.50	0.66	0.82	0.92	0.96	1.00
Computer yes	0.75	0.57	0.50	0.34	0.18	0.08	0.04	0.00
Gardening no	0.86	0.69	0.57	0.52	0.51	0.54	0.65	0.78
Gardening yes	0.14	0.31	0.43	0.48	0.49	0.46	0.35	0.22

We can confirm this result using the variable 'number of activities'. To include this quantitative variable in the analysis, this variable can be divided into classes and included as an illustrative variable. A division such as this can highlight nonlinear effects (for example, the quadratic effect of the variable 'age' on the two-dimensional map). It is also possible to retain the variable in its quantitative state and to project it as supplementary, which is what we have done here. The graph in Figure 11.10 is constructed in the same way as the graph of correlations in PCA: the coordinate of the variable on dimension k corresponds to the coefficient of the correlation between the principal component k and the variable. This graph (Figure 11.10) confirms that the first axis of the MCA separates the individuals, according to a gradient, from the number of activities (the correlation coefficient between the first principal component and the number of hobbies is 0.975).

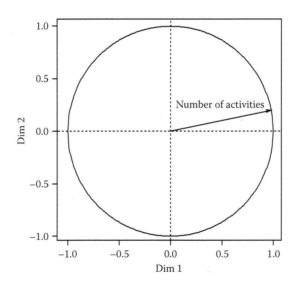

FIGURE 11.10
Map of the quantitative supplementary variable 'number of activities'.

11.6 Discussion

Let us end this introductory chapter to MCA with a number of important additional aspects:

1. Although data analysis methods are used as multidimensional descriptive statistics, it is more and more common to evaluate the stability of the results. In order to do this, methods such as bootstrap techniques have been suggested in order to construct confidence zones (Greenacre, 1984; Lebart et al., 1997; Le Roux and Rouanet, 2010).
2. In the analysis of survey data, one is often faced with problems relating to missing data (individuals who leave out certain questions for whatever reason). A common solution is to create a 'missing' category for each variable. However, this strategy is not always appropriate, and other solutions have been suggested in various works (van der Heijden and Escofier, 2003; Greenacre and Pardo, 2006a; Josse et al., 2012).
3. The principal components of the MCA are synthetic quantitative variables that summarize all the categorical variables. It is thus possible to use MCA as a preprocessing technique and to perform a clustering method on the principal components, for example (Husson et al., 2011), omitting the last components that are considered to be random variation.

12

Structured Data Analysis

Brigitte Le Roux

CONTENTS

12.1 Empirical Example: Lifestyles in the UK 187
12.2 Algebraic Aspects of Analysis of Variance 188
      12.2.1 Cloud of Points 189
      12.2.2 Notations 189
      12.2.3 Contrast and Comparison 190
      12.2.4 Double Breakdown of Inertias 191
12.3 Nesting of Two Factors... 191
      12.3.1 Between and Within Clouds 192
      12.3.2 Example: Lifestyles and Social Classes 192
12.4 Crossing of Two Factors 196
      12.4.1 Main and Within Effects and Clouds 196
      12.4.2 Lifestyle Study: Crossing of Age and Gender 197
            12.4.2.1 Additive and Interaction Effects and Clouds.............. 197
            12.4.2.2 Lifestyle Study: Within, Additive, and Interaction
                     Effects..200
12.5 Conclusion 203

In this chapter, I present the study of Euclidean clouds, taking into account the structures they are equipped with. The motivation is as follows: In *analyse des données*, the basic object of methods is a two-way table. The analysis brings out the relations between the two sets indexing the entries of the table, that is, the structures of the table. Such methods have the remarkable ability to identify these structures—this is one of their most valuable assets—but they do not take into account the structures the two sets themselves may be equipped with.

By *structuring factors*, we mean relevant variables describing the two sets that do not serve to construct the Euclidean clouds. By *structured data*, I designate data tables whose sets are equipped with structuring factors, and by *structured data analysis*, I mean the embedding of structuring factors into the data analysis, in the line of analysis of variance (ANOVA)—including its multivariate extension (MANOVA)—while preserving the principles of the construction of clouds (Le Roux and Rouanet, 1984).

Clearly, structured data constitute the rule rather than the exception. The data analysis based on geometric modeling of data (construction of Euclidean clouds), including structured data analysis and its inductive extensions, is called geometric data analysis (GDA) (Le Roux, 2014). The phrase 'geometric data analysis' was suggested in 1996 by Patrick Suppes of Stanford University; see the foreword of the book by Le Roux and Rouanet (2004).

The traditional technique in GDA, namely, that of supplementary elements (see, e.g., Le Roux and Rouanet, 2010, p. 42; Chapter 9 by Cazes in this book), can be considered as the beginning of structured data analysis; GDA users have widely used this technique both in a predictive and an explanatory perspective. For instance, in *La Distinction*, Bourdieu (1979), when building the *space of lifestyles*, points out that he has put the sociodemographic variables (age, father's profession, education level, and income) not as active variables but as supplementary ones—that is, they contribute neither to the construction of clouds nor to the determination of principal axes—in order to 'give its full strength to the demonstration' (Bourdieu, 1979, p. 294) that differences in lifestyle can be explained by the positions in the social space.

The limitations of this technique become apparent as soon as one realizes that, on the one hand, the categories of supplementary variables induce subclouds of individuals, and that, on the other hand, considering only supplementary categories amounts to confining attention to the mean points of subclouds while leaving aside their dispersion.

Referring again to *La Distinction*, when Bourdieu studied the fractions of the upper class—'the most powerful explanatory factor' (Bourdieu, 1979, p. 294)—this concern led him to examine the corresponding subclouds of individuals, that is, to treat class fractions as a structuring factor (see Diagrams 11 and 12 of *La Distinction*, 1979, p. 296). The examination of subclouds revealed, for example, that the individuals of 'professions libérales' are concentrated in the centre of the space, whereas those of 'cadres' are scattered throughout the whole plane. As supplementary elements, these two class fractions would have been merely represented by two points, both located near the centre of the diagram (see Rouanet et al., 2000).

In the experimental paradigm, there is a clear distinction between *experimental factors* (independent variables) and *dependent variables* (response variables) with a *factorial design* that expresses the relationships between factors and often corresponds to strong structures like balanced design (each combination of categories of the factors of the design has the same number of observations), orthogonal design, etc. Statistical analysis aims at studying the effects of factors in the framework of the design.

Structured Data Analysis

In observational data, even in the absence of prior design, there are also relationships between structuring factors that generate structures, formally equivalent to those of an experimental design, but often much simpler. They are generally not more complex than nesting and crossing. For instance, the individuals in a survey are nested within age groups, since any individual belongs to one and only one age group; gender and age groups are crossed factors whenever there are individuals of both genders in every age group. Therefore, all effects of interest are of the following types: main effects, between effects, within effects (also called conditional effects in regression analysis), and interaction effects. On the other hand, when factors happen to be crossed, the crossing relation is usually *not orthogonal* (as opposed to experimental data, where orthogonality is often ensured by design); that is, structuring factors are typically *correlated*. As a consequence, conditional effects must be carefully distinguished from main effects. For instance, if age and gender are correlated, the effect of gender conditional upon age may differ from the main effect of gender. In this chapter, as will become clear, factors have nothing to do with principal variables (also called factors), produced by principal components analysis (PCA), correspondence analysis (CA), or multiple correspondence analysis (MCA).

12.1 Empirical Example: Lifestyles in the UK

The data source is the project 'Cultural Capital and Social Exclusion: A Critical Investigation' (see Bennett et al., 2009). The data were collected in 2003–2004.

For the construction of the *space of lifestyles*, we used 41 questions of the questionnaire (17 on participation and 24 on taste) covering a wide diversity of the cultural field, from television, film, reading, music, visual art, eating out, and participation in leisure and sport. These 41 questions generate 166 active categories; the rare categories (frequencies less than 4%) as well as 'others' and 'don't know' were excluded from the analysis; to do so, we use the variant of MCA called specific MCA (see Le Roux and Rouanet, 2004, 2010). The data set includes 1,529 individuals since 35 individuals who had failed to respond to several questions were discarded from the analysis (for further details about the construction of the space and the interpretation of axes, see Le Roux et al. (2008); Bennet et al. (2009, chap. 3).

The interpretation of axes is based on the method of contributions of points and deviations (see Le Roux and Rouanet, 1998) and may be summarized as follows:

- Axis 1 ($\lambda_1 = 0.1641$). *Cultural engagement: Involvement vs. disengagement.*
  This axis opposes absence and frequent attendance at legitimate

cultural events; it also opposes differences over taste for legitimate genres.

- Axis 2 ($\lambda_2 = 0.1188$). *Contemporary taste: established vs. emergent.* This axis shows an opposition between tastes for established cultural genres and emergent ones. Established genres cover items linked to traditionally accepted forms of culture, while emergent genres refer to contemporary/commercial cultural practices.
- Axis 3 ($\lambda_3 = 0.0746$). *Expression of inwardly vs. outwardly oriented dispositions.* The former identifies engagement with fictional genres focused on personal concerns; the latter refers to preferences for more factual and physically robust genres.

Using the technique of supplementary variables, we study the sociodemographic variables, namely, age, gender, income, and education level (see Le Roux et al., 2008; Bennett et al., 2009).

For gender, the deviation between men and women is negligible on axis 1, small on axis 2, and large on axis 3 (men are located on the side of outwardly oriented dispositions and women on the side of inwardly oriented dispositions). For age, the categories are ordered along axis 2, and the deviation between extreme categories is very large on this axis, medium on axis 1, and very small on axis 3. The household income categories are ordered along axis 1, and the deviation between extreme categories on axis 1 is large. Education levels are also ordered on axis 1 with a large deviation between 'university' and 'no qualification'.

In summary, the first axis is related to income and level of education, the second axis to age, and the third one to gender. In the following, we will study in detail social class, age, and gender as structuring factors.

12.2 Algebraic Aspects of Analysis of Variance

Analysis of variance, like geometric data analysis, can be tackled with the formal approach. The resulting construction—called *analysis of comparisons*—is founded on the set theory to formalize factors and factorial designs, and on abstract linear algebra to formalize ANOVA concepts, such as sums of squares (SS), etc. Analysis of comparisons was developed by Rouanet and Lépine (1976); originally designed for numerical variables, it was extended to a Euclidean cloud by Le Roux and Rouanet (1983, 1984, 2004).

12.2.1 Cloud of Points

Let us consider a set of n observations indexed by I and m_I a strictly positive measure over I with total mass (sum of coefficients) denoted m_{tot}. In GDA, the observations are the points of a cloud, denoted M^I, in a Euclidean space U of dimensionality p. Each point M^i is weighted by m_i. Let G denote the mean point (or barycentre) of the cloud.

12.2.2 Notations

Taking any Cartesian frame of U, we define the following:

- The vector $\mathbf{x}_i$ of the p coordinates of point M^i.
- The matrix $\mathbf{X}$ $(n \times p)$, the n rows of which are the vectors $\mathbf{x}_i$; the general term of the matrix is x_{ij}, that is, the coordinate of point M^i on the jth dimension ($1 \leq j \leq p$) of the space.
- The vector $\mathbf{m}$ of the n weights, with

$$\sum_{i=1}^{n} m_i = \mathbf{1}^T \mathbf{m} = m_{tot},$$

where $\mathbf{1}$ is a vector of n 1s.

- $\mathbf{D}_m$ $(n \times n)$ the diagonal matrix of the weights m_i.
- The symmetric positive definite matrix $\mathbf{Q}$ $(p \times p)$ defines the inner product, and thus the norm and the metric of U: the scalar product between vectors $\mathbf{u}$ and $\mathbf{v}$ is equal to $\mathbf{u}^T \mathbf{Q} \mathbf{v}$, the squared norm of a vector $\mathbf{u}$ is equal to $\mathbf{u}^T \mathbf{Q} \mathbf{u}$, and the squared distance between $\mathbf{u}$ and $\mathbf{v}$ is equal to $(\mathbf{u} - \mathbf{v})^T \mathbf{Q} (\mathbf{u} - \mathbf{v})$.

The coordinate of point G on the jth dimension is

$$g_j = \sum_{i=1}^{n} m_i x_{ij} / m_{tot},$$

hence the vector

$$\mathbf{g} = \frac{1}{m_{tot}} \mathbf{X}^T \mathbf{m}.$$

When the mean point G is taken as the origin, each vector $\mathbf{x}_i$ is centred with respect to $\mathbf{g}$ and the centred vectors are gathered in the table $\mathbf{X}_0 = \mathbf{X} - \mathbf{1} \mathbf{g}^T$.

By definition, the overall *sum of squares* (or *inertia*) of the weighted cloud (M^I, m_I) is the weighted sum of the squared distances from its points to the mean point, that is,

$$\sum\nolimits_{i=1}^{n} m_i(\mathbf{x}_i - \mathbf{g})^\mathsf{T} \mathbf{Q}(\mathbf{x}_i - \mathbf{g});$$

hence, the total variance of the cloud is denoted Var_{tot}, with

$$\text{Var}_{tot} = \sum\nolimits_{i=1}^{n} m_i(\mathbf{x}_i - \mathbf{g})^\mathsf{T} \mathbf{Q}(\mathbf{x}_i - \mathbf{g}) / m_{tot}$$

(see Benzécri, 1992, p. 36).

In the sequel, we will consider the breakdown of variance along principal axes of the cloud, so it is convenient to write the formulas in the basis of the principal axes. We will denote $\mathbf{Y}$ the matrix ($n \times p$) of the *principal coordinates* of the points of the cloud, verifying the principal variable equation (see Le Roux and Rouanet, 2004, p. 119):

$$\mathbf{X}_0 \mathbf{Q} \mathbf{X}_0^\mathsf{T} \mathbf{D}_m \mathbf{Y} = \mathbf{Y} \Lambda \quad \text{with} \quad \mathbf{Y}^\mathsf{T} \mathbf{D}_m \mathbf{Y} = \Lambda$$

where Λ is the diagonal matrix ($p \times p$) of eigenvalues (λ). Relative to the principal basis, the matrix defining the metric is equal to $n \times n$ identity matrix $\mathbf{I}$.

12.2.3 Contrast and Comparison

By definition, a contrast over I is a measure over I whose total mass is null. A comparison over I is a subspace of contrasts over I (its dimensionality defines the number of degrees of freedom of the comparison).

With any contrast $\mathbf{c}$ over I, with masses c_i ($\Sigma_i c_i = \mathbf{1}^\mathsf{T}\mathbf{c} = 0$), one associates:

1. A vector effect, or effect of the contrast on the cloud M^I, which is the geometric vector $\Sigma_i c_i \mathrm{M}^i$, denoted $\mathbf{e}$ with $\mathbf{e} = \mathbf{Y}^\mathsf{T}\mathbf{c}$.
2. A sum of squares (SS), or inertia, which is, by definition, the ratio of the squared norm of the vector effect (in the vector space associated with the geometric space) to the squared norm of the contrast (in the vector space of measures on I; see Le Roux and Rouanet (2004, pp. 27–28), that is,

$$\frac{\mathbf{e}^\mathsf{T}\mathbf{I}\mathbf{e}}{\mathbf{c}^\mathsf{T}\mathbf{D}_m^{-1}\mathbf{c}} = \frac{\mathbf{c}^\mathsf{T}\mathbf{Y}\mathbf{Y}^\mathsf{T}\mathbf{c}}{\mathbf{c}^\mathsf{T}\mathbf{D}_m^{-1}\mathbf{c}}$$

This SS is also the one of any proportional contrast, and therefore defines the SS of the 1 d.f. (one degree of freedom) comparison generated by the contrast. The SS of a comparison with p d.f. is the sum of the SS of p orthogonal contrasts generating the comparison (see the example in Section 12.4.2.2).

12.2.4 Double Breakdown of Inertias

The double breakdown of inertias according to axes and sources of variation consists in calculating the inertias of each source of variation (factor) on each principal axis, in the line of Le Roux and Rouanet (1984; see also Rouanet, 2006). Nonorthogonality is no obstacle to breaking down sources of variation, even though the breakdown of inertias is no longer additive.

Two sorts of relative contributions follow:

1. The contribution of a source of variation to an axis is obtained by dividing the variance of the source on an axis by the variance of the axis (eigenvalue).
2. The contribution of an axis to the source of variation is obtained by dividing the variance of the source on an axis by the total variance of the source.

The first contribution extends the contribution of a point to an axis; the second extends the quality of representation of a point on an axis (see Benzécri et al., 1973, pp. 37–40; Benzécri, 1992, pp. 99–103).

From the double-breakdown table, one may calculate the variance of a source of variation in a plane, a subspace, etc.; one may seek the axes for which, in the case of *nesting*, the opposition between some groups is important (see *lifestyle study*, Section 12.3), or, in the case of *crossing*, for which the interaction is important (see Section 12.4.2.2).

12.3 Nesting of Two Factors

Consider two factors A and B; if each category (level) of factor A is paired with only one category of factor B, then A is said to be nested within B (see Winer, 2004, p. 184). Such a definition implies a partition of A into B groups; the subset of categories of A that are paired with category b of B is denoted $A <b>$; the partition (and also the nesting) is denoted $A <B>$. In ANOVA, it is classical to break down the source of variation $A <B>$ into two sources: the between B and the A-within-B ones. We denote the former B and the latter $A(B)$ (read 'A-within-B').

In the lifestyle study, there are the factors 'individuals', denoted I, and 'occupational groups', denoted A. Then, from the space of lifestyles we want to reduce the 12 occupational groups to 3, hence, a factor B such that A is nested within B, and we compare our classification to the official one (Le Roux et al., 2008). So we have the following relations: $I <A>$, $A <B>$ and also $I <A < B>>$.

12.3.1 Between and Within Clouds

Consider here two factors I (for instance, a set of n individuals) and A (for instance, a set of A occupational groups) with the nested relation $I <A>$ (as a general rule, we denote by the same letter a factor and its number of categories, with the exception of I, where we use n). With each category a is associated the subset of individuals belonging to category a that we call 'group a', hence a subcloud whose mean point is denoted M^a with weight m_a, which is the sum of the weights of individuals of group a. The A weighted mean points $(M^a, m_a)_{a \in A}$ define the *between cloud*; its variance is the between-A variance.

For a point M^i belonging to group a, the deviation from the mean point M^a to point M^i is the geometric vector $M^i - M^a$. The *within cloud* (residual cloud) with points denoted $M^{i(a)}$ is such that the deviation from point G to point $M^{i(a)}$ is equal to the one from M^a to M^i (see Le Roux and Rouanet, 2004, pp. 100–105).

By construction, the within cloud has mean point G and its variance is the I-within-A variance. According to the rule of parallelograms for adding geometric vectors, the deviation from G to M^i is the sum of the deviations from G to $M^{i(a)}$ and from G to M^a; that is, there is a breakdown of the initial cloud into the between cloud and the within cloud. This breakdown is orthogonal in the sense that, for any axis of the space, the coordinate variable of the between cloud is uncorrelated with the one of the within cloud; as a consequence, there is an additive breakdown of the variance into between variance and within variance. The proportion of the variance due to factor A is the coefficient $\eta^2 = \text{Var } M^A / \text{Var}_{\text{tot}}$.

12.3.2 Example: Lifestyles and Social Classes

For the study of social classes, we use a version (2000) of the National Statistics Socioeconomic Classification (NS-SEC), which distinguishes between 13 occupational categories coded L1 to L13. The 12 categories (L1 and L2 are pooled because of the small counts involved) retained for analysis are

L1/L2	($n = 29$)	Employers in large organizations and higher managerial occupations[a]
L3	($n = 91$)	Higher professional occupations
L4	($n = 237$)	Lower professional and higher technical occupations
L5	($n = 77$)	Lower managerial occupations
L6	($n = 72$)	Higher supervisory occupations
L7	($n = 192$)	Intermediate occupations
L8	($n = 36$)	Employers in small organizations
L9	($n = 68$)	Own account workers
L10	($n = 121$)	Lower supervisory occupations
L11	($n = 53$)	Lower technical occupations
L12	($n = 311$)	Semiroutine occupations
L13	($n = 198$)	Routine occupations

[a] The 41 individuals who have never worked and the 3 whose occupation was not classifiable were excluded from the analysis, which leaves us with $n = 1485$ individuals.

Structured Data Analysis

We will take the set of occupational categories as the structuring factor (factor A): factor I (individuals) is nested within factor A (occupational categories).

First issue: Where are these groups of individuals located in the space of lifestyles?

Relative to the first three principal axes, the coordinates of the mean points and the variances of the 12 subclouds are given in Table 12.1. The 12 mean points are represented in plane 1-2 (Figure 12.1) and in plane 1-3 (Figure 12.2). Principal axes being perpendicular, the variances of each source of variation add up in a plane or in a three-dimensional space. Thus, the between variance of A is equal to 0.0440 in plane 1-2 and to 0.0512 in the three-dimensional space.

From the double breakdown of variances, we notice that the contribution of the first axis to A (see Section 12.2.4) is predominant on axis 1. Moreover, the order of the 12 category mean points corresponds closely to axis 1 (see Figure 12.1). The proportion of the variance of axis 1 accounted for by the 12 category mean points is $\eta^2 = 0.0422/0.1648 = 0.256$; the eta-square coefficients corresponding to the other axes are much smaller. So, we may confine attention to the first axis and conclude that axis 1 is also the axis of occupational groups; for a more detailed study, see Le Roux et al. (2008, pp. 1061–1064).

TABLE 12.1

Coordinates of Mean Points, Variances of the 12 Subclouds, and Double Breakdown of Variances on the First 3 Axes, in Plane 1-2 and Subspace 1-2-3

A	n_a	Coordinates			Variances			1-2	1-2-3
		Axis 1	Axis 2	Axis 3	Axis 1	Axis 2	Axis 3		
L1/L2	29	+0.345	+0.051	0.167	0.098	0.119	0.071	0.217	0.288
L3	91	+0.356	−0.026	−0.095	0.085	0.136	0.058	0.221	0.279
L4	237	+0.266	−0.045	+0.018	0.112	0.139	0.066	0.251	0.317
L5	77	+0.139	+0.066	−0.102	0.157	0.119	0.059	0.276	0.335
L6	72	+0.129	−0.081	+0.014	0.098	0.101	0.093	0.199	0.292
L7	192	+0.065	−0.021	+0.099	0.132	0.132	0.069	0.264	0.333
L8	36	−0.021	−0.039	−0.156	0.114	0.100	0.053	0.214	0.267
L9	68	−0.017	+0.036	−0.106	0.132	0.116	0.068	0.248	0.316
L10	121	−0.114	−0.027	−0.071	0.117	0.099	0.073	0.216	0.289
L11	53	−0.159	+0.119	−0.162	0.128	0.112	0.052	0.240	0.292
L12	311	−0.133	+0.015	0.089	0.139	0.111	0.066	0.250	0.316
L13	198	−0.311	+0.022	−0.032	0.117	0.101	0.084	0.218	0.302
	1,485			Within A	0.1226	0.1173	0.0692	0.2339	0.3091
				Between A	0.0422	0.0018	0.0072	0.0440	0.0512
				Total	0.1648	0.1191	0.0764	0.2839	0.3603
				η^2	0.256	0.015	0.094	0.155	0.166

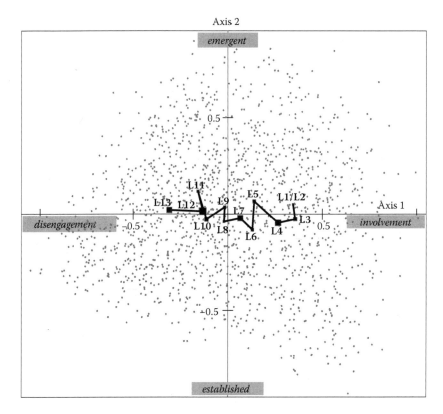

FIGURE 12.1
Cloud of 1,529 individuals with the 12 mean points of occupational groups (L1/L2 through L13) in plane 1-2.

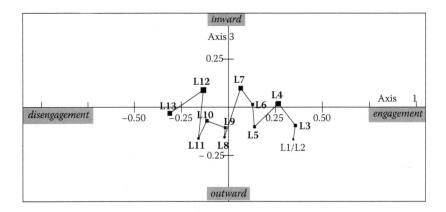

FIGURE 12.2
The 12 occupational category mean points in plane 1-3 (cloud of individuals). The size of squares is proportional to the frequencies of occupational groups.

Second issue: It consists in reducing the 12 occupational groups to 3.

We will study two partitions: the one produced by NS-SEC and the one stemmed from the lifestyle space.

The nested relationship between the 12- and 3-category versions of the NS-SEC is the following: managerial and professional occupations (L1/L2, L3, L4, L5, L6) (n = 506, 34%), intermediate occupations (L7, L8, L9) (n = 296, 20%), and routine and manual occupations (L10, L11, L12, L13) (n = 683, 46%), hence the factor B with $A <B>$.

From the *space of lifestyles*, we determine a partition into three clusters, first resulting from an ascending hierarchical clustering with aggregation according to variance (Ward method; see Roux's Chapter 17 in this book), and second with a reallocation of the 12 occupational category mean points to the three clusters using the K-means procedure. Thus, we obtain the following nested relationship: a small *service class* of professional and higher managers (L1/L2, L3, L4) (n = 357, 24%), an *intermediate class* that includes lower managers (L5, L6, L7, L8, L9) (n = 445, 30%), and a relatively large *working class* that includes lower supervisors and technicians (n = 683, 46%), hence the factor B' with $A <B'>$.

We will compare the two 3-cluster partitions. Because the between variance of the partition into 12 clusters is low on axes 2 and 3, we study the two 3-cluster partitions only on axis 1. For the NS-SEC partition (B), the between variance on the first axis is equal to 0.0366, hence $\eta^2 = 0.0366/0.0422 = 0.866$. For our partition ($B'$), the between variance is equal to 0.0377, hence $\eta^2 = 0.0377/0.0422 = 0.894$, a fit that leads to a slight improvement, but the sizes of the groups are very different (see Figures 12.3 and 12.4). For a more general

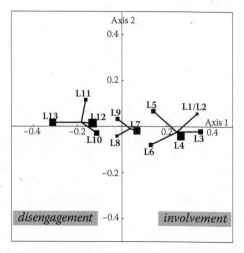

FIGURE 12.3
$A <B>$: 12 occupational categories split up into 3 clusters (NS-SeC classification) in the cloud of individuals (plane 1-2).

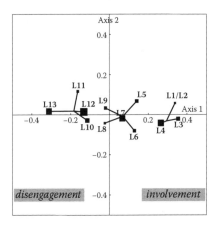

FIGURE 12.4
A <B'>: 12 occupational categories split up into 3 clusters (our classification) in the cloud of individuals (plane 1-2).

discussion about a new model of social class in the UK, see Savage et al. (2013).

12.4 Crossing of Two Factors

Two factors A and B are said to be crossed if each level of A is associated with each level of B; in other words, the Cartesian product $A \times B$ is itself a factor. Let m_{AB} denote the weighting measure over $A \times B$, m_A and m_B the marginal measures (with $m_a = \Sigma_b m_{ab}$, $m_b = \Sigma_a m_{ab}$), and m_{tot} the total mass ($m_{\text{tot}} = \Sigma_a \Sigma_b m_{ab}$).

12.4.1 Main and Within Effects and Clouds

When two factors A and B are crossed, it is customary to break down the source of variation $A \times B$ into three sources: the two main sources A and B and the interaction source $A.B$. Besides, a breakdown of the source of variation $A \times B$ is associated with the nesting of $A \times B$ in A, namely, the breakdown of $A \times B$ into the between source A and the within source (residual) denoted $B(A)$, or also, in a symmetrical way, into B and $A(B)$.

Given a Euclidean cloud defined on the crossing of two factors A and B, we define the main clouds (M^A, m_A) and (M^B, m_B) where M^a is the barycentre of the B points $(\mathrm{M}^{ab}, m_{ab})_{b \in B}$, and similarly, M^b is the barycentre of the A points $(\mathrm{M}^{ab}, m_{ab})_{a \in A}$.

Structured Data Analysis

12.4.2 Lifestyle Study: Crossing of Age and Gender

We will now attempt to answer the following question: Is there a difference between men and women within each age group?

For this, we will consider age and gender variables as structuring factors and study their crossing. The age factor is denoted A with $a1 = 18 - 24$, $a2 = 25 - 34$, $a3 = 35 - 44$, $a4 = 45 - 54$, $a5 = 55 - 64$, and $a6 \geq 65$; the gender factor is denoted B with $b1 =$ men and $b2 =$ women.

In the cloud of individuals, the crossing of age and gender induces a cloud of $6 \times 2 = 12$ mean points (see Table 12.2 and Figure 12.5). This weighted cloud will be taken as the basic data set for the analyses that follow and will be denoted M^{AB}.

- *Age cloud and main effects of age.* For age, one has $6 - 1 = 5$ degrees of freedom. The age cloud (M^A) consists of six points: $a1$ is the barycentre of the two gender points ($a1b1$ and $a1b2$) of age group $a1$, and so on (coordinates of the 12 points are in Table 12.2 (right)). For age groups $a1$ and $a2$, the main effect is the vector joining $a1$ to $a2$. Figure 12.6 shows the five vectors (in gray) representing main effects of age in plane 1-2 (left) and in plane 2-3 (right).
- *Gender cloud and main effect of gender.* The mean point of the six age points for men and that of the six age points for women both define the gender cloud. The mean effect of gender is the vector (black arrow) joining these two mean points (see Figure 12.6).

The variances of clouds $M^{A \times B}$ (12 points), M^A (6 points), and M^B (2 points) are given in Table 12.3.

Comments: As already said, axis 2 is related to age. There is no main effect of gender on axis 1, a small one on axis 2, and a large one on axis 3.

12.4.2.1 Additive and Interaction Effects and Clouds

> *Definition*: A cloud on the crossing of two (structuring) factors A and B is without interaction if the within effects of A are the same for the different levels of B, or equivalently, if the within effects of B are the same for the different levels of A.

Given a category a of A and two categories b and b' of B ($b \neq b'$), the vector $M^{ab} - M^{ab'}$ is the (b, b')-within-a effect (for example, the vector M^{ab}-$M^{ab'}$ is the effect of gender B for age group a). If the cloud is without interaction, for $a' \in A$ with $a' \neq a$, then the (b, b')-within-a effect is equal to the (b, b')-within-a' effect; that is, we have the relation $M^{ab} - M^{a'b} = M^{ab'} - M^{a'b'}$.

TABLE 12.2
Weights and Coordinates of the 12 Age × Gender Mean Points, of the 2 Gender Ones (B) and of the 6 Age Ones (A), for Axes 1, 2, and 3

	men (b1)				women (b2)				Age (A)			
A	n_{ab}	Axis1	Axis2	Axis3	n_{ab}	Axis1	Axis2	Axis3	n_a	Axis1	Axis2	Axis3
a1	51	0.013	0.516	−0.105	63	−0.098	0.380	0.342	114	−0.048	0.441	0.142
a2	120	0.120	0.387	−0.172	169	0.053	0.182	0.265	289	0.081	0.267	0.083
a3	133	0.121	0.239	−0.214	172	0.100	0.010	0.207	305	0.109	0.110	0.024
a4	94	0.126	0.074	−0.207	141	0.096	−0.078	0.126	235	0.108	−0.017	−0.007
a5	118	−0.054	−0.101	−0.247	121	0.069	−0.305	0.050	239	0.008	−0.204	−0.097
a6	144	−0.240	−0.254	−0.223	203	−0.217	−0.353	0.047	347	−0.226	−0.312	−0.065
B	660	0.003	0.095	−0.205	869	−0.003	−0.072	0.156				

Structured Data Analysis

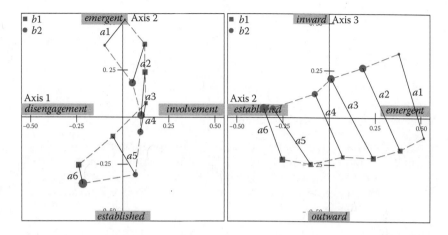

FIGURE 12.5
The 12 mean points of the cloud M^{AB} in planes 1-2 and 2-3 (cloud of individuals).

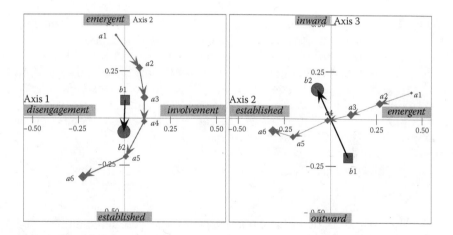

FIGURE 12.6
Main effects of age (gray vectors) and of gender (black vector) in planes 1-2 and 2-3.

TABLE 12.3

Double Breakdown of Variances according to the Three Sources of Variation A, B, and $A \times B$ and the First Three Axes, Variances in Plane 1-2 and in Subspace 1-2-3

	Axis1	Axis2	Axis3	Plane1-2	1-2-3
A	0.01720	0.05904	0.00536	0.07264	0.08160
B	0.00001	0.00690	0.03185	0.00691	0.03876
$A \times B$	0.01830	0.06686	0.03802	0.08516	0.12318

The *geometric interpretation* of this property is the following one:

> A cloud is without interaction if and only if the figure formed by the four points M^{ab}, $M^{a'b}$, $M^{ab'}$, and $M^{a'b'}$ is a parallelogram.

Given a normalized positive measure φ_{AB} over $A \times B$ ($\Sigma_a \Sigma_b \varphi_{ab} = 1$) with marginal measures φ_A and φ_B with $\varphi_a = \Sigma_b \varphi_{ab}$ and $\varphi_b = \Sigma_a \varphi_{ab}$, the following *additive property* can be easily shown.

> *Additive property*: Let P^a be the mean point of the B points $(M^{ab})_{b \in B}$ weighted by $(\varphi_b)_{b \in B}$ and P^b be the mean point of the A points $(M^{ab})_{a \in A}$ weighted by $(\varphi_a)_{a \in A}$; then, for any cloud M^{AB} without interaction, the deviation from G to point M^{ab} is the sum of the deviation from G to point P^a and of the deviation from G to point P^b.

This is why a cloud without interaction is called an *additive cloud*.

> *Property*: Given a cloud (M^{AB}, m_{AB}), there is a unique additive cloud fitted to the cloud with the same main clouds M^A and M^B, called *additive reference cloud* and denoted $M^{A \oplus B}$.

If the basic cloud is without interaction, the additive reference cloud coincides with the cloud; in other words, there are interaction effects only if there are deviations from the observed cloud to the additive reference cloud.

If the crossing is orthogonal, one has $P^a = M^a$ and $P^b = M^b$, and the point $M^{a \oplus b}$ can be directly constructed by adding the deviations from G to point M^a and from G to point M^b. If the crossing is not orthogonal, the construction is not straightforward; terms must be added that depend on the canonical correlations associated with the measure m_{AB} (see Le Roux, 1991; Le Roux and Rouanet, 2010).

The residual cloud $M^{A.B}$ is the *interaction cloud*.

> *Property*: The interaction cloud is doubly centred at point G.

The breakdown of the initial cloud into additive and interaction clouds is *orthogonal*, in the sense that, for any axis of the space, the coordinate variable of the additive cloud and the one of the interaction cloud are uncorrelated.

12.4.2.2 Lifestyle Study: Within, Additive, and Interaction Effects

- *Within effect*. To answer the question, we will study the comparison $B(A)$ (B-within-A) with $12 - 6 = 6$ degrees of freedom.

    For each age group a, the vector joining the two gender points defines the effect of B-within-a (see Figure 12.7).

    The $B(A)$ comparison is generated by the orthogonal basis of the six contrasts over $A \times B$, defined as follows: for each a, the contrast is equal to 1 for $(a, b1)$, to -1 for $(a, b2)$, and to 0 for all other pairs $(a'$,

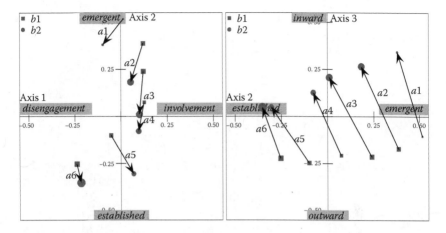

FIGURE 12.7
Six within effects (black vectors) of gender for each age group in planes 1-2 and 2-3.

$b1$), and (a', $b2$) with $a' \neq a$; this contrast is called B-within-a contrast. By definition (see p. 190), the (m_{AB}/m_{tot})-weighted sum of squares (SS) associated with the B-within-a contrast is equal to the squared distance between M^{ab1} and M^{ab2} divided by the squared norm of the contrast, which is equal to

$$1 / \left(\frac{m_{tot}}{m_{ab1}} + \frac{m_{tot}}{m_{ab2}} \right).$$

The basis being orthogonal, the sum of the six SSs is equal to the SS associated with the comparison $B(A)$, that is, the B-within-A variance (see Table 12.4).

Table 12.5 shows that the B-within-A effect is quite small on axis 1, medium on axis 2, and large on axis 3. The detailed study of the variance of B-within-A comparisons shows a larger deviation

TABLE 12.4

SS Associated with the Canonical Contrasts $B(a)$ ($a \in A$) and with the Comparison $B(A)$

	Axis1	Axis2	Axis3
$B(a1)$	0.00022	0.00034	0.00367
$B(a2)$	0.00021	0.00191	0.00875
$B(a3)$	0.00002	0.00256	0.00868
$B(a4)$	0.00003	0.00085	0.00407
$B(a5)$	0.00058	0.00163	0.00346
$B(a6)$	0.00004	0.00053	0.00403
$B(A)$	0.00110	0.00782	0.03266

TABLE 12.5

Breakdown of Variances of $A \times B$ into between (A) and within $(A(B))$ Variances in Plane 1-2 and Subspace 1-2-3

	Axis1	Axis2	Axis3	Plane1-2	1-2-3
$B(A)$	0.00110	0.00782	0.03266	0.00892	0.04158
A	0.01720	0.05904	0.00536	0.07624	0.08160
$A \times B$	0.01830	0.06686	0.03802	0.08516	0.12318

between men and women for age groups $a2$ and $a3$ (see Table 12.4 and Figure 12.7). Together, categories $a2$ and $a3$ contribute 57% to the within variance for axis 2 and 53% for axis 3. Thus, we can conclude that women's and men's lifestyles diverge the most during the 25–44 age period. In terms of life course, this age period coincides with women being predominantly occupied with child rearing and related domestic activities (see Silva and Le Roux, 2011).

- *Interaction and additive effects.* We have seen that the within effects of gender are not the same for the different age groups. Equivalently, the within effects of age are not the same for men and women. In such a situation, it is commonly said that there is an *interaction effect* of the two factors. Figure 12.8 depicts the additive reference cloud in Planes 1-2 and 2-3 fitted to the $M^{A \times B}$ cloud, whose derived clouds are the ones shown in Figure 12.6. This cloud will be compared to the $M^{A \times B}$ cloud represented in Figure 12.5.

By construction, in the additive cloud, the within effects for each factor are equal to one another. For instance, the six gender vector effects have the same coordinates (−0.005, −0.172, +0.358). This vector

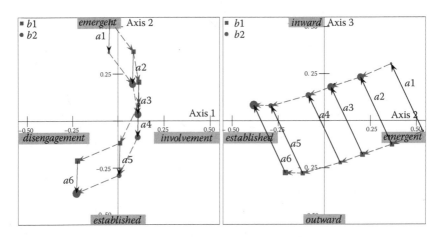

FIGURE 12.8
The additive reference cloud in planes 1-2 and 2-3 (cloud of individuals).

TABLE 12.6

Breakdown of $A \times B$ Variances into Additive ($A \oplus B$) and Interaction ($A.B$) Variances, and Variances in Plane 1-2 and Subspace 1-2-3

	Axis1	Axis2	Axis3	Plane1-2	1-2-3
$A \oplus B$	0.01721	0.06628	0.03680	0.08348	0.12028
$A.B$	0.00109	0.00058	0.00122	0.00167	0.00289
$A \times B$	0.01830	0.06686	0.03801	0.08516	0.12317

effect can be taken as the averaged within effect of B-within-A. By definition, the interaction cloud is the residual cloud. The double breakdown of variances is given in Table 12.6.

Remark: The two factors A and B are not orthogonal (the Cramér's coefficient for the measure m_{AB} is equal to 0.061 ≠ 0). Consequently, the sum of the two variances of the main effects A and B is not the variance of the additive cloud: there is a structure effect. There is also a weak interaction effect: the variance of the additive cloud is slightly different from the variance of the crossing $A \times B$. Structure effect and interaction are two different things.

12.5 Conclusion

The procedures presented in this chapter are purely descriptive in the technical sense that they do not depend on the sample size. For extending descriptive conclusions of existence of effects, significance testing is the natural tool. This can be done in GDA using combinatorial inference, namely, permutation testing that is entirely distribution-free and can even be made free from random assumptions altogether: see test values (Lebart et al., 1995) and typicality and homogeneity tests (Le Roux and Rouanet, 2010; Bienaise, 2013b).

Finally, we will say that structures govern procedures, and we will emphasize that, in structured data analysis, to avoid the crumbling down of procedures, it is sufficient to combine several approaches, namely, analysis of comparisons (for building derived clouds) and geometric data analysis (for searching principal axes). Thus, structured data analysis enriches the analysis of observational data, without having to invent a new method every time a new structure is encountered. To sum up, one avoids two pitfalls: on the one hand, the rigid mold of a standard analysis, and on the other hand, the proliferation of ad hoc procedures.

SOFTWARE NOTE: The analyses of this chapter were performed using SPAD software distributed by SPAD-Coheris (http://www.coheris.fr/en/page/produits/Spad.html).

13

Empirical Construction of Bourdieu's Social Space

Jörg Blasius and Andreas Schmitz

CONTENTS

13.1 History of Jean-Paul Benzécri and Pierre Bourdieu 206
13.2 Bourdieu and His Social Space Approach 207
13.3 Example of Social Space Interpretation 215
13.4 Conclusion 221

Pierre Bourdieu belongs to the most often cited social scientists worldwide, whose books and papers are translated into many languages. His most famous book is probably *La Distinction* (in French, 1979; English version *Distinction*, 1984), which had already been translated into 10 languages by 2002, the year he died (cf. Delsaut and Rivière, 2002), including German, Spanish, Russian, and Japanese. Within the social sciences, the book is appraised as one of the most important contributions to stratification research. Outside of France, it is less known that an essential part of Pierre Bourdieu's work is closely related to his interest in statistics and methodology, and here especially in the work of his friend Jean-Paul Benzécri.

In *La Distinction*, Bourdieu was mainly interested in identifying and distinguishing different groups within the French society. As group indicator he used occupational status. He assigned these groups to three classes: the upper, the middle, and the lower classes. One of his main interests was to show differences within the upper classes, for example, between artists and managers, and how they defend their positions against those groups from the middle classes aspiring to enter the upper classes. To distinguish these groups, he used lifestyle attributes such as preferences for clothes, food, and artists, combining them with elements of traditional stratification research such as age (groups) and educational level. As part of his concept of the *social space*, or in French, *l'espace social* (Bourdieu and de Saint Martin, 1976, p. 45), he visualized the relations between class fractions and their associated lifestyles. In addition to the social sciences, in the following years this concept was applied in many different areas, such as medicine, marketing research, linguistics, and geography.

The first part of this chapter concentrates on describing the influence of a group of Parisian statisticians contributing to the development of *l'analyse des données* and being interested in the work of Bourdieu. Among these statisticians were Brigitte Le Roux, Ludovic Lebart, Henry Rouanet, and especially the founder of the French tradition of correspondence analysis, Jean-Paul Benzécri. A special focus of this chapter, which is related to the work of Bourdieu, is on the differentiation between the application of simple correspondence analysis (CA), using the frequencies of a stacked cross-table as input data, and multiple correspondence analysis (MCA), using an indicator matrix as input data. Thereby, we discuss the theoretical backgrounds for these two kinds of input formats in terms of Bourdieu's underlying methodological implications.

13.1 History of Jean-Paul Benzécri and Pierre Bourdieu

In a personal memoriam to Bourdieu, published four years after Bourdieu's death in January 2002, Benzécri wrote: 'La dernière lettre reçue de Bourdieu me posait une question trop difficile: c'était, ou presque: qu'est-ce que l'Analyse des Données? Je lui fis une réponse élusive'. ('The last letter received from Bourdieu asked me a too difficult question: more or less it was: What is Data Analysis? I gave him an elusive reply'; Benzécri, 2006, p. 1).

Benzécri remarked that they had known each other for half a century, since the time when Bourdieu and he were students at the Ecole Normale Supérieure. Bourdieu was at the Faculty of Humanities, Benzécri at the Faculty of Sciences, and they had been close friends since their student days in the early 1950s. Benzécri continued by describing the progress of computer achievements and how difficult it was in 1963 for a modestly equipped laboratory to diagonalize a 7×7 matrix, and that hierarchical algorithms were even more complex.

At this time, in the early stages of his academic career, Bourdieu was working on an ethnographic study in Algeria. There he utilized complex data using punch cards, which he marked in order to work out all simultaneously occurring or mutually exclusive characteristics, and hence the net of contrast and equivalence relations (Bourdieu, 1990, p. 8). One difficulty was to simultaneously consider more than a certain number of oppositions within the data (Bourdieu, 1990, p. 9). Although he applied several graphical visualization techniques and different statistical approaches, he could not produce a satisfactory representation of dimensionally structured relations, as embedded in his theoretical concept.

With increasing computing power and with more efficient algorithms, the dream of analysing complex data could gradually be realized: 'Mais la puissance des machines croissant avec l'efficacité des algorithmes, notre carrière de statisticien se développa...; en mettant au service d'ambitions croissant sans cesse, des techniques dont le progrès défiait tous les rêves!' (Benzécri, 2006, p. 2). In the social sciences it became possible to include more and more variables as well as an increasing number of cases into a postulated model.

Continuing his personal memoriam to Bourdieu, Benzécri jumped from the early 1960s and the start of his academic career into the year 2000, when a microcomputer such as those offered to customers in any shop was able to classify several thousand individuals within minutes. This involved a dramatic change in analysing any kind of data. At this time it took only a few minutes for clustering algorithms and factor analytic techniques to classify or decompose large data sets, while the design of the data, its formatting, and reviewing the results could take months. Analysing complex data sets is no longer a computational problem, but the problem of data analysis remains—according to Benzécri it became even larger because nowadays the calculation does not involve a terminal point for research. There was no excuse left to stop data collection and meditation. Benzécri (2006, p. 2) concluded that compared to 1960, the ratio of difficulty between intellectual projects and calculations is inverted: 'Relativement à 1960..., le rapport de difficulté, entre projets intellectuels et calculs, est inverse'.

Benzécri continued to give some examples on the significance of data in medicine, physics, and other areas of science, moving into philosophical questions of science and life. He concluded with the remark that the work of his generation was exciting, and that 30 years ago he could not have dreamt about what can be realized today with this new computational tools. At the end of his paper, he referred to his friendship with Bourdieu with the comment: 'A la mémoire de Pierre Bourdieu, je devais présenter ces excuses pour n'avoir pas répondu à sa dernière lettre. Moi qui dois tant à sa familière et indulgente amitié' (Benzécri, 2006, p. 5).

13.2 Bourdieu and His Social Space Approach

Bourdieu (1984, 1988) and his scholars (e.g., Rosenlund, 1996; Le Roux and Rouanet, 1998; Lebaron, 2000, 2009; Rouanet et al., 2000; Rouanet, 2006; Hjellbrekke et al., 2007; Le Roux et al., 2008; Sallaz and Zavisca, 2007; Blasius and Friedrichs, 2008) are embedded in the French tradition of data analysis, called *analyse des données* (Benzécri et al., 1973). The most prominent method within this approach is MCA, an exploratory method that follows the philosophy of its principal investigator, Jean-Paul Benzécri, who pointed out that 'the model should follow the data, not the inverse' (cf. Greenacre and Blasius, 2006, p. 6); the method counterbalances the high attention that is paid to confirmatory modelling in statistics, for example, structural equation modelling.

Central to Bourdieu's social space approach is the concept of taste. Taste permits individuals to distinguish themselves from others in a variety of domains of everyday life, such as music, arts, preferred meals, and styles of clothing. The combinations of preferences in different domains of everyday life constitute different lifestyles, which can be located in the social space. In his empirical examples, Bourdieu and his scholars (Bourdieu and de Saint Martin, 1976; Bourdieu, 1979, 1984) allocated tastes or lifestyles to different population groups (compare also Bourdieu and Passeron, 1977; Bourdieu, 1979; Blasius and Friedrichs, 2008).

Fundamental to Bourdieu's (1979, 1984) lifestyle approach is the focus on the triad of *economic, social,* and *cultural* capital. All three forms of capital are latent variables that are uncorrelated to each other in a factorial space. Economic capital describes all kinds of material property, especially income and the security of income (Bourdieu, 1983). Social capital includes those resources that arise from social networks, from relationships persons have built up and where they can fall back on in case of need. Cultural capital is subdivided into (1) *objective cultural capital*, such as precious artworks (for example, paintings and sculptures); (2) *incorporated cultural capital*, which comprises the cultural skills of a person mainly gained through socialization within the family (e.g., aesthetic preferences, table manners, etc.); and (3) *institutionalized cultural capital*, which includes academic titles and other certificates of qualification. Bourdieu (1979, 1984) also introduced the concepts of *capital volume*, which is the weighted sum of the three forms of capital, and, uncorrelated with capital volume, the *composition of capitals*, for example, the composition of cultural and economic capital.

According to Bourdieu (1984), differences in the composition of the three forms of capital result in different lifestyles, such as different preferences for meals to be served to guests or different preferences with regard to furniture and artists. Although managers and professors may have the same capital volume, their composition of capitals will be quite different: managers have more economic capital, while professors have more cultural capital. Compared to the majority of the population, both have a clearly higher cultural and economic capital; managers usually have a university degree, and professors have salaries that are clearly above average. For more details we refer to Blasius and Friedrichs (2008), who give a detailed description of the different capitals, the alignment of their respective axes in the social space, as well as exemplified locations of social groups in the social space.

In his empirical studies, Bourdieu (1979, 1984) often worked with multiresponse questions, for example, when asking the respondents to name their preferred sources of furniture, their preferred artists, and meals preferably served for guests. Thereby, respondents were asked to select them from a given list (usually up to three). Please note that it is not necessary that the respondents perceive the items in a similar way. For example, the choice of the item 'my dwelling is clean and tidy' does not imply any common understanding of 'cleanliness' among the respondents; this variable has more than one interpretation. Respondents choosing this item may just want to declare that it is important for them to label their dwelling as clean and tidy. However, this choice does not guarantee that their homes are clean and tidy; their dwellings might be even less clean and tidy than the dwellings of respondents not choosing this option. A similar interpretation holds true for the opposite direction: if a respondent does not give priority to this item, it does not imply that his or her dwelling is dirty and untidy. An analogous objection holds for categories such as 'my furniture is stylish', 'my dwelling is warm', and 'when I have guests for dinner, I serve meals that are fine and exquisite'; the respective items can be understood in different ways. As discussed by Bourdieu (1979, 1984), all these response options are related to cultural and economic capital. Thereby, a dwelling labeled as clean and tidy is an indicator for a low cultural capital in most Western countries (assuming that people with a high cultural capital will mention this item relatively seldomly because their dwellings are clean and tidy, so there is no need to stress this point), while 'fine and exquisite meals for guests' is an indicator of high economic capital since people offering these kinds of meals need a certain amount of money.

As early as the mid to late 1960s, Bourdieu interviewed more than a thousand people in Paris and Lille, among others, about their preferences toward the meals they serve their guests, the sources of their furniture, and the artists they like most. Most of these lifestyle attributes admitted multiple responses and were later recoded into dichotomized items with categories 'yes' and 'no'. In addition, he asked for the sociodemographic characteristics of the individuals to combine lifestyle information with stratification indicators.

Essentially, there are two possibilities to analyse this kind of data. In the first case, as is the case in *La Distinction*, and which is the one that needs less computational power, the data can be analysed at the level of cross-tables, relating a set of describing variables and a variable to be described (cf. Greenacre and Blasius, 1994, p. ix), that is, at the level of aggregate data. For example, Bourdieu used lifestyle attributes as describing variables and occupational groups as variables to be described. In the French tradition, and in most applications based on stacked tables, scholars use the so-called symmetric map, sometimes also referred to as the French plot (see Chapter 9 by Cazes in this book; Greenacre, 2007). In *La Distinction* as well as in *Anatomie du Goût* (Bourdieu and de Saint Martin, 1976) the social space has been constructed on the basis of the row-wise concatenated cross-tables of occupational groups (the column variable) multiplied by a set of dichotomized lifestyle attributes and some sociodemographic characteristics (the row variables).

In CA (and in MCA) one can make use of the concept of supplementary (or passive) variables, or in Bourdieu's terminology, 'illustrational variables'. The respective categories do not have any influence on the geometric orientation of the axes, but they can be interpreted together with the active variables. According to Bourdieu's approach, the social space should be constructed using lifestyle attributes only; supplementary variables should be used to confirm the interpretation of the axes estimated on the basis of the active variables. Bourdieu (1979) used the rectangular matrix of the frequencies as input for correspondence analysis, with the lifestyle attributes as active variables and the sociodemographic characteristics as passive variables.

The first two dimensions of the resulting solution can be interpreted as in traditional factor analytic approaches and have been labeled by Bourdieu (1979) as composition of economic and cultural capital and capital volume, or when rotating the axes by 45 degrees, as economic and cultural capital. He used the passive sociodemographic characteristics to confirm the interpretation of these dimensions; among others, he showed that income (in groups) is positively related to economic capital, and that the level of education is positively related to cultural capital.

In this approach the positions of single individuals are not available, but rather aggregated, since the cells of the table of input data contain the frequencies of cross-tabulations. To include the individuals in this approach, one could add the information of each respondent as a supplementary column variable. Assume that we have 12 groups (g1 to g12), N (= 600) respondents, 40 lifestyle attributes (A1 to A40) with two categories each ('yes/mentioned', 'no/not mentioned'), and several sociodemographic characteristics, such as marital status (with categories 'single', 'married', 'divorced', etc.). For reasons of simplification, we assume there are no missing values on the lifestyle attributes. The structure of the respective matrix of input data is given in Table 13.1, which shows 'marital status' with four categories as supplementary rows and 600 respondents as supplementary columns, the latter with cell values of 0 (not mentioned) and 1 (mentioned).

The first 80 rows and the first 12 columns (block upper left) contain the frequencies of the cross-tabulations between the 40 lifestyle attributes (with categories 'yes' and 'no') and the 12 groups. This active part of the table will be used for constructing the social space using CA, noting that in the given example the weight of each lifestyle attribute is constant: $n_{1+} + n_{2+} = n_{3+} + n_{4+} = \ldots = n_{79+} + n_{80+} = N$.

In theory, and maybe even stronger in coherence with the ideas of Bourdieu, one could also use the 'yes/mentioned' categories only. In this case, the weight of each lifestyle attribute would depend on the number of choices, so that the more often a certain category was mentioned, the higher its weight. With a fixed number of choices, each multiresponse variable would have the same weight; in the case of three options, it would be three times the sample size. Another possibility to focus only on the 'yes' categories would be to apply *subset correspondence analysis* (Greenacre and Pardo, 2006a).

Empirical Construction of Bourdieu's Social Space

TABLE 13.1

Matrix of Input Data, Stacked Table, with Supplementary Rows and Columns

	Attribute	Active				Passive				Sum (g1 to g12)
		g1	g2	...	g12	1	2	...	600	
Active	A1_yes	$n_{1,1}$	$n_{1,2}$	...	$n_{1,12}$	1	0	...	0	n_{1+}
	A1_no	$n_{2,1}$	$n_{2,2}$	...	$n_{2,12}$	0	1	...	1	n_{2+}
	A2_yes	$n_{3,1}$	$n_{3,2}$	...	$n_{3,12}$	1	0	...	1	n_{3+}
	A2_no	$n_{4,1}$	$n_{4,2}$	...	$n_{4,12}$	0	1	...	0	n_{4+}
	...	...	...	...	...	...	...	...	...	...
	A40_yes	$n_{79,1}$	$n_{79,2}$	...	$n_{79,12}$	1	0	...	1	n_{79+}
	A40_no	$n_{80,1}$	$n_{80,2}$	...	$n_{80,12}$	0	1	...	0	n_{80+}
Passive	Single	$n_{81,1}$	$n_{81,2}$	...	$n_{81,12}$	1	0	...	1	(s_+)
	Married	$n_{82,1}$	$n_{82,2}$	...	$n_{82,12}$	0	0	...	0	(m_+)
	Widowed	$n_{83,1}$	$n_{83,2}$	...	$n_{83,12}$	0	0	...	0	(w_+)
	Divorced	$n_{84,1}$	$n_{84,2}$	...	$n_{84,12}$	0	1	...	0	(d_+)
	Sum	40 × g1	40 × g2	...	40 × g12	(40)	(40)	(40)	(40)	40 × N

In the French tradition, for visualizing the solution researchers usually use the so-called symmetric map, which is the joint representation of column and row profiles. This map is easy to read; the distances between the rows and between the columns are Euclidean ones, while the distances between row and column profiles are not defined. To avoid misinterpretations resulting from the missing distance function between rows and columns, one can show an asymmetric map with either the rows or the columns in profile format and the others as unit profiles, or vertices (for more details, see Greenacre, 2007; Chapter 9 in this book by Cazes). Alternately the contribution biplot (Greenacre, 2013) visualizes directly which attributes are the most important discriminators of the groups.

In the second case, as it is elaborated for *Homo Academicus* (Bourdieu, 1984 in French, 1988 in English), *Le Patronat* (Bourdieu and de Saint Martin, 1978), and for one of his last publications ('Une revolution conservatrice dans l'édition', Bourdieu, 1999), the columns contain the variable categories, and the rows the single individuals. The cells of this cross-table contain ones and zeros only; thereby, the one is used for the true options and the zero for the false options. In case of lifestyle attributes with categories 'yes/mentioned' and 'no/not mentioned', the ones reflect the 'yes/mentioned' categories from those attributes that have been mentioned and the zeros reflect the 'no/not mentioned' categories from those attributes that have not been mentioned. This is the case of MCA where we analyse the data on individual level instead of aggregate level, as it is the case for CA.

Assuming that the previous example contains $N = 600$ cases, the matrix to be decomposed would be 600 × 80; in the early 1970s this was a sophisticated computational problem compared to the decomposition of an 80 × 12 matrix or, in case only the positive options have been considered, a 40 × 12 matrix. Nowadays, the time difference it takes a standard PC with standard statistical software to decompose a 40 × 12 matrix compared to a 600 × 80 matrix will be some tenths of a second. In contrast, the time it takes to set up the theoretical model is clearly higher than in the 1970s, since the number of options to build such a model has been drastically increased. Nowadays, at least for social science applications, there are no computational limitations, which may be an excuse for operating with a reduced model.

For variables with more than two categories, for example, marital status, the one is used for the true category, for example, 'married', and the zero for all the other categories. Assuming that there are no missing values, it follows that all row sums of the resulting indicator matrix are equal to the sum of active variables, while the column sums contain the frequencies of the single categories. Thereby, the number of columns is equal to the number of variable categories, and the number of rows is equal to the sample size. An example of an indicator matrix is given in Table 13.2.

The hypothetical example of Table 13.2 contains 40 lifestyle attributes with two categories each ('yes, mentioned' and 'no, not mentioned'), and the supplementary variable 'marital status' with four categories. In total there are Q (= 40) active variables with a total of J = 80 categories and N = 600 individuals. The 1st and the 600th person mentioned all three lifestyle attributes (A1, A2, A40), the 1st person is single, and the 600th person is married. The second person did not mention any of the three lifestyle attributes; he or she is widowed. The first lifestyle category was mentioned n_{1+} times and n_{2+} times not mentioned. Since no missing cases are considered in Table 13.2, it yields $n_{1+} + n_{2+} = n_{3+} + n_{4+} = \ldots = n_{79+} + n_{80+} = 600$.

In case of applying MCA, both individuals and variable categories can be visualized in a map, which can be read off like a geographic map. Thereby, associations between indicators (and between individuals) are reflected by short distances and lack of association by large distances, or more generally, the closer two categories (two individuals) in the social space are located to each other, the more similar they are.

Handling missing values in stacked tables (the case of CA) is relatively easy since the inertia of each variable is the product of the marginal times the squared distance to the centroid (on each dimension). If one just neglects them, the mass of the respective variables decreases with the increasing percentage of the missing values (cf. Blasius, 1994). In many cases this might be meaningful from a substantive point of view—variables with many missing values should have a lower weight for constructing the social space (the inertia of each variable category on each dimension is calculated by multiplying the mass of the variable category times the squared distance to the centroid). Whether the percentage of missing values is large and substantively meaningful, as it might be true for the question on party preference (many people will not know or will

Empirical Construction of Bourdieu's Social Space

TABLE 13.2
Indicator Matrix with Supplementary Columns

| | Active | | | | | | | | Passive | | | |
| | A1 | | A2 | | ... | | A40 | | Marital Status | | | |
I	Yes	No	Yes	No	Yes	No	Yes	No	Single	Married	Widowed	Divorced	Sum
1	1	0	1	0	.	.	1	0	1	0	0	0	40
2	0	1	0	1	.	.	0	1	0	0	0	1	40
3	1	0	0	1	.	.	1	0	0	0	1	0	40
4	0	1	1	0	.	.	0	1	0	1	0	0	40
...	.	.	.	.			.	.	.	.	.	.	40
600	1	0	1	0	.	.	1	0	0	1	0	0	40
Sum	n_{1+}	n_{2+}	n_{3+}	n_{4+}	...	...	n_{79+}	n_{80+}	(s_+)	(m_+)	(w_+)	(d_+)	24,000

keep it secret which party they will vote for in the next election), one can add an additional 'missing' category.

If one applies correspondence analysis to an indicator matrix (the case of MCA), on the level of the individuals, neglecting missing values would lead to an unusual distribution of profile values. For example, someone responding to only 20 out of 40 questions with a total of 80 categories would have 20 nonzero profile values of 1/20, whereas respondents replying to all questions would have 40 nonzero values of 1/40. Adding an additional 'missing value' category might work if missing responses are not highly correlated with each other. This assumption might not be very realistic; for example, in survey research missing values are usually highly intercorrelated. In this case the first or the second dimension will have a very trivial solution, contrasting missing responses (or nonsubstantive responses) against substantive responses (Blasius and Thiessen, 2012). A possibility to handle this problem is subset MCA (Greenacre and Pardo, 2006a; Greenacre, 2007; Blasius and Thiessen 2012), also called *specific MCA* (Le Roux and Rouanet, 2004, 2010; see also Chapter 12 from Le Roux in this book) or *missing value passive* (Gifi, 1990).

For the visualization of the results of MCA, scholars usually prepare two plots. The first contains the objects, the individuals—in Bourdieu's and his scholar's terminology this plot is called *cloud of individuals*. The second plot contains the column variables and is called *cloud of categories* or *cloud of variables* (see also Chapter 11 from Husson and Josse in this book). It should be mentioned that in the French tradition the cloud of individuals is also shown when the individuals are a random sample of respondents without having any specific characteristics—in this case the plot simply shows the distribution of respondents in the social space. The cloud of individuals can also carry some substantive information when labeling the respondents, for example, by gender, by occupation, by a specific lifestyle attribute, by the preference for a specific actor or for a specific soft drink (Blasius and Mühlichen, 2010).

However, in neither *La Distinction* nor *Anatomie du Goût* did Bourdieu visualize the single respondents; instead, he showed geometric areas such as squares, rectangulars, and triangles, where he expected the majority of subjects. These geometric figures could have been constructed in a way that the centroids of the respective class fractions were in the centre of the figure. Alternatively, he could have included the individuals as supplementary points, as shown in Table 13.1, and use this information for drawing the geometric forms. Although Bourdieu did not say whether he applied CA or MCA, from his theoretical approach to describe occupational groups by lifestyle attributes, as well by the inertias shown for the first and second dimensions (with respect to the upper classes, 0.070 and 0.034), it can be assumed that he applied CA. This is contrary to the assumption of Rouanet et al. (2000) postulating that Bourdieu and de Saint Martin (1976) as well as Bourdieu (1979) analysed a table 'Individuals × Properties' (Rouanet et al., 2000, p. 6).

However, we agree with Rouanet et al. (2000) that Bourdieu and de Saint Martin (1978) in *Le Patronat* did use MCA, and the same holds for *Homo Academicus* (Bourdieu, 1984). In contrast to *La Distinction* and *Anatomie du Goût*, in these papers the authors showed the cloud of individuals. The basis of analyses also changed—from groups of individuals to single individuals. In *Le Patronat*, Bourdieu and Saint Martin (1978) described (individual) French entrepreneurs by their characteristics, such as the position of their fathers and their academic background, and located each of them in the social space (Bourdieu and de Saint Martin, 1978, p. 10). Later in *Homo Academicus* (Bourdieu, 1984) and in many other studies by Bourdieu and his scholars they investigated the location of single individuals in social space. Furthermore, in *Le Patronat* it is the first time that Bourdieu mentioned that he had used *l'analyse des correspondances multiples*, here with thankful reference to Ludovic Lebart for helping him to analyse the data with his program MULTM (Bourdieu and de Saint Martin, 1978, p. 9).

Bourdieu's way of constructing social spaces has been employed in many areas, such as in medicine (cf., Gatrell et al., 2004; Lengen and Blasius, 2007; Lengen et al., 2008), in marketing (cf., Snelders et al., 1994; López Sintas and García Álvarez, 2002), but especially in the social and political sciences (cf. Bourdieu, 1984, 1988; Le Roux and Rouanet, 1998; Lebaron 2000, 2009; Mochmann and El-Menouar, 2005; Rouanet, 2006; Coulangeon and Lemel, 2007; Hjellbrekke et al., 2007; Le Roux et al., 2008). A large number of empirical examples are given in the book of Robson and Sanders (2009).

13.3 Example of Social Space Interpretation

The data of the following example were collected between 2007 and 2010 within a 'Online-Dating as Mating Market' project via a web survey (for an overview of the survey and further information, see Blossfeld and Schmitz; 2011, see also http://www.partnerwahlforschung.de/page.php?id = 6). Apart from sociodemographic information, 2,113 persons of a German dating platform were asked about their self-assessment, for example, their self-perceived 'mate-value', that is, the subjective chance to find a partner. The question we use for our example is: 'How likely do you think is your chance to find someone for a marriage?' running on a 5-point scale from 'very unlikely' to 'very likely'. As an example, we cross-tabulated this variable with age (five categories), education (four categories), gender, the subjective relevance of education and income (running from 'not important at all' to 'very important'), as well as the relative education and income preferences, i.e., whether the future partner should have lower, same, higher, or much higher education and income, or if it does not matter. The frequencies of these cross-tables

TABLE 13.3

Stacked Table of Frequencies

	Chance to Find Someone to Marry				
	Very Unlikely (cp--)	Unlikely (cp-)	Neither Nor (cp0)	Likely (cp+)	Very Likely (cp++)
Relative income (ri), partner					
Lower (_l)	64	13	20	8	11
Same (_s)	113	26	31	9	16
Higher (–h)	211	37	35	7	7
Much higher (–hh)	103	21	13	5	7
No matter (_dm)	445	96	82	17	53
Relative education (re), partner					
Lower (_l)	34	7	14	3	6
Same (_s)	286	72	64	19	19
Higher (_h)	373	63	50	12	24
Much higher (_hh)	30	4	7	5	3
No matter (_dm)	213	47	46	7	42
Education (edu), partner					
Not important at all (1)	60	11	12	2	18
Not important (2)	53	14	21	7	9
Neither nor (3)	280	63	71	15	27
Important (4)	334	67	49	13	19
Very important (5)	209	38	28	9	21
Income (inc), partner					
Not important at all (1)	265	48	42	9	38
Not important (2)	206	42	44	11	9
Neither nor (3)	309	74	72	18	28
Important (4)	130	24	15	5	8
Very important (5)	26	5	8	3	11
Sex					
Male	549	138	134	38	74
Female	387	55	47	8	20
Education level					
Primary	61	15	26	7	23
Basic	631	126	124	31	55
Secondary	110	26	15	3	3
University	134	26	16	5	13
Age					
29 and younger	151	32	39	8	19
30–39	202	52	46	12	28
40–49	301	58	49	16	26
50–59	205	38	39	7	14
60 and older	77	13	8	3	7

Empirical Construction of Bourdieu's Social Space

and the abbreviations of the variable categories as used for the maps are shown in Table 13.3.

To avoid the effect of different masses (see Blasius, 1994) and to have exactly the same data as those we use for MCA, we applied listwise deletion of cases with missing values, which reduced the data set to 1,450 cases (from 1,716). From all respondents 936 (= 64.6%) answered the question about the chance of getting married with 'very unlikely'. In contrast, 94 (= 6.5%) persons answered the respective question with 'very likely', and an additional 46 (= 3.2%) with 'likely'. In the following we apply both techniques, CA on the stacked table of frequencies (Table 13.3) and MCA on the indicator matrix of the same set of data. Analogous to Bourdieu, we used the sociodemographic indicators 'sex', 'education', and 'age' as supplementary information. For all analyses we applied the R package FactoMineR (Lê et al., 2008).

Using the stacked table as input data (see Table 13.3), CA provides a decomposition of inertia with the first dimension explaining almost 66.3% of the total; the second dimension explains another 24.1%. Using a criterion similar to the eigenvalue criteria in PCA, the threshold for considering a dimension for interpretation would be total inertia divided by the number of dimensions or, when using the explained variance, 100 divided by the number of dimensions (Blasius, 1994). Since there are four dimensions, the threshold would be 25%; hence, the solution would be one-dimensional. However, for aesthetic reasons we show the two-dimensional solution (Figure 13.1). The response categories of the question on chance of marriage mirror their inherent order on dimension 1, running from 'very unlikely' (cp--) on the left

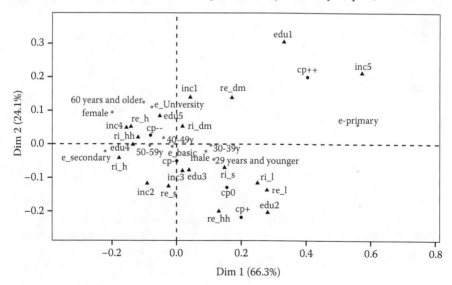

FIGURE 13.1
CA on stacked table, symmetric plot of dimensions one and two.

to 'very likely' (cp++) on the right part of the scale; the second dimension reflects the two extreme values from the three middle categories.

Please note again that the distances between rows (the locations of the active rows are symbolized with a solid triangle, the passive rows with a star) and columns (their locations are symbolized with a solid circle) are not defined. However, it is possible to compare them via their common loadings on the principal axes. The categories positively associated with 'very unlikely to find someone for a marriage' (cp−−) are 'female', the preference that the income of the future partner is much higher than one's own one (ri_hh), as well as '60 years and older'; all these categories are located in the negative part of dimension 1 and in the positive part of dimension 2. In contrast, a relatively high chance of marriage (cp++) is associated with 'primary education' and the statements that education is 'not important at all' and that income is 'very important'. These associations can also be verified by comparing the profile values, which can easily be calculated by dividing the frequencies in the single cells by their corresponding row sums (Table 13.3). For example, the row category 'education of the partner: not important at all' (edu1) has a profile value of 18/103 = 0.175; the profile value of the 'income of the partner: very important' is 11/53 = 0.208. In contrast, the respective value for edu5 is 21/305 = 0.048 and for inc1 is 38/402 = 0.095 (cf. Table 13.3); for further details on the meaning of profiles, see Chapter 10 from Bécue-Bertaut (in this book).

Applying correspondence analysis on the indicator matrix (= MCA) of the five active variables and the three passive variables described above exhibits a two-dimensional solution in which the first dimension explains 11.7% of the variation, and the second another 8.8%. These percentages are heavily underestimated. It is well known that MCA produces a number of artificial eigenvalues, namely, those that are smaller than $1/Q$, with Q = number of variables (cf. Benzécri, 1979; Greenacre, 1988a). To exclude this artificial variation, both Benzécri and Greenacre consider only those eigenvalues of the indicator matrix $\mathbf{Z}$ that are greater than $1/Q$, in the given case $1/5 = 0.2$. For adjusting the eigenvalues they both propose the formula

$$\tilde{\lambda}_k = \left(\frac{Q}{Q-1} \left(\lambda_k - \frac{1}{Q} \right) \right)^2,$$

where λ_k is the kth eigenvalue from the indicator matrix $\mathbf{Z}$. Benzécri and Greenacre differ in how these adjusted eigenvalues are converted to percentages of explained inertia. Benzécri merely expresses them relative to the sum of adjusted eigenvalues (those for which $\lambda_k > 1/Q$), but this is overoptimistic since it implies that one can explain 100% of inertia, which is not correct, as shown by Greenacre (2007). He argues that one should use as total inertia the average inertia of all two-way tables in the MCA based on the Burt table,

Empirical Construction of Bourdieu's Social Space 219

which is defined as $\mathbf{B} = \mathbf{Z}^T\mathbf{Z}$ (for the different input formats in MCA and their relations toward each other, see Greenacre, 2007). Applying this latter criterion, the first dimension explains 54.4% of the total variation, and the second another 17.7%, while adjusting the explained variances according to Benzécri (1979), the first dimension explains 65.4% of the total variation, and the second another 21.2%.

It should be noted that many researchers from the social sciences as well as from other applied areas appreciate high explained variances, which may explain the success of Benzécri's procedure. However, whatever form of adjustment one uses, the resulting figures only differ by scaling factors of the axes; the substantive interpretation of the solution is the same.

As Figure 13.2 shows, in the MCA solution, the previous column variable 'chance of marriage' does not keep its inherent ordinality in the social space; four of the five categories are close to the centroid, which implies that at least in the two-dimensional solution there is no association with the other variables. The only exception is 'chance to find someone for marriage: very likely' (cp++), which is associated with a low education (e_prim), and the responses that income and education of the future partner are 'not at all important' (edu1, inc1). In contrast to this variable, the other four ordinal-scaled variables keep their inherent ordinality along the first dimension. From left to

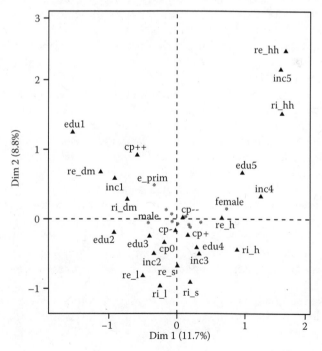

FIGURE 13.2
MCA on indicator matrix, cloud of categories.

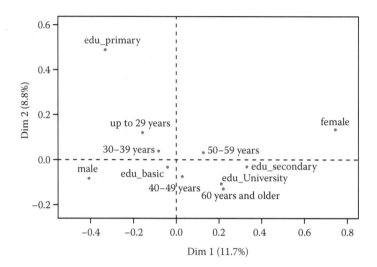

FIGURE 13.3
MCA on indicator matrix, passive variables.

right the importance concerning education and income of the future partner increases, and further, the expected relative income and the expected relative education become higher. On the left part of dimension 1, both education and income of the future partner can be lower than one's own, or it doesn't matter (re_dm, ri_dm), while the right part is associated with the expectation that both are clearly higher than one's own (re_hh, ri_hh). The second dimension reflects the horseshoe, or Guttman, effect in ecology known as arch effect (see Chapter 5 from ter Braak in this book), with the extreme values on the positive part and the middle values in the negative part.

From the passive variables only three categories are easy to distinguish in the two-dimensional social space. However, they are still good to interpret, so we show them on a separate map (Figure 13.3). Age, for example, also keeps its inherent ordinality on dimension one: the farther a category is to the right, the higher the age and the higher the expectations concerning the future partner; education and income become more important, and both are expected to be higher than one's own. Further, men have clearly lower expectations than women on the future partner, while men look more downward, women look more upward. This can be interpreted as the gender-specific character of mate values (Bourdieu, 2001; Schmitz, 2012).

This distinction between men and women becomes visible when showing the cloud of individuals with labeling the single individuals by gender. Figure 13.4 shows that both groups are well separated in the two-dimensional social space. Furthermore, the confidence ellipses indicate that this separation is a substantive one, as the directions of them are almost orthogonal.

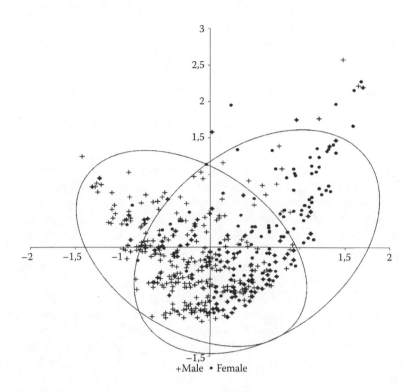

FIGURE 13.4
MCA on indicator matrix, cloud of individuals.

13.4 Conclusion

In this chapter we discussed the elective affinities of the work of Jean-Paul Benzécri and Pierre Bourdieu. While Benzécri gained his scientific background from the Faculty of Sciences, Bourdieu's background is routed in the humanities. Both were already friends in the 1950s when they went to university; since then their works have been quite strongly related to each other. Next to his mathematical and statistical interests, Benzécri also has a strong connection to questions from the philosophical and social sciences. On the other hand, next to his interests in the humanities and social science relations, Bourdieu was highly interested in statistical techniques and especially in the visualization of complex societal phenomena. Already in his time in Algeria in the 1960s, he collected large data sets from different sources and tried to combine them in a meaningful way. Via Benzécri he learned the advantages of CA and later MCA for visualizing complex sets of data. With the help of Benzécri and other friends from statistics working together with Benzécri, he started to construct latent factor spaces. The congruence of both philosophies lies in the epistemological emphasis of the relation: both Bourdieu's theory and Benzecri's method assume that 'the real is relational' (cp. Bourdieu, 1998, p. 3).

Although Benzécri speaks many languages fluently, among others, English and German, he published nearly all of his work in French. Outside the French-speaking world correspondence analysis was almost unknown until Lebart et al. (1984) and Greenacre (1984) published their textbooks, books that are not easy to understand for readers with no statistical training. At the same time, Bourdieu became well known worldwide, especially via his book *La Distinction*, which was published in 1979 in French; the English translation was released in 1984. His readers, who mainly came from the humanities, and, in the beginning, especially from the social sciences, were now confronted with this new way of visualizing data. They had to learn it to understand his ideas and especially his social space approach.

The increasing success of correspondence analysis in France is also related to achievements in computation. When Benzécri started his academic career also for a modestly equipped laboratory, it was already very difficult to diagonalize a 7×7 matrix, and to perform hierarchical algorithm was even more complex. It is no wonder that Bourdieu first started with the application of CA, analysing stacked tables with a limited number of columns and rows. With the increasing power of machines, he could extend his theoretical models to analyse the interrelations of an increasing number of indicators, and the solution that allowed him to do this job was MCA.

14

Multiple Factor Analysis: General Presentation and Comparison with STATIS

Jérôme Pagès

CONTENTS

14.1 Data 224
14.2 Method 225
14.3 Balance between the Groups of Variables 226
14.4 Superimposed Representation 228
14.5 Overall Comparison of the Groups of Variables 231
14.6 Comparison between MFA and STATIS 234
14.7 Qualitative Variables and Mixed Data 235
14.8 Conclusion 237

In this chapter, I will analyse a table detailing a set of individuals according to several groups of quantitative or qualitative variables. Two main methods have been developed in France to process this kind of data: STATIS (from the French *Structuration des Tableaux a Trois Indices de la Statistique*) and multiple factor analysis (MFA).

The first description of STATIS can be found in the thesis by L'Hermier des Plantes (1976), a doctoral study supervised by Yves Escoufier. The thesis extended the work of Escoufier, who is most well known for introducing the RV coefficient (Escoufier, 1973; Robert and Escoufier, 1976). Just after completing this thesis, he published a concise presentation of the method accompanied by an application to real data in *La Revue de Statistique Appliquée* (L'Hermier des

Plantes and Thiébaut, 1977). A comparison between STATIS and the INDSCAL model was the subject of a later thesis supervised by Escoufier (Glaçon, 1981), and a general presentation of STATIS was later written by Lavit (1988).

The first description of MFA appeared in Escofier and Pagès (1982), followed soon afterwards by a comprehensive theoretical presentation in Escofier and Pagès (1984a). MFA is at the heart of Escofier and Pagès (1988), a work that has been republished four times (most recently in 2008). A recent book (Pagès, 2013) is devoted to MFA. A number of extensions of MFA have been suggested. MFA on contingency tables (MFACT), first suggested by Abdessemed and Escofier (1996), has been extended by Bécue-Bertaut and Pagès (2004, 2006). Hierarchical MFA (HMFA) extends MFA to a set of nested partitions defined on the variables (Le Dien and Pagès, 2003a, 2003b). A number of applications of MFA have been published, particularly in the field of sensory analysis (e.g., Pagès, 2005a; Morand and Pagès, 2006, 2007; Pagès et al., 2010). Finally, remaining within the field of sensory analysis, specific tests of validity have also been introduced (Pagès and Husson, 2005). This chapter first presents MFA through an application. It is then compared with STATIS.

14.1 Data

Here we will reuse data that have been analysed previously, most notably in Pagès and Husson (2005). We will study six pure orange juices (see Table 14.1) of two different kinds: refrigerated (*fr* in the graphs; these juices, which are pasteurized at a lower temperature, must be keep refrigerated in the store shelves) and ambient (*amb*; these juices are presented at room temperature in the store shelves). Three of these juices are made with Florida oranges (both of the Tropicana juices and the Fruivita brand).

These six juices were subjected to eight chemical measurements and seven sensory descriptors (mean of the ratings on five-point scales given by a panel of 96 tasters). Table 14.2 shows in the columns the six orange juices as well as their mean values, and in the rows the eight plus seven quantitative variables to which we add two qualitative variables, each with two categories: 'origin' (Florida/other) and 'type' (ambient/refrigerated). In the following, the data are centred and reduced.

TABLE 14.1

The Six Orange Juices Studied

No.	Brand	Origin	Type	No.	Brand	Origin	Type
P1	Pampryl	Other	Ambient	P4	Joker	Other	Ambient
P2	Tropicana	Florida	Ambient	P5	Tropicana	Florida	Refrigerated
P3	Fruivita	Florida	Refrigerated	P6	Pampryl	Other	Refrigerated

TABLE 14.2

Chemical and Sensory Data

	P1	P2	P3	P4	P5	P6	Mean
Chemical							
Glucose (g/l)	25.32	17.33	23.65	32.42	22.70	27.16	24.76
Fructose (g/l)	27.36	20.00	25.65	34.54	25.32	29.48	27.06
Sucrose (g/l)	36.45	44.15	52.12	22.92	45.80	38.94	40.06
Raw pH	3.59	3.89	3.85	3.60	3.82	3.68	3.74
pH after centrifugation	3.55	3.84	3.81	3.58	3.78	3.66	3.70
Acidity	13.98	11.14	11.51	15.75	11.80	12.21	12.73
Citric acid	0.84	0.67	0.69	0.95	0.71	0.74	0.77
Vitamin C	43.44	32.70	37.00	36.60	39.50	27.00	36.04
Sensory							
Odour intensity	2.82	2.76	2.83	2.76	3.20	3.07	2.91
Odour typicity	2.53	2.82	2.88	2.59	3.02	2.73	2.76
Pulpy	1.66	1.91	4.00	1.66	3.69	3.34	2.71
Taste intensity	3.46	3.23	3.45	3.37	3.12	3.54	3.36
Sourness	3.15	2.55	2.42	3.05	2.33	3.31	2.80
Bitterness	2.97	2.08	1.76	2.56	1.97	2.63	2.33
Sweetness	2.60	3.32	3.38	2.80	3.34	2.90	3.06

14.2 Method

There are two groups of quantitative variables ('chemical' and 'sensory') mainly, in addition we will use the two qualitative variables for interpretation. Many different factorial methods can be used to analyse a table such as this. The most common is to conduct a principal components analysis (PCA) by introducing one group of variables as active and the other as supplementary. Thus, for example, we can first highlight the principal dimensions of chemical variability and then identify the sensory descriptors related to these dimensions. This approach is quite straightforward and is very useful in many situations (Lebart et al., 1977).

It is desirable to give the two groups of variables an equal role in a PCA where both sets of variables are active. With this in mind, there can be a number of consequences to taking into account the structuring of variables in groups. First, it is necessary to balance the groups of variables in this overall analysis. Then a number of new issues also arise, such as: Do the structures on the individuals, as defined by each of the groups, resemble one another?

14.3 Balance between the Groups of Variables

Table 14.3 contains the inertias of the PCA separated for the two groups of variables. The first principal dimension of the chemical data is associated with a slightly higher inertia than its corresponding inertia for sensory data (6.212 > 4.744). This is due both to the greater number of variables for this group (8 vs. 7) and to the relative importance of the first axis, which is greater for the chemical data (77.7% > 67.8%). By 'mixing' the two groups of variables in a PCA, the first axis of this overall PCA would automatically be more greatly influenced by the chemical data. This example illustrates the need to account for both the total variance (here the number of variables, because they are standardized) and its distribution in space—the aim being to balance the two groups of variables. This is why we decided to standardize the maximum axial inertias for each group. In MFA, each variable of group q is attributed a weight inverse to the first eigenvalue of the separate analysis for group q (denoted λ_1^q). If we conduct a separate PCA of the group q with the weight $1/\lambda_1^q$, we obtain a first eigenvalue of 1 for each group by default.

Table 14.4 contains the decomposition of the inertia, associated with the MFA, by axis and by group. It shows a first axis that

- Is predominant (64.6%) and thus a first plane with sufficient inertia (77.8%) to limit ourselves to this simple comment.
- Is associated with high inertia (1.79; not far from its maximum value of 2, which is equal to the number of active groups).

TABLE 14.3

Inertias of the PCA Separated for the Two Groups of Variables

	Total Inertia	F1	F2	F1 (%)	F2 (%)
PCA chemical	8	6.212	1.100	77.7	13.7
PCA sensory	7	4.744	1.333	67.8	19.1

TABLE 14.4

Inertias of the MFA by Axis and by Group

	Total Inertia	F1	F2
MFA	2.763	1.785	0.365
		(64.6%)	(13.2%)
Chemical	1.288	0.891	0.099
	(46.6%)	(49.9%)	(27%)
Sensory	1.476	0.894	0.266
	(53.4%)	(50.1)	(73%)

- Has an inertia divided almost equally between the two groups of variables. In this case, the balancing in MFA worked perfectly: the share of inertia for each group (0.89) is close to its maximum value of 1. The first axis corresponds to high direction of inertia for each group.

The representations of individuals and variables (Figure 14.1) should be interpreted as those of a PCA. In particular, the representation of variables can be read as in PCA: coordinates are coefficient of correlation; the closeness of the vector to the unit circle measures its quality of representation. The first axis relates to both measured (titre, citric acid, pH) and perceived (sourness) acidity. On the chemical level, this acidity is accompanied by the opposition between sucrose on the one hand and fructose and glucose on the other. This opposition is related to the hydrolysis of sucrose in glucose and fructose, hydrolysis that is accentuated in acidic environments. On the sensory level, sourness is linked to bitterness and opposed to the sweetness, which evokes the notion of balance in taste. Finally, this axis opposes:

- On the one hand, juices 2, 3, and 5, with, on the chemical level, a high pH and a strong ratio of sucrose/(glucose + fructose) and, on the sensory level, a 'sweet' profile (i.e., sweetness, low sourness, and low bitterness)
- On the other hand, juices 1, 4, and 6 with the opposite characteristics

This opposition, which is based on the two groups of variables, illustrates the dimensions that we might expect from a factor analysis conducted on multiple groups of variables. It corresponds to the variable 'origin of the juice', as it separates the Florida juices from the others.

The interpretation of the second axis is not so clear. However, the second bisector, i.e., the direction oriented from lower right to upper left, relates to the characteristics 'pulpy' and 'odour intensity'. It opposes juices 3 and 5 with juices 1 and 4. This opposition corresponds to the variable 'type of juice', separating the ambient juices, perceived as not pulpy and weak smelling,

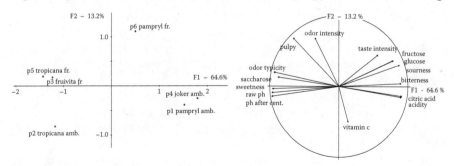

FIGURE 14.1
Representation of the individuals and variables on the first plane of the MFA.

from the refrigerated juices, perceived as pulpy and strong smelling. This dimension is not related to the chemical group (vitamin C is slightly related to this direction, but as there is no explanation, it is not kept in the interpretation). The first bisector, upper left to lower right, clearly separates juices 2, 3, and 5 from juices 1, 2, and 6, i.e., other juices from Florida juices. This analysis accurately illustrates a common situation in the analysis of several groups of variables: some factors that are common to all groups and the others that are specific to only one group.

14.4 Superimposed Representation

Let us consider the data from the group of variables q within the table $\mathbf{X}_q$ with I rows (I: number of individuals) and J_q columns (J_q designates both the number of variables in group q and the set of these variables). In table $\mathbf{X}_q$, we can associate the two clouds of points involved in its PCA:

- The cloud N_J^q of variables for group q only. This cloud lies within space $\mathbb{R}^I$.
- The cloud N_I^q of individuals, considered from the point of view of group q only. This cloud lies within space $\mathbb{R}^{J_q}$ generated by the variables of group q alone. The elements of this cloud are denoted i^q and are called *partial individuals*.

The simultaneous analysis of the Q clouds N_J^q aims to examine the relationships between the variables both within and between groups. All these clouds lie within the same space, and together they make up cloud N_J. MFA, which analyses cloud N_J, is therefore a simultaneous analysis of the N_J^q.

However, the simultaneous analysis of clouds N_I^q is not as easy since these clouds do not lie within the same space. Such an analysis aims to compare the shapes of the clouds N_I^q using questions such as: Are two individuals, which are close according to one group of variables, also close according to the other groups? Is an individual that is unusual according to one group (i.e., far away from the origin/group mean) also unusual according to the other groups?

The MFA includes a superimposed representation of N_I^q. This representation is based on the fact that the space $\mathbb{R}^J$ is the direct sum of $\mathbb{R}^{J_q}$. It is therefore possible to represent all of N_I^q in the space $\mathbb{R}^J$ (see Figure 14.2, left): each N_I^q is the projection of N_I on the subspace generated by the columns of group q alone (isomorphic to $\mathbb{R}^{K_q}$), which is why they are called partial clouds.

Clouds N_I^q are different from N_I. However, in some ways they are also the constituent elements as N_I brings together their coordinates. From this point

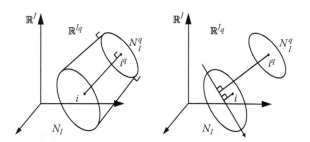

FIGURE 14.2
Clouds of individuals in $\mathbb{R}^J$. $\mathbb{R}^{K_q}$: Subspace generated by the columns of group q only. N_I^q: Partial cloud of individuals from the point of view of group q only; construction (left) and representation (right).

of view, the analysis of N_I conducted in the MFA can be seen as a simultaneous analysis of (the data in) N_I^q. By accepting this point of view, we are led to represent N_I^q by projecting them on the principal axes of N_I, as if they were supplementary elements. By working in this way, we obtain a superimposed representation that presents a crucial property when interpreted: the *partial transition relationship*.

Let us thus rewrite the usual transition relationship that, in PCA, expresses the coordinates of an individual according to those of the variables (Escofier and Pagès, 2008, p. 20). By denoting f_{ik} (and g_{jk}, respectively), the coordinate of individual i (and of variable j, respectively) along axis k and the inertia λ_k associated with that axis, this relationship is expressed, where J is the total number of variables whatever the group and λ_k is the kth eigenvalue of the PCA:

$$f_{ik} = \frac{1}{\sqrt{\lambda_k}} \sum_{j \in J} x_{ij} g_{jk}$$

By applying this relationship to the MFA, that is, taking into account the group structure of the variables and the weighting by $1/\lambda_1^q$ of the variables in group q, we obtain

$$f_{ik} = \frac{1}{\sqrt{\lambda_k}} \sum_q \frac{1}{\sqrt{\lambda_1^q}} \sum_{j \in J_q} x_{ij} g_{jk}$$

We denote f_{ik}^q, the coordinate of the partial point i^q along the axis k. Given the construction of N_I^q, we obtain

$$f_{ik} = \sum_q f_{ik}^q.$$

The result is thus:

$$f_{ik}^q = \frac{1}{\sqrt{\lambda_k}} \frac{1}{\sqrt{\lambda_1^q}} \sum_{j \in J_q} x_{ij} g_{jk}$$

This partial transition relationship expresses that, along axis k and discarding a coefficient (the same for all individuals), the partial individual i^q is found on the side of the variables (of the group q) for which it carries high values and opposite the variables for which it carries low values. This relationship therefore makes it possible to interpret the representation of a cloud N_I^q in the same way as for a traditional PCA.

Remark: In practice, all of the partial points are dilated with the coefficient Q in order to make point i appear at the barycentre of i^q, $q = 1, Q$. This is why N_I is known as the average cloud.

When applied to the six orange juices, this representation yields Figure 14.3. This figure features the representation of the individuals from Figure 14.1, completed by the partial points. Overall, the partial points relating to the same individual are close together. Thus, the first axis confronts individuals {1, 4, 6} with individuals {2, 3, 5} both from chemical and sensory points of view. In terms of the partial individuals, we observe the nature of the axis that is common to the two groups of variables.

Along the second axis, the partial points of the sensory groups vary more than those of the chemical group. At the individual level, we can therefore observe the nature of this axis, which is specific to the sensory group.

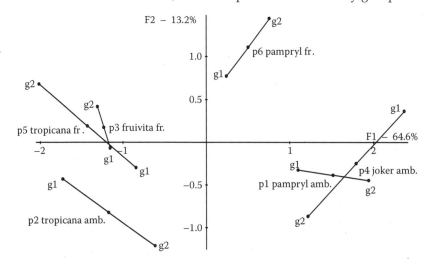

FIGURE 14.3
Superimposed representation of partial clouds: g1, chemical; g2, sensory.

As well as these main traits, this representation allows for further interpretation. For example, let us compare juices 1 and 4. On a chemical level, juice 4 is more atypical than juice 1: its citric acid content and its titre are greater, and its sucrose content is lower, etc. However, this extreme characteristic is not reflected in it sensory profile: in comparison, juice 1 is perceived as more sour, more bitter, and less sweet. This relative mismatch between the chemical and sensory images of juices 1 and 4 is clearly identified by the MFA.

Remark: The simultaneous representation of clouds of corresponding points is also the goal of Procrustes analysis. A comparison of MFA with this method is in Pagès (2005b).

14.5 Overall Comparison of the Groups of Variables

The superimposed representation presented in the previous section is used to compare the groups of variables accurately in terms of the structure that they induce on the individuals. However, particularly when they are numerous, it can be useful to have an overall view of the similarities/relationships between the groups of variables.

This was first suggested by Escoufier (1973), whose work is the basis for STATIS. The classical presentation of STATIS holds for individuals of any weight. However, to simplify, we here assume that all of the individuals have the same weight.

The STATIS method is based on four principles.

First, the group of variable q is represented by the matrix of scalar products (denoted $\mathbf{W}_q$) between individuals induced by this group, so $\mathbf{W}_q = \mathbf{X}_q \mathbf{X}_q^T$. This decision is justified by the fact that the eigendecomposition of $\mathbf{W}_q$ allows us to perfectly reconstruct the shape of cloud N_I^q.

Second, each group is therefore represented by a matrix $\mathbf{W}_q$ with dimensions $I \times I$. $\mathbf{W}_q$ can be considered an element of a vectorial space with I^2 dimensions denoted $\mathbb{R}^{I^2}$. The set of $\mathbf{W}_q$ makes up the cloud of groups of variables denoted N_Q.

Third, space $\mathbb{R}^{I^2}$ has the usual metric. Thus, $\langle \mathbf{W}_q, \mathbf{W}_{q'} \rangle = \text{trace}(\mathbf{W}_q \mathbf{W}_{q'})$.

$\mathbf{W}_q$ are standardized to have a length 1. The cosine of $\mathbf{W}_q$ and $\mathbf{W}_{q'}$, denoted $RV(J_q, J_{q'})$ (Escoufier, 1973), can be written (the formula includes a reminder of the standardization of the $\mathbf{W}$):

$$RV(J_q, J_{q'}) = \left\langle \frac{\mathbf{W}_q}{\|\mathbf{W}_q\|}, \frac{\mathbf{W}_{q'}}{\|\mathbf{W}_{q'}\|} \right\rangle$$

The *RV* coefficient is always between 0 and 1. It is worth 0 if each variable in group q does not correlate with each variable in group q'. It is worth 1 if clouds N_I^q and $N_I^{q'}$ are homothetic (they differ by a simple scale factor or a constant term). Thus, the *RV* coefficient can be considered an indicator of the relationship between groups of variables (Escoufier, 1973).

Fourth, a representation of the cloud N_Q is obtained by projecting it on its principal axes. As the cosine between two $\mathbf{W}_q$ is always positive, all of the coordinates of $\mathbf{W}_{q'}$ on the first principal axis of N_Q, are also positive. The unit vector of this first axis, denoted $\mathbf{W}$, is thus the linear combination of $\mathbf{W}_q$, which is the most closely related to all of the groups. In this sense, it is a compromise between all of the $\mathbf{W}_q$. The eigendecomposition of $\mathbf{W}$ (as vector of $\mathbb{R}^{I^2}$, $\mathbf{W}$ is a matrix) provides a representation of individuals known as compromise representation.

In MFA, cloud N_Q is projected on the elements of $\mathbb{R}^{I^2}$ induced by the principal components of the weighted PCA. Let $\mathbf{v}_k$ be the standardized principal component of rank k. With $\mathbf{v}_k$, we associate $\tilde{\mathbf{W}}_k = \mathbf{v}_k \mathbf{v}_k^T$, the element of $\mathbb{R}^{I^2}$ said to be of rank 1 (associated with only one dimension of $\mathbb{R}^I$). It can be shown that $\hat{\mathbf{w}}_s$ is standardized and that, for two distinct principal components $\mathbf{v}_k$ and $\mathbf{v}_{k'}$, $\hat{\mathbf{w}}_k$ and $\hat{\mathbf{w}}_{k'}$ are orthogonal. Finally, the coordinate of the projection of $\mathbf{W}_q$ on $\hat{\mathbf{w}}_k$ is

$$\left\langle \frac{\mathbf{W}_q}{\lambda_1^q}, \tilde{\mathbf{W}}_k \right\rangle = \frac{1}{\lambda_1^q} \operatorname{trace}\left(\mathbf{W}_q \mathbf{v}_k \mathbf{v}_k^T\right) = \frac{1}{\lambda_1^q} \operatorname{trace}\left(\mathbf{v}_k^T \mathbf{X}_q \mathbf{X}_q^T \mathbf{v}_k\right)$$

$$= \frac{1}{\lambda_1^j} \text{inertia of } N_K^j \text{ projected onto } \mathbf{v}_s$$

The coordinate of $\mathbf{W}_q$ on $\tilde{\mathbf{W}}_k$ can be clearly interpreted. It is always between 0 and 1. It is 0 if each variable of group J_q is not correlated with $\mathbf{v}_k$. It is 1 if $\mathbf{v}_k$ coincides with the first principal component of J_q. Thus, this coordinate can be interpreted as a measurement of the relationship between J_q on the one hand and $\mathbf{v}_k$ on the other. Moreover, as the groups are not standardized in MFA, they appear within a square (see Figure 14.4, left), known as the *relationship square*, with a side of length 1 and the points (0, 0), (0, 1), (1, 0), and (1, 1) as vertices.

By bearing in mind the meaning of the axes, the relationship square (of the MFA; Figure 14.4, left) shows:

- A strong similarity between the two active groups from the point of view of axis 1. The two groups separate the 'smooth' juices (2, 3, and 5) from the 'hard' juices (1, 4, and 6). This opposition is a direction of high inertia for each group (coordinates near 1).

Multiple Factor Analysis

- A difference between the two active groups from the point of view of the second axis. This axis is not related to the chemical group. It is related to the sensory aspect but corresponds to a direction with rather weak inertia in this group (i.e., related to few variables).

This representation does not give new ideas in this example, which has only two active groups. It is, however, sufficient to show it works, which stems from the fact that each axis of $\mathbb{R}^{I^2}$ on which N_Q is projected, being linked to one single direction of $\mathbb{R}^I$ (tangibly a principal component), is interpreted as such.

The same does not apply for the representation provided by STATIS (Figure 14.4, right). The first STATIS axis (in $\mathbb{R}^{I^2}$) highlights the equal role played by the two groups in constructing the compromise, which is automatic when there are only two active groups. As expected, the second axis separates the two groups but fails to indicate that which differentiates the two. This inability to interpret a STATIS axis is due to the fact that, unlike MFA, they do not correspond to a unique direction in the variables' space $\mathbb{R}^I$. In making up for this absence of constraint, the representation quality of N_Q in STATIS is better than that of MFA. Here, with two groups, it is automatically perfect with two axes, whereas, in this example, with MFA the percentage of projected inertia in N_Q is only 78%.

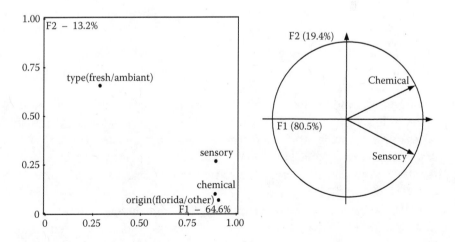

FIGURE 14.4
Representations of groups of variables resulting from MFA (relationship square, left) and from STATIS (right).

14.6 Comparison between MFA and STATIS

The previous section identified a big difference in principle between the representation of the groups resulting from MFA and those from STATIS. However, when representing individuals and variables, the solutions of the two methods are very similar. They both lead to a weighted PCA of the complete table, with the difference between the two methods residing in the choice of weights for the variables.

In STATIS, the weight of variable j in group q can be expressed: $\langle \mathbf{W}_q, \mathbf{W} \rangle / \|\mathbf{W}_q\|$. This yields two coefficients:

- The denominator corresponds to standardization in $\mathbb{R}^{I^2}$. It plays a role up to a certain point analogous to the weighting in MFA. In particular, it leads to giving the same importance to two groups associated with homothetic clouds of individuals.

- The numerator corresponds to the coordinate of the projection of $\mathbf{W}_q$ on the principal axis of N_Q. It reinforces a group J_q as this group resembles the others (or, symmetrically, it weakens a group that is different from the others). This aspect is entirely absent from MFA.

No one balancing system can claim to be better than another across all applications. When compared to that of STATIS, MFA's balancing presents the following characteristics:

- It is clearly interpreted in spaces $\mathbb{R}^J$ and $\mathbb{R}^I$, which are major spaces in the interpretations.

- A group that is different from the others is not 'diminished' but rather expressed on specific axes.

However, in practice, both methods often lead to very similar representations of individuals. This is particularly true for the example of the six orange juices, for which the two methods yield almost exactly the same first plane (correlation coefficients between homologous rank factors greater than 0.999; very similar percentages of inertia that are, for the compromise of STATIS, 65.6 and 18.9%).

Finally, it must be noted that, originally, STATIS included a superimposed representation of partial clouds. However, this representation is currently not available in software for STATIS (see Section 14.8). A complete comparison between the two methods can be found in Pagès (1996, 2013) and in Dazy et al. (1996).

14.7 Qualitative Variables and Mixed Data

Let us summarize the three issues associated with multiple tables that we have dealt with: balance between the groups of variables, superimposed representation of partial clouds, and representation of groups of variables. These three aspects, previously examined within the context of quantitative variables, are of course valid for all types of variables.

For qualitative variables, the standard factorial method is multiple correspondence analysis (MCA). Yet, in the same way as we previously defined a weighted PCA, MCA can also be weighted; in doing so, the weight of each category of usual MCA (proportional to its frequency) is multiplied by the weight of the variable to which it belongs.

It is therefore possible to extend MFA to qualitative variables by replacing the weighted PCA with a weighted MCA. Balancing is the same as for the quantitative variables in such cases: to ensure that the maximum axial inertia of the partial clouds is equal to 1, we attribute the weight $1/\lambda_1^q$ to each variable of group q, where λ_1^q is the first eigenvalue of the separate MCA of group q. This method is described in further detail in Pagès (2002, 2013).

Finally, the nature of the balancing, which is applied in the same way to both quantitative and qualitative variables, means that both types of groups of variables can be introduced as active elements within the same analysis. It also means that mixed groups can be introduced, using factor analysis for mixed data (FAMD) (Escofier, 1979a, 2003; Saporta, 1990; Pagès, 2004) as a reference method for analysing a mixed table.

In the orange juice example, there are two qualitative supplementary variables: the origin and the type of juices. However, introducing each of these two variables as a supplementary group illustrates the possibilities of MFA. Figure 14.5 represents the categories of the qualitative variables on the first plane of the individuals from Figure 14.2. Much like in PCA, a category lies at the barycentre of the individuals possessing that category. Figure 14.5 illustrates:

- The relationship between the first axis and the origin of the juices (opposition between Florida and other origins)
- The relationship between the second bisector and the type of juice (opposition between refrigerated and ambient)

This representation of categories, which is classic in MCA, is complemented by a representation of partial categories obtained as follows. A partial category (of group q) is at the barycentre of the partial individuals i^q possessing that category. In the example, Figure 14.5 shows:

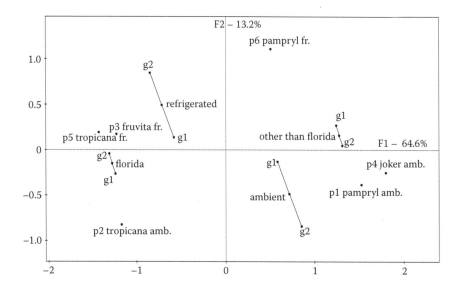

FIGURE 14.5
Representation of categories, average and partial, superimposed onto those of the individuals (Figure 14.2): g1, chemical; g2, sensory.

- Along axis 1, great proximity between the two partial points related to the same category (example: first coordinate of Florida g1 is very similar of the one of Florida g2); the mismatches (difference between the points g1 and g2 related to the same juice) observed for some juices (mainly juices 2, 4, and 5; see Figure 14.3) do not seem to be related to the factors 'origin' and 'type'.
- Along axis 2, greater variability in the types of juice from a sensory perspective, an aspect that has already been identified for the juices themselves in relation to the fact that axis 2 is specific to the sensory group.

As a group of variables, each qualitative variable leads to the calculation of a matrix of scalar products between individuals ($\mathbf{W}_q$). This matrix can be situated in the cloud N_Q (in $\mathbb{R}^{I^2}$) and projected as supplementary on the axes of the MFA (in $\mathbb{R}^{I^2}$). It can be shown that, in the specific case of a group q reduced to one single qualitative variable, the coordinate, in $\mathbb{R}^{I^2}$, of $\mathbf{W}_q$ along the axis k (of the MFA) is equal to the square of the correlation ratio between the variable and the factor (of the MFA) k (see representation of variables in MCA). This possibility is used in Figure 14.4, which shows:

- The close relationship between the first axis and the origin of the juices
- The strong link between the second axis and the type of juice

This new representation illustrates the relevance of this methodology in more complex data analysis.

14.8 Conclusion

Taking into account groups of variables within the factorial analysis of a individuals × variables table poses a technical problem: that of balancing the influence of the different groups of variables. Moreover, taking this into account means the analysis is enriched: in particular, in terms of the overall comparison of groups of variables and the comparison of clouds of individuals induced by each group. MFA offers a solution to each of these issues and does so within a unique frame of reference. The result is a powerful tool for analysing tables with a very general structure, mixing quantitative and qualitative variables, analysing variation both between and within groups. On some points, STATIS also brings an interesting point of view, but it is less general. STATIS is less useful in the application, but it deserves to be mentioned because of its historical role.

From a practical point of view, MFA is available in different software packages: SPAD, XLSTAT, Statgraphics (Uniwin), and in the R packages ade4 (Dray and Dufour, 2007) and FactoMineR (Husson et al., 2013). A SAS macro is available (Gelein and Sautory, 2013). One specific comment must be made about FactoMineR, which is the only package to contain, along with a very comprehensive version of MFA, an extension for contingency tables (multiple factor analysis of contingency tables (MFACT)) and another one for a hierarchy defined on the variables (hierarchical multiple factor analysis (HMFA)). STATIS is available in SPAD and ade4.

15

Data Doubling and Fuzzy Coding

Michael Greenacre

CONTENTS

15.1 Doubling of Ratings 240
15.2 Doubling of Preferences and Paired Comparisons 245
15.3 Fuzzy Coding of Continuous Variables 247
15.4 Explained Inertia 252
15.5 Discussion 253

An important aspect of Benzécri's French school of data analysis is the recoding of data prior to visualization by correspondence analysis (CA), a theme treated in detail in the book by Murtagh (2005). The method of CA is seen as a universal tool for visualizing data of any kind, once recoded into a suitable form. To understand what makes a data set suitable for CA, one has to consider the elements of a frequency table, which is the primary object on which CA is applicable, and which needs no pretransformation:

- Each cell of a frequency table is a count.
- The row or column frequencies are expressed relative to their marginal totals as profile vectors.
- The marginal frequencies of the table provide masses that weight the row and column profiles in the measure of variance and in the dimension reduction.
- The chi-square distance between profiles inherently standardizes the profile elements.

The idea of recoding data prior to CA is illustrated nicely in the original definition of multiple correspondence analysis (MCA), where all categorical variables are recoded as dummy variables (see Chapter 3 by Lebart and

Saporta in this book). Variants of this idea are the so-called *doubling* of data and *fuzzy coding*, which are the topics of this chapter.

Doubling (*dédoublement* in French) can be considered a simple generalization of the MCA of dichotomous variables, that is, when each categorical variable has only two categories. But, instead of the coding [0 1] or [1 0] used in MCA, there is a pair of nonnegative values between 0 and an upper bound, for example, 10, and the two values have a sum equal to the upper bound: for example, [3 7] or [9 1], or even [0 10] where the bounds are included. This type of coding arises in the analysis of ratings, preferences, or paired comparisons, as I will show later.

Doubling can be considered a special case of fuzzy coding (*codage flou* in French), for which a set of fuzzy categories is defined, often between two and five in number. In fuzzy coding the data are nonnegative values between 0 and 1 that sum to 1, like a small profile vector: for example, examples of fuzzy coding with three categories are [0.2 0.3 0.5] and [0.012 0.813 0.175]. Since the doubling described previously could just as well be coded in CA between 0 and the upper bound of 1 rather than between 0 and 10 in that case, by dividing by the upper bound, the three examples of doubling mentioned above can equivalently be considered two-category fuzzy coding, with values [0.3 0.7], [0.9 0.1], and [0 1], respectively. In what follows the idea of doubling and fuzzy coding will be further developed in the specific contexts in which they are applied.

15.1 Doubling of Ratings

Rating scales are ubiquitous in the social sciences, for example, the five-point Likert scale of agreement: 1 = strongly agree, 2 = somewhat agree, 3 = neither agree nor disagree, 4 = somewhat disagree, and 5 = strongly disagree. As an example of such data, I reconsider the Spanish data subset from the International Social Survey Program (ISSP, 2002) on attitudes to working women, previously analysed by Greenacre (2010, chap. 9) and available at http://zacat.gesis.org. This data set includes eight statements to which 2,107 respondents answered on this 5-point scale (cases with missing data have been excluded). The statements are

> *A*: A working mother can establish a warm relationship with her child.
>
> *B*: A preschool child suffers if his or her mother works.
>
> *C*: When a woman works the family life suffers.
>
> *D*: What women really want is a home and kids.
>
> *E*: Running a household is just as satisfying as a paid job.

F: Work is best for a woman's independence.

G: A man's job is to work; a woman's job is the household.

H: Working women should get paid maternity leave.

In order to recode the data into its doubled form, the 1–5 scale has first to be transformed to a 0–4 scale by subtracting 1, so that the upper bound is 4 in this case. This variable is then given its label and a negative sign, for example, *A*–, since it quantifies the level of disagreement. The doubled version of the data is then computed by subtracting that value from 4, creating a variable with label *A*+, which codes the level of agreement. Table 15.1 shows the doubled data for the first respondent. The value of 2 ('somewhat agree') for statement *A* becomes 3 and 1 in the doubled coding for *A*+ and *A*–, while the 4 ('somewhat disagree') for statement *B* becomes 1 and 3 for *B*+ and *B*–.

There is a connection between the doubled coding and the counting paradigm inherent in CA. If one thinks of the original scale points 1 to 5 written from right to left, so that strong agreement (= 1 on the original scale) is to the right and strong disagreement (= 5) to the left, a response of 2 ('somewhat agree') has 3 scale points to the left of it and 1 scale point to the right, hence the coding of [3 1], indicating more weight toward the positive pole of agreement. A response of 4 has 1 scale point to the left and 3 points to the right, hence [1 3]. A middle response of 3 has 2 scale points on either side; hence, it is balanced and coded [2 2], giving an equal count to both poles. The 'strongly agree' response to statement *F* is coded as a count of 4 for agreement and a count of 0 to disagreement.

With this doubling of the variables there is a perfect symmetry between the agreement and disagreement poles of the attitude scale. Since there are some statements worded positively and some worded negatively toward women working (statements *A* and *B* are good examples), one would expect these to be answered in a reversed way, as indeed respondent 1 does, agreeing to *A* and disagreeing to *B*. The use of integer values from 0 to 4 is, in fact, arbitrary, since each row of data will sum to a constant row margin, $8 \times 4 = 32$ in this case. So values from 0 to 1 could have been used to give the coding a probabilistic flavour: a [3 1] would thus be [¾ ¼] and the row sums

TABLE 15.1

Original 5-Point Scale Data for Respondent 1, and the Coding in Doubled Form

Original Responses

	A	B	C	D	E	F	G	H
	2	4	3	3	4	1	4	1

Doubled Data

A+	A–	B+	B–	C+	C–	D+	D–	E+	E–	F+	F–	G+	G–	H+	H–
3	1	1	3	2	2	2	2	1	3	4	0	1	3	4	0

Note: For each variable the negative pole of the doubled value is 1 less than the original response code, and the positive pole is 4 minus the negative pole.

would sum to a constant 8 (the number of variables) for each respondent. The coding then looks very much like the coding in MCA for dichotomous variables, but in MCA only the extreme points [0 1] and [1 0] are observed.

The doubled data matrix, with 16 columns in this example, has a dimensionality of 8, exactly the same as if one had analysed the original response data by principal component analysis (PCA). The CA of doubled data has a very close similarity to the PCA of the original (undoubled) data, and the only methodological difference is in the metric used to measure distances between case points. In PCA, because all variables are on the same five-point rating scale, no standardization would be performed and the metric would be the regular unweighted Euclidean distance. Greenacre (1984) showed that the chi-square metric for a matrix of doubled data implied a standardization of the original variables similar to that of a Bernoulli variable, which has variance $p(1-p)$, where p and $(1-p)$ are the probabilities of 'success' (coded as 1) and 'failure' (coded as 0). In the present example, the means across the 2,107 cases of the doubled variables are given in the first row of Table 15.2 (each pair also adds up to 4). The second row gives the same means divided by 4, so that each pair adds up to 1. The third row computes the product of the pair of values of each variable, an estimation of the Bernoulli variance, while the fourth and last row is the square root of the variance estimate, in other words, an estimate of standard deviation.

The result shown by Greenacre (1984, p. 183) is that the chi-square distance between cases, based on the doubled matrix, is—up to an overall scaling constant—the same as the Euclidean distance between cases, based on the original undoubled matrix, after standardizing the variables using the above estimates of standard deviation. Hence, a PCA using this standardization would yield the same result. Compared to the regular unstandardized PCA that would usually be applied to these data, CA, using the chi-square distance, would boost the contribution of variable H slightly, because its standard deviation is somewhat smaller than the others. In other words, compared to the Euclidean distance, the chi-square distance increases the contributions of questions with attitudes toward the extremes of the scale, where the variance is lower.

The CA asymmetric map of these doubled data is shown in Figure 15.1. The geometry of the joint map is very similar to that of an MCA for dichotomous data. Figure 15.2 shows examples of two cases, the first of which (on the left-hand side of the map) has only extreme responses to the eight statements: $A = 1, B = 5, C = 5, D = 5, E = 5, F = 1, G = 5, H = 1$, strongly agreeing to the statements in favour of working women, and strongly disagreeing to those against. This case lies at the average position of the corresponding endpoints of the variables and is consequently the leftmost case in the map. The case on the right, on the other hand, has some intermediate response categories, with the response pattern $A = 4, B = 1, C = 1, D = 2, E = 3, F = 4, G = 2, H = 1$. Some of the response categories thus lie at intermediate positions on the line segments of the variables. For example, the response $D = 2$ is indicated by a

TABLE 15.2
Means of Original Doubled Data, Their Rescaled Values, and Estimates of Variance and Standard Deviation for Each Variable

A+	A−	B+	B−	C+	C−	D+	D−	E+	E−	F+	F−	G+	G−	H+	H−
2.45	1.55	2.14	1.86	2.24	1.76	1.90	2.10	1.90	2.10	2.91	1.09	1.38	2.62	3.47	0.53
0.613	0.387	0.534	0.466	0.560	0.440	0.474	0.526	0.476	0.524	0.728	0.272	0.344	0.656	0.867	0.133
0.237		0.249		0.246		0.249		0.249		0.198		0.226		0.115	
0.487		0.499		0.496		0.499		0.499		0.445		0.475		0.340	

Note: Row 1: The means of the eight doubled variables on a 0–4 scale. Row 2: The means of doubled variables on a 0–1 scale (first row divided by 4), denoted by p_j and $1 - p_j$, $j = 1,\ldots, 8$. Row 3: Values of the product $p_j(1 - p_j)$, an estimate of variance. Row 4: Values of $\sqrt{p_j(1 - p_j)}$, an estimate of standard deviation.

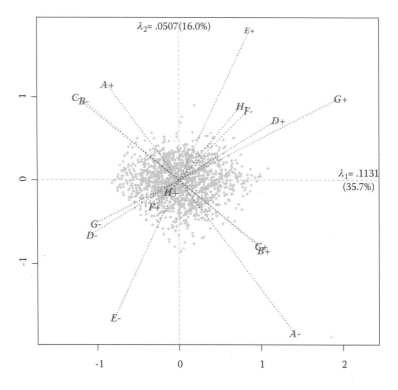

FIGURE 15.1
CA map of the doubled matrix of women working data, showing the agreement (+) and disagreement (–) poles of the rating scales and the 2,107 respondents. The scaling is 'row-principal', that is, variables in standard coordinates and cases in principal coordinates. 51.7% of the total inertia is explained by the two-dimensional solution.

cross at a position on the *D* segment corresponding to the 'somewhat agree' category, as if the segment was cut into four equal intervals between *D–* and *D+*. Similarly, the middle response *E* = 3 corresponds to the midpoint of the *E* segment. Given the positions on the eight segments of this case's responses, the case lies at the average position, as shown.

As for MCA, the category points for each variable have their centroid at the origin. This means that the average rating is exactly displayed by the position of the origin on each segment. For example, the average response to statement *G* is toward disagreement, because the centre is closer to the *G–* pole, while for *F* and *H* the average attitude is closer to agreement. If one calibrates each segment linearly from 1 at the + end to 5 at the – end, the corresponding mean can be read off exactly. Notice, however, that there is an alternative calibration possible for the biplot that allows prediction of the values of each case; see, for example, Greenacre (2010, chap. 3).

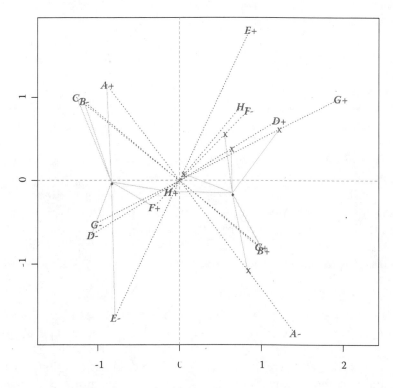

FIGURE 15.2
Examples of the barycentric relationship between cases and variables in the CA map of Figure 15.1. The case on the left has response pattern 1 5 5 5 5 1 5 1, while the case on the right has responses 4 1 1 2 3 4 2 1, where 1 ('strong agreement') corresponds to the + pole of the scale and 5 ('strong disagreement') to the − pole.

15.2 Doubling of Preferences and Paired Comparisons

The same doubling strategy can be used in the CA of preferences, or rank orderings, and paired comparison data. Rank orderings can be thought of as a special case of ratings: for example, if m objects are being ranked, then this is equivalent to an m-point rating scale where the respondent is forced to use each scale point only once. If six objects, A to F, were ranked, from 1 (most preferred) to 6 (least preferred), then doubling would again entail reversing the scale and subtracting 1, giving the + pole and creating the − pole as 5 minus the + pole. Suppose object C, for example, was ranked second; then its doubled coding would be $C+ = 4$ and $C− = 1$, indicating that the respondent preferred C over four other objects, but preferred one object to C. For a sample of n cases ranking m objects the doubled matrix can be constructed as before as an $n \times 2m$ matrix, with each object doubled. But in this case there is an alternative way to set up the matrix: the doubled columns can be stacked on top of each other to form a $2n \times m$ matrix so that the cases are doubled, not the objects. The original data set (after subtracting 1) then receives case labels with − appended, while the doubled matrix has case labels with +. Torres and Greenacre (2002) showed that the CA of this latter data structure is equivalent to Nishisato's dual scaling of preferences (Nishisato, 1994). In this CA map each object appears as a single point, while each case appears as two points, a + case point representing the case's preferences with respect to the six objects and a − point exactly opposite at the same distance from the origin representing the case's 'dispreferences'. Clearly, the − points are redundant to the display, but are nevertheless displayed by analogy with the doubled rating scales described previously.

Paired comparisons can be coded in a similar doubled fashion. Suppose respondents were asked to compare the six objects pairwise. There are 15 unique pairs (in general, $\frac{1}{2}m(m-1)$ pairs for m objects), and a typical response might be $A > B$, $C > A$, $A > D$, $A > E$, $F > A$, $B > C$, $B > D, \ldots$, and so on, where > stands for 'is preferred to'. The doubled variables are now real counts of how many times an object is preferred and dispreferred, respectively. For example, in the above list A is preferred to three others and two are preferred to A, so the data pair would be $A+ = 3$ and $A- = 2$. This type of coding can also be used for incomplete, but balanced, paired comparison designs. Notice that doubled values being coded here are the margins of an objects-by-objects preference matrix for each respondent (6×6 in this illustration)—these margins are not necessarily equivalent to the paired comparisons, when there are inconsistencies in the judgements, whereas in the case of a ranking these margins are equivalent to the ranking information.

Before explaining fuzzy coding let me point out a nonparametric analysis of a table of continuous data, which is related to the CA of preferences. Suppose that a data matrix consists of observations by n cases on p continuous variables. Rank order each variable in ascending order, so that the lowest value is replaced by 1, the second lowest by 2, and so on, up to n for the highest value. Notice that these 'preference orderings' are down the columns of the data matrix, not across the rows as discussed above where objects are ordered. Then double the variables as before, defining the positive pole as the rank minus 1, and the negative pole as $(n-1)$ minus the positive pole. The CA of this matrix gives identical results, up to a scaling factor, to the PCA of the (undoubled) matrix of ranks (Greenacre, 1984, p. 132).

15.3 Fuzzy Coding of Continuous Variables

The idea in fuzzy coding is to convert a continuous variable into a pseudo-categorical (i.e., fuzzy) variable using so-called *membership functions*. I illustrate this with the simplest example of triangular membership functions, shown in Figure 15.3, defining a fuzzy variable with three categories. On the horizontal axis is the scale of the original variable and three *hinge points*, chosen as the minimum, median, and maximum values of the variable. The three functions shown are used for the recoding, and on the left a particular value of the original variable is shown to be recoded as 0.3 for category 1, 0.7 for category 2, and 0 for category 3. This coding is invertible: given the fuzzy observation [0.3 0.7 0.0] the value of the original variable is

$$\text{Original variable} = 0.3 \times \text{minimum} + 0.7 \times \text{median} + 0.0 \times \text{maximum} \quad (15.1)$$

Fuzzy coding in CA first appeared in the thesis of Bordet (1973) and was subsequently used by Guittonneau and Roux (1977) in a taxonomic study where continuous variables characterizing plants were coded into fuzzy categories, thus enabling them to be analysed jointly with several categorical variables. The data set in question consists of 82 examples from a taxonomic study of 75 different species, which can be combined into 11 groups of the plant genus *Erodium* (see Figure 15.4): for example, *Erodium plumosa* and *Erodium romana*. Apart from the botanical objective of seeing how these groups differ in terms of their characteristics, the interest from a data analytic point of view is the mix of 33 categorical and five continuous variables, and the challenge of analysing them jointly. Like the original authors, I will apply fuzzy coding to the continuous variables in order to analyse them jointly with the categorical variables, recoding each variable into four categories using four hinge points: minimum, first tercile, second tercile, and maximum.

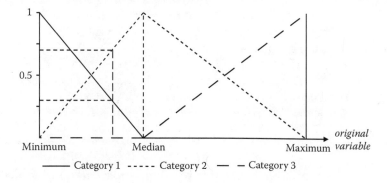

FIGURE 15.3
Triangular membership functions that convert a continuous variable into a fuzzy variable with three categories. The minimum, median, and maximum are used as hinge points.

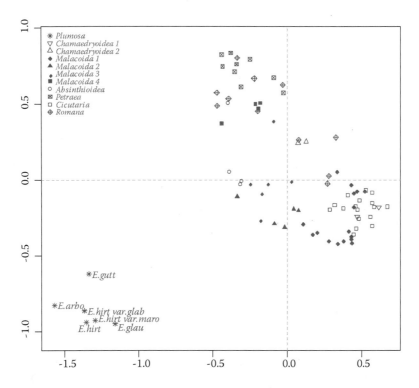

FIGURE 15.4
Correspondence analysis of recoded *Erodium* data, showing species only in principal coordinates, with clear separation of the six species forming the subgenus *Plumosa*. For species names, see Guittonneau and Roux (1977).

Of the categorical variables 15 are dichotomous, 13 are trichotomous, four have four categories, and one has six. Guittonneau and Roux (1977), who published the data, recoded the 13 trichotomous variables into two categories by assigning values [½ ½] for the second category when it occurred, arguing that it was always an intermediate category between categories 1 and 3. In my reanalysis, I prefer to keep all three categories. The 33 categorical variables thus generate $15 \times 2 + 13 \times 3 + 4 \times 4 + 1 \times 6 = 91$ dummy variables. Each of the five continuous variables is transformed to four fuzzy categories, using triangular membership functions as described above. Thus, $5 \times 4 = 20$ additional variables are generated, so that the grand total is 111 variables instead of the original 38. (It is interesting to notice how times have changed. In their data recoding, which extends the number of variables to be analysed by introducing many dummy variables and fuzzy categories, Guittonneau and Roux (1977) say that an inconvenience of the recoding is that 'after transformation, the table requires much more memory in the computer; moreover, the computation time is increased by an important proportion'. Their table, after transformation, was only of size 82×98, very small by today's standards.)

Before passing these data through CA, there is an issue of weighting of the dummy and fuzzy variables, which authors Guittonneau and Roux also considered. The argument was, and still is, that the more categories a variable has, the higher the variance created in the generated dummy variables. Also, crisp (0/1) and fuzzy coded variables with the same number of categories do not have the same inherent variances. In their analysis Guittonneau and Roux (1977) applied an algorithm proposed by Benzécri (1977a) and implemented by Hamrouni (1977) to reweight the groups of variables, each considered to be homogeneous internally (i.e., on the same measurement scale), but each having different number of categories. In my reanalysis, I have followed this idea, reweighting the four groups of variables—15 dichotomous, 13 trichotomous, five polychotomous, and five fuzzy—so that their variances (i.e., inertias) are equalized in the analysis. This amounted to multiplying the four sets of columns by the following respective coefficients: 0.228, 0.132, 0.201, and 0.439. Figure 15.4 shows the display of the samples in the CA of the full 82 × 111 recoded and reweighted matrix, with symbols indicating the 11 different groups of species. There is a clear separation of one group of six species at the bottom left, forming the subgenus *Plumosa*. Figure 15.5 shows

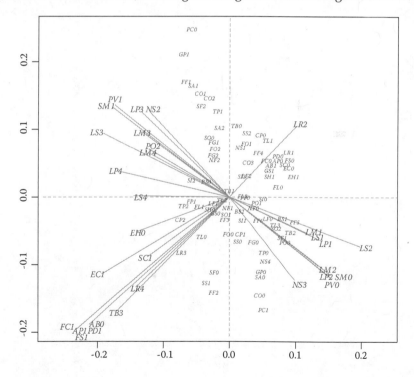

FIGURE 15.5
Correspondence analysis of recoded *Erodium* data, showing in standard coordinates the categories of the variables that contribute more than average to the two-dimensional solution. For a description of the variable names, see Guittonneau and Roux (1977).

the corresponding plot of the categories in the contribution biplot scaling (see Greenacre, 2013), with variable categories that contribute more than average to the axes of the two-dimensional solution shown in larger font. The categories *AB0*, *AP1*, *FS1*, *PD1*, and *FC1*, for example, are associated exclusively, or almost exclusively, with subgenus *Plumosa* and are its main botanical indicators. Five out of these six species also fall into category *TB3*, for example, the highest category of the four-category variable *TB*, which is also stretching far in the direction of these species.

In order to visualize the variation within the other species, either the *Plumosa* species should be removed, or the variables associated with them, but this is unsatisfactory in both cases. If the species are removed, then some variables have zero observations in the remaining species and have to be removed anyway; if we remove the variables, then some information is lost where there are some observations of these variables in other species. A good compromise here is to use subset CA (Greenacre and Pardo, 2006a), retaining all the variables but excluding the six species of subgenus *Plumosa* (subset CA has been used in multiple correspondence analysis as well—see Greenacre and Pardo (2006b)—to exclude missing value categories, where

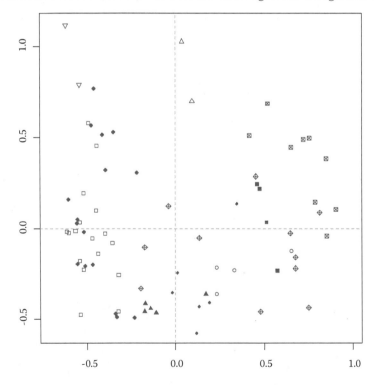

FIGURE 15.6
Subset correspondence analysis of recoded *Erodium* data, omitting species in subgenus *Plumosa*. See Figure 15.4 for key to group symbols.

it is also called missing data passive (Gifi, 1980) or specific multiple correspondence analysis; see Chapter 12 by Le Roux in this book). In subset analysis the column margins of the original table are maintained; hence, the centre of the space as well as the metric from the original analysis are preserved. Figure 15.6 shows the species in the subset excluding those in subgenus *Plumosa* (see Figure 15.4), but maintaining exactly the same space (i.e., same origin, same distance metric, and same point masses). Figure 15.7 shows the categories, in standard coordinates, of the most contributing variables—this is different from Figure 15.5, which showed the highly contributing categories, in contribution coordinates with lines from the origin. In Figure 15.7 the contributions of each variable were aggregated, and the variables with higher than average contribution are plotted, showing all their categories, connected to show their trajectories. The second principal (vertical) axis shows the trajectories of the variables **LP, LS, LM, LR**, and **NS**, all lining up from low to high in their expected order. However, there is an interesting mirror-image pattern of these trajectories along the first principal (horizontal) axis, contrasting low values of variable **NS** and high

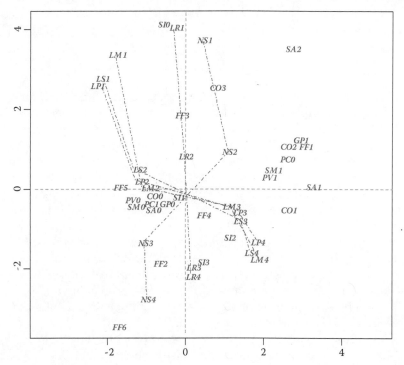

FIGURE 15.7
Subset correspondence analysis of recoded *Erodium* data, showing in standard coordinates the categories of the variables that contribute more than average to the two-dimensional solution. The fuzzy-coded variables, among the top contributors, are shown with their consecutive categories connected.

values of ***LM, LP,*** and ***LS*** against the high values of ***NS*** and low values of the other three, with variable ***LR*** not following this contrast. While the pattern is clear, it has no obvious interpretation.

15.4 Explained Inertia

Up to now I have generally refrained from reporting inertias and percentages of explained inertia. With respect to the data matrix entering the respective analyses, the relevant results are as follows:

- Analysis of reweighted data, Figures 15.4 and 15.5: Total inertia = 1.642, of which 27.7% is explained by the two-dimensional solution.
- When doing a subset analysis excluding subgenus *Plumosa*, the inertia of 1.642 splits into two parts: 0.0256 due to *Plumosa* and 1.386 due to the rest, the latter being decomposed in the subset analysis.
- Subset analysis excluding *Plumosa*, Figures 15.6 and 15.7: Total inertia = 1.386, of which 26.0% is explained by the two-dimensional solution.

It is well known that in the CA of data coded in an indicator matrix, for example, in a multiple correspondence analysis (MCA), the percentages of inertia seriously underestimate the explained variance. Greenacre (1995) demonstrated an easy way to adjust the total inertia as well as the parts of inertia to obtain improved estimates of the overall quality of the solution (for a recent account, see Greenacre, 2007, chap. 19). The same phenomenon appears in the analysis of fuzzy-coded data, and Aşan and Greenacre (2011) showed how estimates from a CA of fuzzy data can be transformed back to the original continuous scales of the data, a process called defuzzification, in order to obtain improved parts of variance explained. In the present example, however, there are two additional aspects that make this strategy more complicated: first, there is a mixture of measurement scales, including crisp (zero/one) and fuzzy coded data; and second, the groups of variables have been reweighted. Estimating more realistic overall percentages of explained inertia in this situation remains an open problem.

For the present situation, therefore, it is not possible to obtain a single adjusted estimate of the global quality of the solution, but one can make separate computations of the adjusted inertia explained for the crisply coded categorical data (variables 1 to 33) and fuzzy coded data (variables 34 to 38), in both the analysis reported in Figures 15.4 and 15.5 and the subset analysis of Figures 15.6 and 15.7. The adjusted percentages of each group as well as their original percentages are reported in Table 15.3. In all cases the adjusted

TABLE 15.3

Adjusted Percentages of Inertia Explained by Crisp and Fuzzy Variables in the Two Analyses of the *Erodium* Data, Before and After Adjustment

	Crisp Variables		Fuzzy Variables	
	(Original)	Adjusted	(Original)	Adjusted
Global analysis, 2D solution	(27.4%)	71.2%	(28.5%)	43.8%
Subset analysis, 2D solution	(19.0%)	35.8%	(44.9%)	67.5%

percentages are higher, especially for the categorical data coded crisply in indicator form.

15.5 Discussion

Doubling and fuzzy coding are part of the philosophy of Benzécri's French school of data analysis, enabling ratings, preferences, paired comparisons, and continuous data to be coded in a way that fits the correspondence analysis paradigm. Correspondence analysis is thus considered to be a versatile dimension reduction and visualization technique, where the data are pre-transformed in various ways prior to analysis. Fuzzy coding transforms continuous data to a form that is comparable to categorical data, and so enables analysis of mixed measurement scales. Doubling can be considered a special case of fuzzy coding when there are only two fuzzy categories. Reweighting the variables can help to compensate for different numbers of categories in the various measurement scales. Explained variance is still an open issue: here it was necessary to consider separately the parts of explained variance for the crisply coded (zero/one) variables from that of the fuzzy coded ones, and it would be preferable to develop a global measure.

16

Symbolic Data Analysis: A Factorial Approach Based on Fuzzy Coded Data

Rosanna Verde and Edwin Diday

CONTENTS

16.1 Symbolic Input Data 258
16.2 Generalized Canonical Analysis Strategy for Symbolic Data 259
    16.2.1 Coding of the Symbolic Variables 260
    16.2.2 Symbolic Generalized Canonical Analysis under a
           Cohesion Constraint 263
    16.2.3 Factorial Representation of Symbolic Data: An Example
           on Developing Countries Data 265
16.3 Conclusion 270

In recent years symbolic data analysis (SDA) has assumed an important role in the framework of Benzecri's French school of *analyse des données*, developing standard exploratory techniques of multivariate data analysis for the study of more general and complex data, called symbolic data.

Since the first papers by Diday (1988, 1989, 1992), SDA has known considerable developments and become a new field of research gathering contributions from different scientific communities: statistics, knowledge extraction, machine learning, data mining. Three European projects supported the systematic developments of the main methodological contributions, and a wide collection of SDA methods are available in three reference books (Bock and Diday, 2000; Billard and Diday, 2006b; Diday and Noirhomme-Fraiture, 2008).

The very first book on SDA, *Analysis of Symbolic Data*, published by Bock and Diday (2000), contains a comprehensive overview on data analysis methodologies. Since then, the development of SDA has continued at an intensive rate involving more and more research teams. Currently, the number of publications in international journals (e.g., Billard and Diday, 2003; Duarte Silva and Brito, 2006; De Carvalho et al., 2009a) is approximately 100. A special issue of the journal *Statistical Analysis and Data Mining* was published containing an overview of the recent symbolic data analysis developments and analytic issues (Billard, 2011).

SDA is a new paradigm based on the transition from standard individual observations to higher-level observations described by symbolic data. Symbolic variables can assume values in the form of intervals, multicategories, or sets of categories with an associated probability, frequency, or percentage value.

We can define three domains of SDA application. First, *standard data tables* where observations are described by standard variables. In this case, numerical or categorical variables induce sets of categories (or fuzzy categories) that can be considered higher-level observations. Second, *native symbolic data* represented by tables where the observations are of higher level and the ground level of observations is not known (for example, ranges of values considered to be normal by medical experts for parameters such as cholesterol and glucose levels in the blood). Third, the case of *complex data*, which covers all types of data that cannot be reduced to a standard data table. In practice, complex data are usually based on several kinds of observations described by numerical and categorical data contained in several tables from different sources. In this case the higher-level units are obtained by a fusion process.

As an example of complex data (provided by the UN and World Bank and completed by some indicators on socioeconomic health[*]) we consider a study on developing countries, where for each country, three standard data tables of different observations and different variables are considered. The observations of the first data table are related to agriculture, with variables such as added value, rural population, and so on. In the second data table the observations are relative to fertility rates, and mortality rates for several levels of age. In the third data table, inequality is expressed by a set of indices (inequality in education, inequality in income, etc.). In order to compare typologies of countries according to a geographical classification, a fusion process is achieved. This aggregation is performed on the three data tables so that the new sampling units are geographical regions described by multivalued variables. A graphical illustration of the kind of symbolic data table obtained after the fusion process is given in Figure 16.1. The interval value variables are represented from left to right by the minimum of the lower bound values of all the intervals associated to this symbolic variable, the mean of the lower bound values, the mean of the means, the mean of the upper bound values, and the maximum of the upper bound values. The shaded part is the interval value of the associated sampling units. The *histogram variables* are built using

---

[*] Kindly made available by Frédéric Lebaron.

Symbolic Data Analysis: A Factorial Approach Based on Fuzzy Coded Data 257

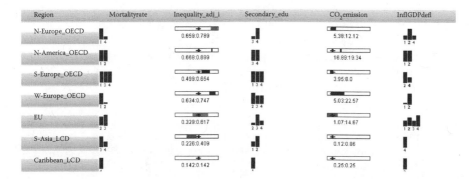

FIGURE 16.1
Symbolic data table that combines different kinds of symbolic variables such as histograms and intervals (rows have been sorted from the highest to the lowest frequency of category 4 of level of secondary education). The bars of the histogram are proportional to the intensity of character on the intervals of values in which the continuous variable is shared. The intervals of values observed for the different regions of the interval variables are drawn as grey strips; a cross indicates the median value of the interval variables.

the domain of the numerical variables in nonoverlapping intervals, with the histogram bars showing the intensity of the phenomenon in each geographical area. Moreover, in order to use symbolic data to summarize large data sets, Stéphan et al. (2000) developed an approach to extract raw detailed data from relational databases and to construct high-level data (symbolic data) by a classification process.

For a univariate and a bivariate analysis, Bertrand and Goupil (2000) and Billard and Diday (2006a) proposed different basic statistics according to the nature of the data (interval, multicategorical, modal). Dependence measures between bivariate symbolic interval variables were introduced by Billard (2004). Among the exploratory tools, visualization techniques of symbolic and complex data were presented by Noirhomme-Fraiture and Rouard (2000). An extension of exploratory factorial techniques was developed for symbolic data; according to the type of the multivalued symbolic variables, principal component analysis (PCA) was proposed for quantitative interval data (Cazes et al., 1997; Lauro et al. 2000), generalized canonical analysis for categorical multivalued data (Verde, 1997; Lauro et al. 2008c), factorial discriminant analysis for quantitative or mixed (interval, multicategorical, modal) predictors, and a single categorical predicted variable (class variable) (Lauro et al., 2000, 2008a). Different methods of PCA were proposed representing the intervals by their centres (*centres method*) or representing each individual in a p-dimensional variable space as a hypercube given by the 2^p combination of the bounds of the intervals (*vertices method*). A different approach based on the transformation of each variable by the midpoints and radii of its interval values was proposed by Palumbo and Lauro (2003). An extension of PCA to histogram variables was also introduced by Rodriguez at al. (2001, 2004) and Ichino (2008). The peculiarity of such methods is that each observation, described by a histogram, is represented by nested intervals on reduced factorial spaces.

An important set of alternative methods is offered by cluster analysis techniques, which aim to perform groups or clusters of homogeneous elements according to a suitable dissimilarity measure.

In the context of SDA, cluster analysis is highly relevant because the new clusters can be suitably interpreted as higher-level symbolic data. The clustering techniques proposed for symbolic data can be shared in agglomerative and partitioning ones. Some contributions to the first set of techniques were given by Brito (1994, 1995), Brito and De Carvalho (2002), and Gowda and Ravi (1995); a large contribution to the extension of the classical partitioning techniques to symbolic data was furnished by De Carvalho et al. (1999, 2009b, 2012) and Chavent et al. (2002, 2006).

Among the most recent developments must be mentioned several models for representing symbolic data distributions (Arroyo and Maté, 2009; Brito and Duarte Silva, 2012; Irpino and Verde, 2010) and for analysing relationships between symbolic variables using suitable dissimilarity and distance measures (Irpino and Verde, 2006; Verde and Irpino, 2007; Irpino and Verde, 2010). New investigating fields of research in SDA are actually about inferential aspects in symbolic data models (Arroyo and Maté, 2009).

In this chapter we propose an extension of generalized canonical analysis (GCA) as a general method for representing symbolic data, described by several kinds of symbolic variables, in reduced factorial subspaces. An example on synthetic and real data sets will allow us to justify the proposed GCA procedure. The results of the analysis show graphically the associations between symbolic descriptors and the influence of the different internal variability of the symbolic data in reduced subspaces. An essential step in this approach is the fuzzy coding of all symbolic variables, which results in a homogeneous system of coding and enables the application of a variant of multiple correspondence analysis (MCA) for fuzzy coded categorical data (see Chapter 15 by Greenacre in this book). The analyses have been supported by the academic free software SODAS 2 (a free version of this software is available at http://www.info.fundp.ac.be/asso/) and by the commercial software package SYR Co.

16.1 Symbolic Input Data

Symbolic data can be characterized to be described by multivalued symbolic variables Y of different types. A multivalued symbolic variable Y is one that takes one or more values from the list of values in its domain D (Bock and Diday, 2000, chap. 3). The complete list of possible values in D is finite, and values may be categorical or quantitative. For example, if D_{empl} = {teacher, professor, manager} is the domain of the categorical multivalued variable Y_{empl} 'employees', the observation i 'educational institution' is characterized by the following categories of 'employees': $Y_{i,empl}$ = {teacher, professor}.

An interval-valued symbolic variable Y is one that takes values in an interval $[l, u] \subset \Re$ with the lower bound l and upper bound u of the interval. Hence, the domain D is $\Re$ and the set of possible values that Y may take are subsets of $\Re$. For example, the *income* for the category i of employees at educational institutions is $Y_{i,income}$ = [45,000–100,000].

A *modal variable* Y is one that takes a subset of values in its domain D that can be a list of nominal or finite numerical values, with associated nonnegative measures (*modes*), such as probabilities, frequencies, or weights. That is, a particular observation takes the form $Y_i = \{(k_1, r_1), \ldots, (k_s, r_s)\}$, with $\{k_1, \ldots, k_s\}$ the list of nominal values in D corresponding to i and $\{r_1, \ldots, r_s\}$ the respective associated modes. When the associated k_l's are nominal values (or numerical discrete values), Y is also defined as a *bar-chart variable*, while if the k_l's are continuous and nonoverlapping intervals, Y is defined as a *histogram variable*. In the case where modes are measures of probability or relative frequencies, they sum up to 1. For example, considering the distribution of professions at educational institutions, the modal (bar-chart) variable Y assumes categorical values with associated relative frequencies $Y_{i,empl\ dist}$ = {(teacher, 0.7), (professor, 0.3)}. Similarly, taking into account the distribution of classes of income, the modal (histogram) variable Y_{inc} for education institutions takes values Y_{inc} = {([45,000–60,000], 0.4), ([60,000–80,000], 0.4), ([80,000–100,000], 0.2)}. The symbolic description of an individual i, with respect to a set of symbolic variables Y_j, is given by a symbolic assertion, for example: $a(i = Ed\ Inst) = [Y_{empl}$ = {teacher, professor}] $\wedge$ [Y_{inc} = {([45,000–60,000], 0.4), ([60,000–80,000], 0.4), ([80,000–100,000], 0.2)}] $\wedge$ [Y_{gender} = {M (0.4), F(0.6)}] $\wedge \ldots$

Such data descriptions for a set of individuals are then assembled in a symbolic data table **X**, with rows containing the individuals and the columns the multivalued symbolic variables. Each cell of the table collects different types of data: a set of categories for the categorical multivalued variable, an interval of values for the interval variables, and a frequency, probability, or belief distribution for the modal variables.

16.2 Generalized Canonical Analysis Strategy for Symbolic Data

Generalized canonical analysis (GCA) for symbolic data (Verde 1997, 1999; Lauro et al. 2008c) is a general factorial method able to analyse data whose descriptors are of different natures (interval, categorical multivalued, modal), leading to a visualization of relationships between symbolic variables in reduced subspaces. It is based on an intermediate *coding phase* in order to homogenize different kinds of symbolic variables using, for instance, a *fuzzy* or a *crisp* (complete disjunctive) coding system (see Chapter 3 by Lebart and Saporta and Chapter 15 by Greenacre, both in this book). GCA on symbolic data is based on an extension of a classical method based on the juxtaposition of complete disjunctive coding tables of categorical variables (Volle, 1985), where the decomposed criterion in the analysis is the squared multiple correlation index computed on coded symbolic variables.

GCA on symbolic data is performed in the following main steps:

1. Symbolic data coding process
2. Application of GCA to the coded symbolic data
3. Symbolic interpretation of the results according to the nature of the original symbolic data

16.2.1 Coding of the Symbolic Variables

Let n be a set of symbolic observations described by Q symbolic variables $\{Y_1, \ldots, Y_q, \ldots, Y_Q\}$ with domains in D_q ($q = 1, \ldots, Q$). The first phase consists in a numerical coding of the symbolic descriptors Y_q (for $q = 1, \ldots, Q$). The example described above consists of three symbolic variables: $\{Y_{gender}, Y_{empl}, Y_{inc}\}$.

Different coding systems for these variables are achieved as follows:

- For a *categorical multivalued variable*, an observation is coded according to the presence/absence of each category in the domain of Y_j with values 1 or 0. For example, the ith symbolic observation that assumes values {teacher, professor} in the domain D_{empl} = {teacher, professor, manager} of the symbolic variable Y_{empl} is coded by two rows of a binary coding table as follows:

$$\begin{bmatrix} \text{teacher} & \text{professor} & \text{manager} \\ 1 & 0 & 0 \\ 0 & 1 & 0 \end{bmatrix}$$

- For a *modal variable*, an observation is coded according to the mode values that it assumes for the categories of Y_j; in the above example, considering the domain of the variable *gender* and the proportions of male and female, the description of Y_{gender} = {M(0.4), F(0.6)} is coded as follows:

$$\begin{bmatrix} M & F \\ 0.4 & 0.6 \end{bmatrix}$$

- For an *interval variable*, a fuzzy coding system is performed using B-spline functions that preserve the numerical information of the original numerical variables. These functions are used in order to assign, with values in [0, 1], the membership of an observation to the pseudocategories (for example: three categories, corresponding to *low, medium*, and *high* levels). To code the interval description for income in the above example, we use triangular membership functions with *knots* (also called *hinge points*) chosen as the lower bound, the middle value, and the upper bound of the range of the interval variable Y_j.

However, a fuzzy coding of interval bounds with respect to three categories *low-medium-high* does not allow recovering all the points of the original interval; that is an important property of the B-spline transformation (De Boor, 1978). In order to overcome this deficiency, we also code each interval with respect to the middle point of the domain of the variable. In this way, if the middle value is between the lower and upper bounds, then the interval is split in two consecutive intervals [lower bound, *middle* value], [*middle* value, upper bound], and these are coded according to fuzzy values, respectively, on two rows.

For example, given the general domain of income is D_{inc} = [30,000–150,000], the observed intervals are coded with fuzzy values according to triangular membership functions with hinges {30,000, 90,000, 150,000}, where 90,000 is the middle point. Thus, the observed interval Y_{inc} = [45,000–100,000] is shared in the two subintervals [45,000–90,000] and [90,000–100,000], which are coded with respect to the three semilinear functions shown in Figure 16.2.

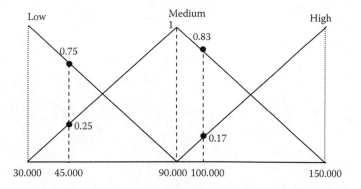

FIGURE 16.2
Coding functions of the interval symbolic variable by B-spline functions (*low, medium, high*) of first order with three knots. For example, the income value of 45,000 is coded as (0.75 0.25 0) and 100,000 as (0 0.83 0.17).

The coding values corresponding to the membership to the three pseudo-categories (*low, medium, high*) of *income* are laid on two different rows of the coding table, respectively, as

$$\begin{bmatrix} 45{,}000; 90{,}000 \end{bmatrix} \Rightarrow \begin{array}{c} 45{,}000 \Rightarrow \\ 90{,}000 \Rightarrow \end{array} \quad \begin{bmatrix} & \text{Income} & \\ \text{low} & \text{medium} & \text{high} \\ \hline 0.75 & 0.25 & 0 \\ 0 & 1 & 0 \\ 0 & 1 & 0 \\ 0 & 0.83 & 0.17 \end{bmatrix}$$

$$\begin{bmatrix} 90{,}000; 100{,}000 \end{bmatrix} \Rightarrow \begin{array}{c} 90{,}000 \Rightarrow \\ 100{,}000 \Rightarrow \end{array}$$

The tables $\mathbf{Z}_q$ contain the coding of the set of values of the symbolic variable Y_q presented by the ith assertion that, in the example, describes the characteristics of an education institution for the gender, and composition, of employees [Y_{gender} = {M(0.4), F(0.6)}], the category of the employees [Y_{empl} = {teacher, professor}], and their income [Y_{inc}—expressed in the range (45,000–100,000)].

The rows of the coding tables $\mathbf{Z}_q$ (for $q = 1, \ldots, Q$), related to the ith unit, are concatenated so that all combinations of the rows are represented, giving the complete coding of the symbolic description of the ith observation. In the case of the above example, the complete coding is the following:

$$\mathbf{Z}_i = \begin{bmatrix} \mathbf{Z}_{i1} \mid \mathbf{Z}_{i2} \mid \mathbf{Z}_{i3} \end{bmatrix}$$

$$\mathbf{Z}_i = \begin{bmatrix} \begin{array}{ccc|ccc|cc} \multicolumn{3}{c|}{\text{income}} & \multicolumn{3}{c|}{\text{employees}} & \multicolumn{2}{c}{\text{gender}} \\ \text{low} & \text{medium} & \text{high} & \text{teacher} & \text{professor} & \text{manager} & M & F \\ \hline 0.75 & 0.25 & 0 & 1 & 0 & 0 & 0.4 & 0.6 \\ 0 & 1 & 0 & 1 & 0 & 0 & 0.4 & 0.6 \\ 0 & 1 & 0 & 1 & 0 & 0 & 0.4 & 0.6 \\ 0 & 0.83 & 0.17 & 1 & 0 & 0 & 0.4 & 0.6 \\ 0.75 & 0.25 & 0 & 0 & 1 & 0 & 0.4 & 0.6 \\ 0 & 1 & 0 & 0 & 1 & 0 & 0.4 & 0.6 \\ 0 & 1 & 0 & 0 & 1 & 0 & 0.4 & 0.6 \\ 0 & 0.83 & 0.17 & 0 & 1 & 0 & 0.4 & 0.6 \end{array} \end{bmatrix}$$

Concatenating vertically all the above Q matrices (for all $i = 1, \ldots, n$) of dimension $N_i \times J$, where the number of columns is

$$J = \sum_q J_q,$$

Symbolic Data Analysis: A Factorial Approach Based on Fuzzy Coded Data 263

J_q being the number of columns of the qth variable ($J_1 = 3$, $J_2 = 3$, $J_3 = 2$ in this example), we obtain a global coding matrix $\mathbf{Z}$ ($N \times J$), where

$$N = \sum_i N_i,$$

and where each N_i depends on the number of variables of different types that influence, in several ways, the number of coding rows for each individual.

16.2.2 Symbolic Generalized Canonical Analysis under a Cohesion Constraint

Symbolic generalized canonical analysis (SGCA) aims to visualize on reduced subspaces, usually Cartesian bidimensional planes, both the relationships among a set of symbolic data, described by p multivalued variables suitably coded as shown in the previous section, and the association between the categories of these symbolic variables. Furthermore, the geometric representation of the data allows us to compare the symbolic data with respect to shape, size, and location of their positions. SGCA looks for factorial axes, as a linear combination of the coded symbolic descriptors.

The method is performed on the global coding matrix $\mathbf{Z}$ ($N \times J$) that can also be considered the juxtaposition of Q fuzzy or disjunctive coding matrices: $\mathbf{Z} = [\mathbf{Z}_1| \ldots |\mathbf{Z}_q| \ldots |\mathbf{Z}_Q]$. In order to keep the cohesion among the N_i coding rows of the subtables of $\mathbf{Z}$ corresponding to the coded symbolic description of each individual, we introduce a cohesion constraint matrix $\mathbf{A}$ on the rows of the global matrix of coding $\mathbf{Z}$ (obtained by the vertical concatenation of the submatrices $\mathbf{Z}_i$). Therefore, $\mathbf{A}_{N \times n}$ is an indicator matrix that identifies the membership of the rows of $\mathbf{Z}$ to the n symbolic data (individuals):

$$\mathbf{A} = \begin{bmatrix} i_1 & i_2 & \cdots & i_n \\ \hline 1 & 0 & \cdots & 0 \\ 1 & 0 & \cdots & 0 \\ \vdots & \vdots & \vdots & \vdots \\ 1 & 0 & \cdots & 0 \\ 0 & 1 & \vdots & 0 \\ 0 & 1 & \cdots & 0 \\ \vdots & \vdots & \vdots & \vdots \\ 0 & 1 & \cdots & 0 \\ \vdots & \vdots & \vdots & \vdots \\ 0 & 0 & \cdots & 1 \\ 0 & 0 & \cdots & 1 \end{bmatrix}$$

The factorial axes are obtained as the solution to the following characteristic equation:

$$\frac{1}{N}\mathbf{A}^T\mathbf{Z}\boldsymbol{\Sigma}^{-1}\mathbf{Z}^T\mathbf{A}\mathbf{u}_k = \lambda_k \mathbf{u}_k$$

under the classical orthonormality constraints: $\mathbf{u}_k^T\mathbf{u}_k = 1$ and $\mathbf{u}_k^T\mathbf{u}_{k'} = 0$ (for $k \neq k'$), where $\boldsymbol{\Sigma}^{-1}$ is a block diagonal matrix of dimension $J \times J$ as follows:

$$\boldsymbol{\Sigma}^{-1} = \begin{bmatrix} (\mathbf{Z}_1^T\mathbf{Z}_1)^{-1} & 0 & \cdots & 0 \\ 0 & (\mathbf{Z}_2^T\mathbf{Z}_2)^{-1} & \cdots & 0 \\ \vdots & \vdots & \ddots & \vdots \\ 0 & 0 & \cdots & (\mathbf{Z}_Q^T\mathbf{Z}_Q)^{-1} \end{bmatrix}$$

The maximum number of nontrivial eigenvalues is $(J - Q + 1)$, provided that the total number of all categories J is less than N.

The coordinates of the categories of the Q transformed variables on each factorial axis are

$$\boldsymbol{\psi}_k = \boldsymbol{\Sigma}^{-1/2}\mathbf{Z}^T\mathbf{A}\mathbf{u}_k \quad \text{for } k = 1, \ldots, J - Q + 1$$

The eigenvectors of dual space are computed as follows:

$$\mathbf{v}_k = \frac{1}{\sqrt{\lambda_k}}\boldsymbol{\Sigma}^{-1}\mathbf{Z}^T\mathbf{A}\mathbf{u}_k.$$

The vector of the coordinates of the symbolic description of the individual i on each factorial axis k, corresponding to the set of N_i rows of the coding table $\mathbf{Z}$ for individual i, is

$$\boldsymbol{\varphi}_{ik} = [\mathbf{Z}_{i1} \ldots \mathbf{Z}_{iQ}]\boldsymbol{\Sigma}^{1/2}\mathbf{v}_k.$$

Moreover, the contribution of each category j and each individual i to an axis k is given by the ratio between the squared coordinates and the eigenvector norm, both with respect to the axis k. These are

$$\text{CTA}_{jk} = \frac{\psi_{jk}^2}{\lambda_k};$$

$$\text{CTA}_{ik} = \frac{1}{\lambda_k}\boldsymbol{\varphi}_{ik}^T\boldsymbol{\varphi}_{ik} \quad \text{for } k = 1, \ldots, K,$$

where $K \leq J - Q + 1$ is the number of the nontrivial eigenvectors.

Symbolic Data Analysis: A Factorial Approach Based on Fuzzy Coded Data 265

The contribution of the *individual i* to axis k is expressed by CTA_{ik} that is a global measure of the contribution of the *individual i* to the axis obtained as a sum of the N_i contribution measures

$$CTA_{sk} = \frac{1}{\lambda_k} \varphi_{sk}^2 \quad (\text{for } s = 1, \ldots, N_i)$$

of the set of coded symbolic description s of individual i. The relative contribution is a measure of the quality of representation of the coded symbolic description of each individual i on the factorial axis k; it is a global measure and it is given by the sum of the N_i elements

$$CTR_{sk} = \frac{\varphi_{sk}^2}{\sum_{h=1}^{K} \varphi_{sh}^2} \quad (\text{for } s = 1, \ldots, N_i).$$

The relative contributions of the categories are computed according to their correlation with respect to the new axes:

$$CTR_{jk} = \frac{\psi_{jk}^2}{\sum_{k=1}^{K} \psi_{jk}^2} \quad \text{for } k = 1, \ldots, K$$

16.2.3 Factorial Representation of Symbolic Data: An Example on Developing Countries Data

The application of the proposed method is performed on a data set on comparative international data on human development derived from world development indicators produced by the World Bank (UNDP, 2010). The Human Development Index (HDI) is used to distinguish whether a country is a developed, a developing, or an underdeveloped country, and also to measure the impact of economic policies on quality of life (United Nations, 2010). Two other categorical variables have been added to the data: the geographic regions and whether the country belongs to the Organization for Economic Cooperation and Development (OECD) or to the low developing countries (LDCs), and countries are aggregated according to these variables. In total, 35 new sampling units are formed, shown in Table 16.1, representing conceptual regions and described by histograms and interval data. For the histograms four intervals are chosen as split points that are most discriminating of the different concepts.

The graphical representation of symbolic data on a factorial plane allows us to interpret the variability of the multivalued variable that describes the

TABLE 16.1

List of the 35 Selected Regions (Concepts) and the Related Short Labels

Labels	Concept-Regions	Labels	Concept-Regions	Labels	Concept-Regions
NAf	North Africa	SEAs	Southeast Asia	WEu	West Europe
SSA	Sub-Saharan	SEAs_L	Southeast Asia, LDC	WEu_O	West Europe, OECD
SSA_L	Sub-Saharan, LDC	SEAs_O	Southeast Asia, OECD	NEu_O	North Europe, OECD
CAm	Central America	SAs	South Asia	SEu	South Europe
CAm_O	Central America, OECD	SAs_L	South Asia, LDC	SEu_O	South Europe, OECD
NAm_O	North America, OECD	EU	European Union	ME	Middle East
LA	Latin America	Ca	Caribbean	ME_O	Middle East, OECD
LA_O	Latin America, OECD	Ca_L	Caribbean, LDC	ME_L	Middle East, LDC
CAs	Central Asia	CE	Central Europe	Oc	Oceania
CAs_L	Central Asia, LDC	CE_O	Central Europe, OECD	Oc_O	Oceania, OECD
As	Asia	EEu	East Europe	Oc_L	Oceania, LDC
EAs	East Asia	EEu_O	East Europe, OECD		

concept-regions: the higher the variability of the symbolic description of a region, the larger is the dimension of the rectangle; the length of the interval (a side of the rectangle) along an axis is also proportional to the variability of those symbolic variables in the description of the individual that have contributed more to the axis. This kind of factorial representation of symbolic data by a rectangle has been largely proposed in several factorial methods on symbolic data, especially in PCA (see Cazes et al., 1997).

In our GCA approach, having transformed the different symbolic variables in a table of numerical coding, the combination of the rows that described each individual can be geometrically interpreted as the hypercubes coordinated in a multidimensional space, so to justify the choice to represent symbolic data on factorial plans by rectangles (for different representations by convex hulls proposed in this context of analysis, see Verde, 1997; Lauro et al., 2008b).

According to this interpretation of the rectangle representation, along the first axis we notice (Figure 16.3) the opposition between East Asia (EAs), European regions (West Europe, OECD (WEu_O); North Europe, OECD (NEu_O); Central Europe (CE); East Europe, OECD (EEu_O)), North America (NAm) and Asia regions (Central Asia, LDC (CAs_L); Southest Asia (SEAs)), Caribbean, LDC (Ca_L), and Africa Sub-Saharan, LDC (SSA_L). It is interesting to observe the contraposition of regions constituted by countries belonging to OECD from these regions of low developing countries (LDCs). In contrast,

TABLE 16.2
Selected Variables, Type (H, Histogram; I, Interval), Categories, Short Labels

Symbolic Variables	Type	Category Labels	Short Labels
Population 2009	H	Pop < 20M, Pop 20–100M, Pop 100–270M, Pop > 270M	Pop1, Pop2, Pop3, Pop4
GDP per capita	H	GDPp <4, GDPp 4–13, GDPp 13–26, GDPp >26	GDPp1, GDPp2, GDPp3, GDPp4
HDI value	H	HDIv <0.5, HDIv 0.5–0.7, HDIv 0.7–0.8, HDIv >0.8	HDIv1, HDIv2, HDIv3, HDIv4
Mean years of schooling	H	MYSc <6, MYSc 6–8.5, MYSc 8.5–10, MYSc >10	MYSc1, MYSc2, MYSc3, MYSc4
Inflation GDP deflator	H	IGDP <−0.9, IGDP 0.9–3.5, IGDP 3.5–5.0, IGDP >5.0	IGDP1, IGDP2, IGDP3, IGDP4
Employment ratio	H	Empl <51.5, Empl 51.5–57.5, Empl 57.5–63.5, Empl >63.5	Empl1, Empl2, Empl3, Empl4
GDP growth	H	GDPg <−3, GDPg 3–0, GDPg 0–2, GDPg >2	GDPg1, GDPg2, GDPg3, GDPg4
Industry value added	H	InVA −22, InVA <22–27, InVA <27–32, InVA >32	InVA1, InVA2, InVA3, InVA4
Agriculture value added	H	AgVA <3, AgVA <3–6, AgVA 6–21.0, AgVA >21.0	AgVA1, AgVA2, AgVA3, AgVA4
Labour participation rate	H	Labr <60, Labr 60–65, Labr 65–70, Labr >70	Labr1, Labr2, Labr3, Labr4
Mortality rate under five years	H	Mort <7, Mort 7–20, Mort 20–55, Mort >55	Mort1, Mort2, Mort3, Mort4
Fertility rate total	H	Fert <1.8, Fert 1.8–2.3, Fert 2.3–3.4, Fert >3.4	Fert1, Fert2, Fert3, Fert4
Enrolment ratio in education	I	EnEd-LOW, EnEd-MED, EnEd-HI	EnEd_L, EnEd_M, EnEd_H
GNI per capita	I	GNIp-LOW, GNIp-MED, GNIp-HI	GNIp_L, GNIp_M, GNIp_H
Human Development Index value	I	HDIv-LOW, HDIv-MED, HDIv-HI	HDIv_L, HDIv_M, HDIv_H
Inequality-adjusted education	I	IHDe-LOW, IHDe-MED, IHDe-HI	IHDe_L, IHDe_M, IHDe_H
Life expectancy at birth	I	Life-LOW, Life-MED, Life-HI	Life_L, Life_M, Life_H
Secondary education	I	SEdu-LOW, SEdu-MED, SEdu-HI	SEdu_L, SEdu_M, SEdu_H

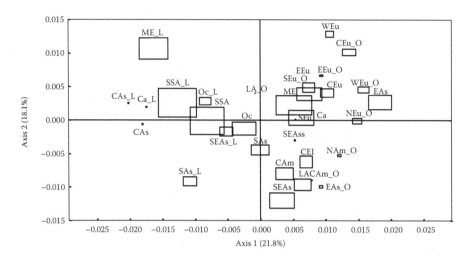

FIGURE 16.3
Graphical representation of the 35 concept-regions on the first factorial plan (39.9% of explained inertia).

along the second factorial axis West Europe (WEu), Central Europe (CEu) (in the first quadrant), and Middle East, LDC (ME_L) are opposed to South America (SAm), Latin America (LA), and Southeast Asia, LCD (SEAs_L).

The rectangles are the projections of the hypercube representations. Their area is proportional to the variability of the regions according to the histogram variables and the categorized interval ones especially. Some hypercubes collapse to a point (such as Caribbean, LDC (Ca_L); Central Asia (CAs) and Central Asia_L (CAs_L); Southeast Asia (SEAs); Central America, OECD (CAm_O), and Latin America, OECD (LA_O)) that means their description does not present variability for the interval variables, since the lower and upper bounds are coincident. The histogram categories and the pseudocategories of the coded interval variables, selected in the analysis, are visualized on the first factorial plan (Figure 16.4; for variable descriptions and abbreviations, see Table 16.2).

Following the trajectory of the categories of each variable, it is evident to observe the nonlinearity, with 'crossing' (see 'mean years of schooling' categories) and 'horseshoes' (see 'fertility rate total' categories) effects in the connection of the ordinal categories, due to the fuzzy coding of data.

Moreover, it is worth observing that according to the opposition of OECD from LDC regions along the first factorial axis and considering the nonlinearity of the trajectories of the categories of both the histogram and interval variables represented on the factorial plane, we prefer to interpret their association in each of four quadrants. Especially the top right quadrant is characterized by the high levels of mean years of schooling (MYSc4), the highest ratio of labour participation ratio (Labr4), the highest values of the GDP per

Symbolic Data Analysis: A Factorial Approach Based on Fuzzy Coded Data 269

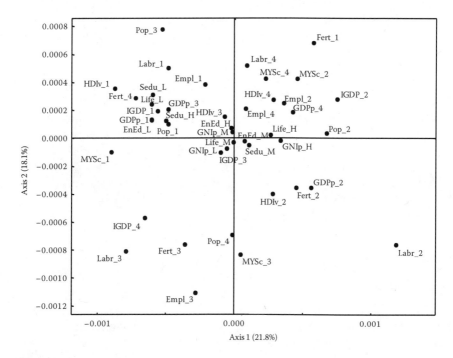

FIGURE 16.4
Graphical representation of the categories of the 18 coded variables on the first factorial plan (39.9% of explained inertia).

capita (GDPp4), of the employment ratio (Empl4), and of the low values of fertility (Fert1), high life expectancy at birth and high GNI per capita (GNIp_H), population contained between 20 and 100 million (Pop2), and globally, the highest HDI values (HDIv4). Those correspond to the characteristics of the OCDE regions, like European countries, as well as East Asia, corresponding to an area of great economic development, that are located in the same quadrant in the factorial representation of the regions.

The bottom right quadrant is characterized by the association of lower values of GDP per capita (GDPp2), higher fertility ratio (Fert2), labour participation rate between 60 and 65% (Labr2), and mean years of schooling quite high (MYSc3). Those characteristics are typical of North America (NAm), Central America, OECD (CAm_O), Latin America (LAm), EU, East Asia, OECD (EAs_O), and Southeast Asia (SEAs), observed on the same quadrant on the factorial plan of individual description.

The bottom left quadrant shows the association between the following categories: high levels of fertility (Fert4), of employment ratio (Empl3), of labour participation ratio (Labr3), and the highest level of IGDP (IGDP4), as well as by the lowest values of mean years of schooling (MYSc1) and the highest intensity of population (Pop4). These categories are characteristic of countries belonging to LCD of South Asia (SAs_L) and Southeast Asia (SEAs) regions.

The top left quadrant shows a high association among the categories mostly related to lowest HDI value (HDIv1); especially, we find the highest values of fertility (Fert4), the lowest values of labour participation ratio (Labr1), of the employment ratio (Empl1), of the GDP per capita (GDPp1) (but we also observe the category GDPp3), low levels of secondary education (SEdu_L), life expectation at birth (Life_L), and enrollment ratio in education (EnEd_L). We observe on the top high-intensity level of population (Pop3) as well as low-intensity population (Pop1) represented in the bottom. On the factorial individual representation we find in this quadrant Caribbean, LCD (Ca_L), and Africa Sub-Saharan (SSA) that especially explain the highest values of fertility in these regions, but not the high intensity of population that is a characteristic of Middle East, LCD (ME_L) and Central Asia, LCD (CA_L).

16.3 Conclusion

In this chapter we have proposed an extension of generalized canonical analysis (GCA) for symbolic multivalued variables (interval, categorical multivalued, modal). The single observations, symbolic data, represent concepts or classes, and they are characterized by a set of values for each variable, so they express an internal variability due to the complexity of their description. We have shown symbolic data can be represented by histograms, intervals of values, and sets of multicategories (with or without weights associated). Especially when symbolic data are described by continuous variables (interval), the internal variability of the data is retained and represented by hypercubes.

GCA is the most general factorial analysis approach. Traditionally, GCA is performed on several sets of quantitative variables observed on the same set of individuals, but this method was extended to the analysis of categorical variables with a disjunctive (0/1) coding. In this case, GCA generalizes the MCA on disjunctive tables (Lebart et al., 1995). Moreover, GCA can be applied to mixed (qualitative and categorical) sets of variables, homogenizing them in categorical ones by splitting numerical variables in intervals; however, in order to not lose the numerical information, B-spline functions are proposed as membership functions for a fuzzy coding of the data. Then, categorical variables are coded in disjunctive tables, whereas numerical categorized variables are set in fuzzy coding tables. The factorial reduction is so performed on the mixed (disjunctive and fuzzy) data tables.

According to this approach, the GCA has been extended to multivalued symbolic variables as we have detailed in this chapter. The results consist especially in a factorial representation of the variability of the individuals (concepts) on factorial plans. That has been realized through the projection of hypercubes in the reduced subspaces, and in their reconstruction as rectangles in 2D space.

17

Group Average Linkage Compared to Ward's Method in Hierarchical Clustering

Maurice Roux

CONTENTS

17.1 Group Average Linkage Algorithm (UPGMA) 272
17.2 Ward's Clustering Algorithm 277
17.3 Comparing the Algorithms 281
    17.3.1 Which Test Examples to Work Out 282
    17.3.2 How to Evaluate the Results 282
    17.3.3 Benchmark Test of Random Data Tables 283
    17.3.4 Results with UPGMA, Ward's, and Some Other Methods 283
    17.3.5 Weighting Considerations 284
17.4 Conclusion 285
Appendix: An Example of Inconsistent Aggregation with Ward's Method 286

The publication of Sokal and Sneath's book (1963) gave rise to the testing and use of the three basic agglomerative procedures in hierarchical clustering: single linkage, complete linkage, and average linkage. The third one was coined UPGMA for unweighted pair group method of aggregation (Sokal and Michener, 1958), *construction ascendante hiérarchique selon la distance moyenne* in French. In many practical cases it appeared useful to associate such methods with correspondence analysis (CA).

Benzécri suggested the use of the inertia as a criterion for joining two groups of observations (Benzécri et al., 1973, pp. 185–191). He proposed three variants of an agglomerative algorithm: (1) joining those two clusters that maximize the *moment d'ordre 2 d'une partition*, that is, the second-order moment of inertia, or between-cluster inertia; (2) joining those two clusters such that the inertia of the newly formed group is at a minimum; and (3) joining those two clusters such that the variance of the newly formed group is at a minimum. The last two methods differ by the weightings when recomputing the dissimilarities after an aggregation step in the algorithm. Later it appeared that the first variant was almost exactly the algorithm descripted by Ward (1963), which worked better than variants 2 and 3. The only difference with Ward's algorithm was that different weights had to be assigned to the objects being clustered.

From various comparisons it turned out that UPGMA was preferable to both single-linkage and complete linkage agglomerative methods (e.g., Benzécri et al., 1973, pp. 375–380). But good results were obtained with Ward's method as well. However, if one applies both methods to the same data set, the resulting hierarchical trees are usually different. Then the question is: Which method is the best? Trying to answer this question brings up a series of preliminary questions, such as how to evaluate a clustering method and whether we should evaluate the methods on mathematical grounds or on a practical bench test. Both approaches are discussed in the present chapter.

17.1 Group Average Linkage Algorithm (UPGMA)

Clustering is usually applied to multidimensional data, but for the sake of simplicity I start with an artificial one-dimensional example inspired from Benzécri et al. (1973, p. 200). The objects are called a, b, c, d, e, f, and g; their values are listed in Table 17.1 (column x_i), which is represented as a single axis (Figure 17.1).

TABLE 17.1

A Simple Data Set

Objects	x_i	$x_i - \bar{x}$	w_i	$w_i(x_i - \bar{x})^2$
a	0	−7	1	49
b	1	−6	1	36
c	4	−3	1	9
d	7	0	1	0
e	8	1	1	1
f	12	5	1	25
g	17	10	1	100
Total	49	0	7	220

Note: Column x_i: List of the coordinates on the single dimension; the mean value of these coordinates is $49/7 = 7$. Column $x_i - \bar{x}$: Deviations from the mean $\bar{x}$; w_i: weight. Column $(x_i - \bar{x})^2$: Weighted squared deviations from the mean. The total of this last column is called the second-order moment of inertia, or inertia, for short.

FIGURE 17.1
A simple one-dimensional data set.

The UPGMA starts with the computation of the distances between all the objects (Table 17.2). At each step the two closest objects or groups are joined to form one new group, and the distance table is updated according to the classical formula

$$d_{i \cup j, k} = \frac{w_i d_{ik} + w_j d_{jk}}{w_i + w_j} \qquad (17.1)$$

where i and j are the objects aggregated and w_i and w_j are their weights, $i \cup j$ is the newly formed group, and d is the distance between single objects or groups of objects. If i and j are single objects and are not differentially weighted, then their weights are equal to 1, and the weights of objects are the numbers of objects in a cluster. Here we assume equal weights.

In this example there are two lowest values of interobject distances: $d_{a,b} = 1$ and $d_{d,e} = 1$. In such a case the usage is to select the first one, or at random, but this choice may result in different clusterings if an object to be merged is common to both choices.

Let us start with the aggregation of a and b; the next step is to recompute the distances between the group {a, b} and the other objects, according to the above formula (17.1) (Table 17.3).

The next step is to aggregate points d and e, the mutual distance of which is again 1. This result is shown in Table 17.4. If this pair of objects would have been aggregated in the first step (before pair {a, b}), the resulting distance matrix would be the same. This is because {a, b} and {d, e} have no common elements, and the elements of each pair are nearest points to each other; they are called mutual nearest neighbours. This property may be profitably used

TABLE 17.2

Distance Matrix between the Seven Points a, b, c, d, e, f, g of Figure 17.1

	a	b	c	d	e	f	g
a	0	1	4	7	8	12	17
b	1	0	3	6	7	11	16
c	4	3	0	3	4	8	13
d	7	6	3	0	1	5	10
e	8	7	4	1	0	4	9
f	12	11	8	5	4	0	5
g	17	16	13	10	9	5	0

TABLE 17.3

After the Aggregation of Elements a and b Only the First Row and the First Column Are Modified to Represent the Mean Values of Distances between the Group {a, b} and the Other Elements

	{a, b}	c	d	e	f	g
{a, b}	0	3.5	6.5	7.5	11.5	16.5
c	3.5	0	3	4	8	13
d	6.5	3	0	1	5	10
e	7.5	4	1	0	4	9
f	11.5	8	5	4	0	5
g	16.5	13	10	9	5	0

TABLE 17.4

Aggregation of Elements d and e

	{a, b}	c	{d, e}	f	g
{a, b}	0	3.5	7	11.5	16.5
c	3.5	0	3.5	8	13
{d, e}	7	3.5	0	4.5	9.5
f	11.5	8	4.5	0	5
g	16.5	13	9.5	5	0

to accelerate the operation of this algorithm and Ward's algorithm as well (de Rham, 1980).

The next aggregation concerns the point c and the group {a, b} at distance 3.5, the lowest value in Table 17.4. Note that another aggregation would be possible: point c and group {d, e} with the same distance value. Continuing with the first aggregation leads to Table 17.5, where the value 5.833, measuring the distance between {a, b, c} and {d, e}, comes from the mean value of 7 and 3.5.

The distance of 7 between {a, b} and {d, e} should be weighted by a coefficient 2 because there are two elements in group {a, b}, while the distance 3.5 between c and {d, e} is to be weighted by a coefficient 1:

$$d_{\{ab\}\cup c,\{de\}} \frac{2\times d_{\{ab\},\{de\}} + 1\times d_{c,\{de\}}}{2+1} = 17.5/3 = 5.833$$

TABLE 17.5

Aggregation of Group {a, b} with Object c

	{a, b, c}	{d, e}	f	g
{a, b, c}	0	5.833	10.333	15.333
{d, e}	5.833	0	4.5	9.5
f	10.333	4.5	0	5
g	15.333	9.5	5	0

This distance recomputation is identical to averaging the six pairwise distances between the sets {a, b, c} and {d, e}: (7 + 8 + 6 + 7 + 3 + 4)/6 = 5.833. The next aggregation concerns groups {f} and {d, e} at distance 4.5, resulting in the distance matrix shown in Table 17.6.

Then the groups {a, b, c} and {d, e, f} are fused at distance 7.333, and the last aggregation is obviously the point g with the group {a, b, c, d, e, f} at distance 11.667. The whole process can be summed up in a single table (Table 17.7) and in a hierarchical tree, also called a dendrogram (Figure 17.2). In Table 17.7, the column 'nodes' represents the aggregations of groups or objects. The levels are the mean values of the intergroup distances. The columns 'oldest' and 'youngest' designate the two groups (or objects) aggregated, as if they were the children of the node. The column 'weights' includes the number of objects at the respective nodes. The initial objects (a, b, c, ...) are considered nodes as well, but without children (zero values in the corresponding columns).

A remarkable property of such a construction is that the weighted average of the levels is equal to the average of the initial distances. In the present case the sum of the initial distances is 154 (subdiagonal values in Table 17.2); therefore, the average distance is 154/21 = 7.3333. The weights to use in the computation of the level average correspond to the number of between-group

TABLE 17.6

Aggregation of Group {d, e} with f

	{a, b, c}	{d, e, f}	g
{a, b, c}	0	7.333	15.333
{d, e, f}	7.333	0	8
g	15.333	8	0

TABLE 17.7

UPGMA: Summary of the Successive Aggregation

	Nodes	Oldest	Youngest	Weights	Levels
a	1	0	0	1	0
b	2	0	0	1	0
c	3	0	0	1	0
d	4	0	0	1	0
e	5	0	0	1	0
f	6	0	0	1	0
g	7	0	0	1	0
{a, b}	8	1	2	2	1
{d, e}	9	4	5	2	1
{a, b, c}	10	8	3	3	3.5
{d, e, f}	11	6	9	3	4.5
{a, b, c, d, e, f}	12	10	11	6	7.333
{a, b, c, d, e, f, g}	13	12	7	7	11.667

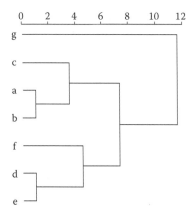

FIGURE 17.2
Dendrogram associated to the successive aggregations of the group average linkage.

distances. For instance, in node 8 there is only one between-group distance, but in node 12 there are nine between-group distances, since there are three elements in the oldest child and three elements in the youngest child. Finally, the average value of the levels results in the following computation:

$$(1 \times 1 + 1 \times 1 + 2 \times 3.5 + 2 \times 4.5 + 9 \times 7.333 + 6 \times 11.667)/$$
$$(1 + 1 + 2 + 2 + 9 + 6) = 7.333$$

It is worth noting that the term *unweighted* in the acronym UPGMA refers to the above computation: all distances weigh equally in the level values of the nodes.

Having obtained the dendrogram (Figure 17.2), one can reconstruct so-called ultrametric distances u_{ij}, where the distance between two objects is equal to the joining level of the branches of the objects (Figure 17.2 and Table 17.8). Such a matrix **U** has the following property, for any triple of elements $\{i, j, k\}$:

$$u_{ij} \leq \max\{ u_{ik}, u_{kj}\} \tag{17.2}$$

TABLE 17.8
Ultrametric Distances associated to the Average Linkage Dendrogram

	a	b	c	d	e	f	g
a	0	1	3.5	7.333	7.333	7.333	11.667
b	1	0	3.5	7.333	7.333	7.333	11.667
c	3.5	3.5	0	7.333	7.333	7.333	11.667
d	7.333	7.333	7.333	0	1	4.5	11.667
e	7.333	7.333	7.333	1	0	4.5	11.667
f	7.333	7.333	7.333	4.5	4.5	0	11.667
g	11.667	11.667	11.667	11.667	11.667	11.667	0

Group Average Linkage Compared to Ward's Method

This inequality is called *ultrametric*, and the distance itself is often called *ultrametric* for short. As a consequence of the ultrametric inequality, all triangles of distances u_{ij} are isosceles with the two equal sides greater than or equal to the third side.

Another interesting property of the average link construction is that, given the shape of the tree, the ultrametric distance is the best one according to the least-squares criterion (Benzécri et al., 1973, p. 181). This criterion is a way to evaluate the quality of the result by comparing the ultrametric distances with the initial distances:

$$\sum_i \sum_{j<1} (d_{ij} - u_{ij})^2 \qquad (17.3)$$

A relative value could be used by dividing each term with the initial value d_{ij} of the distance. The problem of evaluating the results is developed later, in Section 17.3.

17.2 Ward's Clustering Algorithm

Ward's method is based on a generalization of the usual analysis of variance equation:

$$TSS = WSS + BSS = \sum_g WSS_g \; BSS \qquad (17.4)$$

where *TSS* stands for total sum of squares, *WSS* for within-group sum of squares, WSS_g for within-group sum of squares relative to the group g, and *BSS* for between-group sum of squares.

The above equation means that the total sum of squares is equal to the sum, for all groups of a partition G, of the within-group sum of squares plus the between-group sum of squares. Consider a set I of individuals for which only one quantitative variable x is observed. Let $\bar{x}$ denote the mean of x over this sample and $\bar{x}_g$ denote the mean value of the variable x over the subset g of this sample. Then the analysis of variance equation may be rewritten in a more familiar way as

$$\sum_{i \in I} (x_i - \bar{x})^2 = \sum_{g \in G} \sum_{i \in g} w_i (x_i - \bar{x}_g)^2 + \sum_{g \in G} w_g (\bar{x}_g - \bar{x})^2 \qquad (17.5)$$

where w_g is the number of observations in group g. When there is more than one variable, say J variables, the symbol x_i becomes a vector x_i, that is, a set of J quantitative scores associated to each and every variable, and the mean value $\bar{x}$

is replaced by another vector, noted $\bar{x}$, the components of which are the means of the variables. The above equation still holds, provided that the sum of squares is replaced throughout by the sum of squared distances between vectors:

$$\sum_{i \in I} d^2(x_i, \bar{x}) = \sum_{g \in G} \sum_{i \in g} w_i \, d^2(x_i, \bar{x}_g) + \sum_{g \in G} w_g \, d^2(\bar{x}_g, \bar{x}) \quad (17.6)$$

In this equation it is assumed that the squared distances are evaluated according to a Euclidean formula. Let us denote by x_{ij} the value of the jth variable for the ith observation. Then, the jth component of the sample mean is

$$\bar{x}_j = \frac{1}{w} \sum_{i \in I} x_{ij} \quad (17.7)$$

where w is the total weight of the observations. Likewise, the group means $\bar{x}_g$ are vectors, the components of which are the group means of the variables:

$$\bar{x}_{gj} = \frac{1}{w_g} \sum_{i \in g} x_{ij} \quad (17.8)$$

With these notations the usual Euclidean formula for computing the distances is as follows (for squared distances):

$$d^2(x, y) = \sum_{j \in J} (x_j - y_j)^2 \quad (17.9)$$

Using this familiar Euclidean distance, the formula (17.6) is easy to prove. Formula (17.5) holds for each variable j. Then the sum of the similar formulas for all variables j of the set J leads to formula (17.6). Obviously, if there do exist some groups in the sample, these groups must show more internal homogeneity than the whole sample. For a given group of observations, the WSS value, also called the within-group inertia in the French literature, is a measure of heterogeneity. That is, the WSS values should be small, whereas the BSS value should be large, since the sum of both figures is the TSS, which does not depend on any partition. One or the other expression may serve as a criterion for judging the quality of the partition G.

Within the general framework of agglomerative clustering, we may apply the above considerations to the grouping of two clusters. Suppose that at some stage of the hierarchical construction we have two clusters called s and t, which are candidates to be merged into a cluster denoted by $s \cup t$, the fusion of these two groups. We may then apply formula (17.6) to $s \cup t$ ($\bar{x}_{s \cup t}$ being the centre of gravity of $s \cup t$):

$$\sum_{i \in s \cup t} w_i d^2(x_i, \bar{x}_{s \cup t}) = \sum_{i \in s} w_i d^2(x_i, \bar{x}_s) + \sum_{i \in t} w_i d^2(x_i, \bar{x}_t) + \ldots$$
$$+ w_s d^2(\bar{x}_s, \bar{x}_{s \cup t}) + w_t d^2(\bar{x}_t, \bar{x}_{s \cup t})$$

The first two terms on the right side of this equation are the WSS quantities for groups s and t. Should the two groups be aggregated, the sum on the left-hand side represents the WSS value for the newly formed group. In other words, the last two terms in the above equation,

$$\Delta WSS = w_s d^2(\bar{x}_s, \bar{x}_{s \cup t}) + w_t d^2(\bar{x}_t, \bar{x}_{s \cup t}) \quad (17.10)$$

represent the increase of the within-group sum of squares that would result if groups s and t were aggregated (hence the notation ΔWSS). The principle of Ward's method is to select for aggregation the pair made of the two clusters for which this increase is the lowest possible. Taking into account that $\bar{x}_{s \cup t}, \bar{x}_s$ and $\bar{x}_t$ are on a straight line $\bar{x}_{s \cup t}$, being at the centre of gravity of $\bar{x}_s$ and $\bar{x}_t$, there is another way of expressing (17.10):

$$\Delta WSS = \frac{w_s w_t}{w_s + w_t} d^2(\bar{x}_s, \bar{x}_t) \quad (17.11)$$

where $\bar{x}_{s \cup t}$, the centre of gravity of $s \cup t$, no longer appears. Therefore, one can slightly modify the UPGMA algorithm to do Ward clustering by storing the increases ΔWSS instead of the true distances. It can be shown that, after an aggregation of two clusters s and t, it is easy to recompute the new increases that would occur if aggregating the new cluster $s \cup t$ with any other cluster, or object, u. In this formula w (without subscript) stands for the sum $w_s + w_t + w_u$:

$$d^2(s \cup t, u) = \frac{w_s + w_u}{w} d^2(s, u) + \frac{w_t + w_u}{w} d^2(t, u) - \frac{w_u}{w} d^2(s, t) \quad (17.12)$$

In general, at the beginning of the agglomerative procedure, the groups are just single objects, associated with weights equal to 1. Therefore, the initial distances must be replaced by the quantities:

$$\Delta WSS(i, j) = \frac{1}{2} d_{ij}^2 \quad (17.13)$$

where the pseudodistance $\Delta WSS(i, j)$ is the increase of WSS when aggregating objects i and j.

In the next step we apply Ward's method to the above one-dimensional artificial data set. The initial pseudodistances in (17.13) are shown in Table 17.9,

TABLE 17.9

Initial Pseudodistances as Input in Ward's Procedure

	a	b	c	d	e	f	g
a	0	0.5	8	24.5	32	72	144.5
b	0.5	0	4.5	18	24.5	60.5	128
c	8	4.5	0	4.5	8	32	84.5
d	24.5	18	4.5	0	0.5	12.5	50
e	32	24.5	8	0.5	0	8	40.5
f	72	60.5	32	12.5	8	0	12.5
g	144.5	128	84.5	50	40.5	12.5	0

TABLE 17.10

Numerical Results of Ward's Agglomerative Procedure

	Nodes	Oldest	Youngest	Weights	Levels	Percent Inertia
a	1	0	0	1	0	
b	2	0	0	1	0	
c	3	0	0	1	0	
d	4	0	0	1	0	
e	5	0	0	1	0	
f	6	0	0	1	0	
g	7	0	0	1	0	
{a, b}	8	1	2	2	0.5	0.23
{d, e}	9	4	5	2	0.5	0.23
{a, b, c}	10	8	3	3	8.167	3.71
{f, g}	11	7	6	2	12.5	5.68
{a, b, c, d, e}	12	10	9	5	40.833	18.56
{a, b, c, d, e, f, g}	13	12	11	7	157.5	71.59
Total					220	100

and the numerical results are shown in Table 17.10. The dendrogram is shown in Figure 17.3, the main feature of which is that the clusters appear sharply separated. But this is mostly due to an artefact: the node levels are the increases ΔWSS of the within-group sum of squares, which are weighted squared distances. This leads to a flattening of the low levels while stretching the highest ones. Thus, these levels cannot be compared with those of the UPGMA hierarchy, which are just average distances.

However, an interesting property of this method is that the sum of the level values is equal to the whole inertia, which is the total sum of squared distances (*TSS*). In the present example this sum of the levels values is 0.5 + 0.5 + 8.167 + 12.5 + 40.833 + 157.5 = 220, which is the value of *TSS* obtained by direct calculation in the last column of Table 17.1. This property allows for the computation of percentages of inertia associated with the nodes (Benzécri et

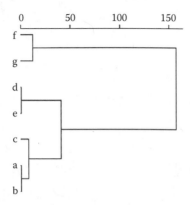

FIGURE 17.3
Dendrogram resulting from Ward's method.

al., 1973, p. 188). In the present example these percentages may be added as an additional column of Table 17.10. Of course, the main contributions to the inertia are given by the upper nodes (17.12 and 17.13).

Both group average linkage and Ward's method operate on distance matrices. It is an advantage of Ward's method to offer the possibility of treating the initial observations-by-variables data table, which may be smaller than the distance matrix. However, this option supposes that all the variables are quantitative. The corresponding version of the program is based on formula (17.11), which operates on the centroïds of the clusters (Benzécri et al., 1973, p. 227; Jambu, 1978, chap. 3).

A classical requirement for a good clustering algorithm is that, applied to an ultrametric distance, it should result in the same ultrametric, or at least with a dendrogram showing the same shape (Fisher and van Ness, 1971). Unfortunately, Ward's method does not meet this apparently logical requirement. The above calculations involve a Euclidean framework, but some researchers have given mathematical support to extend the validity of Ward's formulas, in such a way the algorithm may be applied to any given dissimilarity (Batagelj, 1988; Székely and Rizzo, 2005). Murtagh and Legendre (2011) detailed these variants and the related software packages.

17.3 Comparing the Algorithms

It should be clear from the previous sections that both UPGMA and Ward's method are based on reasonable properties of their corresponding algorithms. Since none of these algorithms have a mathematical superiority over the other, only practical experience can enlighten us as to which may be better. Then two new questions arise: Which data sets should be used for comparison, and which criterion is appropriate for evaluating the results of the algorithms?

17.3.1 Which Test Examples to Work Out

The first idea, which has widely been used before, is to create artificial data sets with a known partition of the objects into a few clusters. High-dimensional data sets can be generated, with or without a real cluster structure, to use as test examples. I followed this idea by computing distance matrices from random data tables. This allows for creating benchmarks as large as needed.

17.3.2 How to Evaluate the Results

The results of a hierarchical clustering can be transformed into an ultrametric distance matrix $\mathbf{U}$ where each value u_{ij} represents the distance between objects i and j, as indicated by the dendrogram (see the end of Section 17.1). Due to the ultrametric constraint, these values differ from the initial distance d_{ij}. Then the evaluation of the methods can be based on the deviation between u_{ij} and d_{ij}. The least-squares criterion (17.3) can be normalized in order to compare the results of various data sets; this leads to the stress criterion currently used in multidimensional scaling (see also Chapter 7 from Borg and Groenen in this book):

$$\text{Stress} = \frac{\sum_i \sum_{j<i} (d_{ij} - u_{ij})^2}{\sum_i \sum_{j<i} d_{ij}^2}$$

However, neither the least-squares criterion nor stress works for evaluating Ward's results because the node levels are not true distances but rather weighted squared distances. Another classical evaluation measure is the so-called cophenetic correlation coefficient (Sokal and Sneath, 1963), an extension of the usual correlation coefficient. Unfortunately, this coefficient has been rightly criticized (Farris, 1969; Gordon, 1996). Finally, I turned to the Goodman–Kruskal coefficient (Goodman and Kruskal, 1954), defined as

$$\gamma = (S^+ - S^-)/(S^+ + S^-)$$

where S^+ (respectively S^-) is the number of concordant (respectively discordant) pairs $\{(d_{ij}, d_{kl}), (u_{ij}, u_{kl})\}$. Such a double pair is said to be concordant if the following logical expression is true:

$$[d_{ij} < d_{kl} \text{ and } u_{ij} < u_{kl}] \text{ or } [d_{ij} > d_{kl} \text{ and } u_{ij} > u_{kl}]$$

and is discordant if

$$[d_{ij} < d_{kl} \text{ and } u_{ij} > u_{kl}] \text{ or } [d_{ij} > d_{kl} \text{ and } u_{ij} < u_{kl}]$$

In words, if the ordering of dissimilarities d_{ij} and d_{kl} is identical to the ordering of ultrametric distances u_{ij} and u_{kl}, then the quadruple is concordant; it is discordant in the opposite case. Only the orders are taken into account in this formula, so that this coefficient uses only the shape, or topology, of the dendrogram, rather than the ultrametric distances. Ties are not taken into account and are simply omitted from the counting. As it is akin to a correlation coefficient, a high value indicates a good hierarchy.

17.3.3 Benchmark Test of Random Data Tables

The random tables consist of 40 observations and four variables. These variables result from independent uniform drawings over the interval [0, 100], so that there are a priori no groups of observations. The complete bench test consists of a hundred 40 × 40 distance matrices.

17.3.4 Results with UPGMA, Ward's, and Some Other Methods

Table 17.11 shows the γ values for the two methods in comparison together with the γ values of some other popular aggregative construction of hierarchies, namely, the single-linkage and complete linkage algorithms (respectively *agrégation selon le saut minimum* and *agrégation par le diamètre* in French). These last two methods are given as reference points.

In Table 17.11 UPGMA appears as slightly better than Ward's method, with both being better than single linkage and complete linkage. In addition, these last two methods show a greater dispersion of their results, as evaluated by the standard deviation.

For a better and fairer comparison I constructed a scatterplot (Figure 17.4) where each data set is located by its UPGMA and Ward's γ values. If all criterion values were equal, then all points would be on the diagonal of this graph. Points above the diagonal indicate better solutions for the UPGMA method, while points under the diagonal correspond to better solutions for

TABLE 17.11

Mean Values and Standard Deviations for the GK Criterion and Four Aggregation Methods

	UPGMA	Ward's	Single Linkage	Complete Linkage
Mean values	0.4808	0.4467	0.3375	0.4082
Standard deviation	0.0477	0.0579	0.0669	0.0662

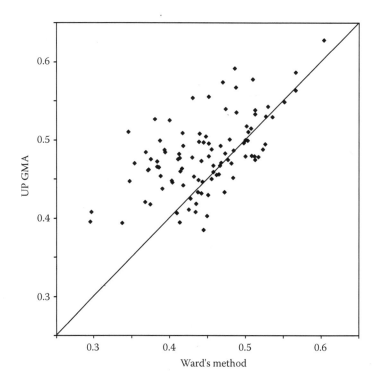

FIGURE 17.4
Scatterplot of the values of γ criterion for the two hierarchical algorithms. Each point represents one data set, located according to its core for both algorithms.

Ward's method. Figure 17.4 shows that UPGMA provides a better solution more often than Ward's does.

More precisely, UPGMA has a better γ value for 70 cases, out of 100, while Ward's has a better γ value for 30 cases. None of the cases shows exactly the same γ values for both methods. My opinion is that the weakness of Ward's method comes from the aggregation criterion based on the within-group sum of squared distances (formulas (17.10) and (17.11)). It tends to build up balanced groups (see appendix at the end of the chapter for a mathematical approach). But such balanced clusters are not frequent in real data sets.

17.3.5 Weighting Considerations

With some data sets it may happen that, from the outset, the observations have natural weights. For instance, an observation may result from repeated experiences, and shows the mean values obtained during these experiences. In such a case it is natural to weight the observations with the number of repetitions. Another similar case would be when a preliminary clustering with k-means, or another partitioning method, was performed prior to the hierarchical algorithm, in order to reduce either the data size or the computing time.

Both UPGMA and Ward's method are able to handle such initial weights. In the usual case the table of weights is initialized with unit values. If the observations are weighted differently, then the weight table should be initialized with the given weights. In addition, in Ward's method, the pseudo-distances should be initialized with a modified version of formula (17.13):

$$\Delta WSS(i,j) = \frac{w_i w_j}{w_i + w_j} d_{ij}^2 \qquad (17.14)$$

where w_i and w_j are the given weights of observations i and j. The result is used by Greenacre (1988b) for clustering the rows and columns of a contingency table.

17.4 Conclusion

I have compared the two classical clustering algorithms, UPGMA and Ward's method, first from a theoretical point of view, and second on a practical bench test. On theoretical grounds these two approaches are based on quite different ideas. UPGMA relies upon the mean values of the between-group distances, while Ward's method is based on the compactness of the groups, represented by their inertia (or generalized variance).

It follows that their hierarchical representations, the dendrograms, have different properties. In the UPGMA, the node levels of the tree are between-group average dissimilarities, while in Ward's method these levels represent the increase of within-group inertia. In UPGMA the weighted average of all the levels is equal to the average of the given dissimilarities. With Ward's method the sum of all levels is equal to the inertia of the whole set of points.

To evaluate the results of both algorithms on the same data sets, I chose the Goodman-Kruskal coefficient. Taking solely into account the ranks of the dissimilarities, this coefficient seems adapted to the problem where the node levels are not comparable between the two algorithms. With the artificial benchmark test of distance matrices based on random data sets UPGMA outperforms Ward's method in 70% of cases. For the other 30 cases out of 100 Ward's is better than UPGMA. In such a situation my recommendation, when dealing with a particular data set, is to apply both algorithms, evaluate the results, and then retain the dendrogram associated with the best evaluation.

Appendix: An Example of Inconsistent Aggregation with Ward's Method

Here is an artificial construction of a data set in order to show how Ward's method behaves in a counterintuitive manner when the clusters are different in sizes.

Suppose the data set is made of three rectangular subsets called A, B, and C. The height of A and B is equal to $2a$. Subset C is a square of dimension $a \times a$. The width of subset B is $2ax$, while the width of subset A is $2a - 2ax = 2a(1 - x)$. Subsets A and B are contiguous, while there is a gap, of length $a/2$, between B and C (Figure 17.5). The density of all parts is supposed to be homogeneous and equal to 1; therefore, the weights of these parts are equal to their area:

$$w_A = 2a \times 2a(1 - x) = 4a^2(1 - x); \quad w_B = 2a \times 2ax = 4a^2x; \quad w_C = a^2$$

The distances d_{AB} and d_{BC} between the centroïds of A and B and between the centroïds of B and C are

$$d_{AB} = a; \quad d_{BC} = ax + a = a(x + 1)$$

Let $f(A \cup B)$ be the pseudodistance between subsets A and B

$$f(A \cup B) = \frac{w_A w_B}{(w_A + w_B)} d_{AB}^2 = \frac{4a^2(1-x)4a_x^2}{4a^2}$$

$$f(B \cup C) = \frac{w_B w_C}{(w_B + w_C)} d_{BC}^2 = \frac{4a^2 \cdot a^2}{(4a_x^2 + a^2)} a^2(x+1)^2 = \frac{4a^4 x}{(4x+1)}(x+1)^2$$

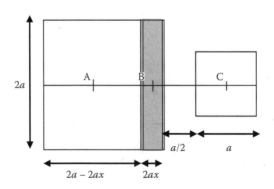

FIGURE 17.5
A three-part data set.

Is it possible to have $f(B \cup C) < f(A \cup B)$? The answer necessitates the solution of

$$\frac{4a^4 x}{(4x+1)}(x+1)^2 < 4a^4 x(1-x)$$

which reduces to

$$(x+1)^2 < (1-x)(4x+1) \Leftrightarrow x^2 + 2x + 1 < -4x^2 + 3x + 1 \Leftrightarrow$$
$$5x^2 - x < 0 \Leftrightarrow x(5x-1) < 0$$

If $0 < x < 0.2$, then Ward's method would aggregate B and C, though the aggregation of A and B seems to be more sensible.

18

Analysing a Pair of Tables: Coinertia Analysis and Duality Diagrams

Stéphane Dray

CONTENTS

18.1 The Duality Diagram 290
    18.1.1 Definition 290
    18.1.2 Properties 293
18.2 Playing with Correspondence Analysis 294
18.3 Relating Two Diagrams 297

In many fields (e.g., ecology, psychometrics, social science, and marketing), researchers are faced with the challenge of summarizing the information contained in large data sets. In this context, multivariate analysis provides efficient tools for identifying the relationships between variables and the similarities between statistical units/individuals. Due to the natural boundaries between disciplines or schools of thought, several multivariate methods have been invented and reinvented by different groups in different countries for different purposes. This situation has resulted in a variety of apparently different methods that actually lead to the same equations for analysing the same data. For instance, Greenacre (1984, Section 1.3) detailed the history of correspondence analysis (CA) and showed how this method has been rediscovered several times in biometrics, psychometrics, and linguistics. This process can be explained by the diversity of viewpoints adopted by researchers to describe a method (e.g., geometrical versus numerical or individual centred versus variable centred). Several authors have tried to provide a unifying mathematical framework to summarize the different properties of a given method, and thus to identify analogies between existing methods. The *duality diagram* theory was first presented in Cazes (1970) and popularized by Cailliez and Pagès (1976) in a French book entitled *Introduction à l'Analyse des Données*. Several French authors adopted this theory, but I believe that it remains poorly known by statisticians outside France. Daniel Chessel, my PhD advisor, used the duality diagram as a formal way to develop new multivariate methods in ecology (e.g., Dolédec and Chessel, 1994; Dolédec et al., 1996). He implemented this framework in the ADE-4 software (Thioulouse et al., 1997) and several years later in the R package ade4 (Chessel et al., 2004; Dray and Dufour, 2007; Dray et al., 2007). Hence, similarly to Obelix (Goscinny and Uderzo, 1989), I fell into the magical duality diagram when I was a little boy, and I have used it as a central framework in my further works.

Seven years after Cailliez and Pagès's book came out, Ramsay and de Leeuw (1983) wrote a dithyrambic review and concluded that they 'hope it will not be long before an English language counterpart appears'. The book has never been translated into English, which is probably a major reason for its low impact on non-French readers. This lack is partially addressed by Escoufier (1987), Holmes (2006), and Dray and Dufour (2007), who provided a general overview of the duality diagram theory in English. More recently, a special section on modern multivariate analysis published in the *Annals of Applied Statistics* demonstrated the power of the duality diagram approach for analysing data of different formats (De la Cruz and Holmes, 2011), including spatial (Dray and Jombart, 2011), temporal (Thioulouse, 2011), or phylogenetic (Purdom, 2011) information. To date, the most convincing application of the duality diagram is probably that found in Tenenhaus and Young (1985), which provided an overview of multiple correspondence analysis and related methods to quantify categorical multivariate data.

This chapter presents the duality diagram theory and its application to the analysis of a contingency table by correspondence analysis. Subsequently, I show how the framework can be generalized to the analysis of a pair of tables focusing on coinertia analysis (Dolédec and Chessel, 1994), and I conclude with several extensions.

18.1 The Duality Diagram

18.1.1 Definition

Let **X** be a matrix containing data for p variables (columns) collected from n individuals (rows). From a geometrical viewpoint, we can consider this information either as p points (the variables $\mathbf{x}^1, \cdots, \mathbf{x}^p$) in $\mathbb{R}^n$ or as n points (the individuals $\mathbf{x}_1, \cdots, \mathbf{x}_n$) in $\mathbb{R}^p$. These two viewpoints suggest two related objectives:

- The comparison of the variables. To conduct this study, it is necessary to define $\mathbf{D}$, an $n \times n$ positive symmetric matrix used as an inner product in $\mathbb{R}^n$ allowing the computation of relationships between the p variables.
- The comparison of individuals. In this case, a $p \times p$ positive symmetric matrix $\mathbf{Q}$ used as an inner product in $\mathbb{R}^p$ allows the quantification of the resemblances (distances) between the n individuals.

Multivariate analysis considers both questions simultaneously and leads to the definition of the triplet $(\mathbf{X}, \mathbf{Q}, \mathbf{D})$ represented in the following diagram:

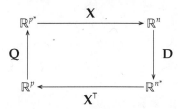

In this diagram, we can see that four spaces are associated to the data table:

- The individual space $\mathbb{R}^p$ which contains the n individuals;
- The variable space $(\mathbb{R}^n)$, which contains the p variables.
- The variable coefficient space $(\mathbb{R}^{p*})$: An element $\mathbf{g} = \begin{bmatrix} g_1, \cdots, g_p \end{bmatrix}^T$ of $\mathbb{R}^{p*}$ is used to define a new synthetic variable

$$\sum_{i=1}^{p} g_i \mathbf{x}^i.$$

It is the space of linear functions on $\mathbb{R}^p$ and may be considered as the dual space of $\mathbb{R}^p$.
- The individual coefficient space $(\mathbb{R}^{n*})$: An element

$$\mathbf{f} = \begin{bmatrix} f_1, \cdots, f_n \end{bmatrix}^T \text{ of } \mathbb{R}^{n*} \text{ is used to define a new individual}$$

$$\sum_{i=1}^{n} f_i \mathbf{x}_i.$$

It is the dual space of $\mathbb{R}^n$.

If a researcher is mainly interested in the first objective (relationships between variables), the analysis of the diagram in $\mathbb{R}^n$ consists of the eigendecomposition of $\mathbf{XQX^TD}$:

$$\mathbf{XQX^TDA} = \mathbf{A}\Lambda_r \text{ and } \mathbf{A^TDA} = \mathbf{I}_r$$

The r nonzero eigenvalues $\lambda_1 > \lambda_2 > \cdots > \lambda_r > 0$ are stored in the diagonal matrix Λ, and $\mathbf{A} = \begin{bmatrix} \mathbf{a}^1, \cdots, \mathbf{a}^r \end{bmatrix}$ is an $n \times r$ matrix containing the associated eigenvectors (in columns). These eigenvectors are typically known as the *principal components* onto which the columns of $\mathbf{X}$ are projected to obtain scores for the variables ($\mathbf{C} = \mathbf{X^TDA}$).

In contrast, if the study aims to compare individuals, the analysis of the diagram in $\mathbb{R}^p$ consists of the eigendecomposition of $\mathbf{X^TDXQ}$. Left multiplying the previous equation by $\mathbf{X^TD}$ leads to the following:

$$(\mathbf{X^TD})\mathbf{XQX^TDA} = (\mathbf{X^TD})\mathbf{A}\Lambda_r$$

If $\mathbf{B} = \mathbf{X^TDA}\Lambda_r^{-1/2}$, we obtain:

$$\mathbf{X^TDXQB} = \mathbf{B}\Lambda_r \text{ and } \mathbf{B^TQB} = \mathbf{I}_r$$

The $p \times r$ matrix $\mathbf{B} = [\mathbf{b}^1, \cdots, \mathbf{b}^r]$ contains eigenvectors (in the columns) that are usually known as the *principal axes*. The rows of $\mathbf{X}$ are then projected onto the principal axes to produce scores for the individuals ($\mathbf{L} = \mathbf{XQB}$).

From this diagram, we can define two other operators, $\mathbf{QX^TDX}$ and $\mathbf{DXQX^T}$, that can be diagonalized in $\mathbb{R}^{p^*}$ and $\mathbb{R}^{n^*}$, respectively. These decompositions produce the same eigenvalues, and the associated eigenvectors are the *principal factors* ($\mathbf{G}$) and the *principal cofactors* ($\mathbf{F}$), respectively. There are several close relationships between the four eigendecompositions; therefore, only one system of axes is required to compute the three others. For instance, we have the following transition formulas:

$$\mathbf{G} = \mathbf{QB}, \mathbf{A} = \mathbf{XG}\Lambda_r^{-(1/2)}, \mathbf{F} = \mathbf{DA} \text{ and } \mathbf{B} = \mathbf{X^TF}\Lambda_r^{-(1/2)}$$

Using these transition formulas, the product $\mathbf{A}\Lambda_r^{1/2}\mathbf{B^T}$ can be rewritten as

$$\mathbf{A}\Lambda_r^{1/2}\mathbf{B^T} = \mathbf{AA^TDX}$$

Left multiplication by $\mathbf{A^TD}$ leads to

$$\mathbf{A^TDA}\Lambda_r^{1/2}\mathbf{B^T} = \mathbf{A^TDX}$$

$$\mathbf{A}\Lambda_r^{1/2}\mathbf{B^T} = \mathbf{X}$$

The diagonalization of a duality diagram is thus similar to the generalized singular value decomposition of **X** (Eckart and Young, 1936). The singular values are contained in the diagonal matrix $\Lambda_r^{1/2}$, and the singular vectors stored in the matrices **A** and **B** are orthonormalized with respect to **D** and **Q**, respectively ($\mathbf{A}^T\mathbf{DA} = \mathbf{B}^T\mathbf{QB} = \mathbf{I}_r$).

18.1.2 Properties

There are several properties linked to the diagonalization of a duality diagram:

- The vectors $\mathbf{a}^1, \mathbf{a}^2, \ldots, \mathbf{a}^r$ successively maximize, under the **D**-orthogonality constraint, the quadratic form $\|\mathbf{X}^T\mathbf{Da}\|_Q^2$.
- The vectors $\mathbf{b}^1, \mathbf{b}^2, \ldots, \mathbf{b}^r$ successively maximize, under the **Q**-orthogonality constraint, the quadratic form $\|\mathbf{XQb}\|_D^2$.
- The vectors $\mathbf{f}^1, \mathbf{f}^2, \ldots, \mathbf{f}^r$ successively maximize, under the $\mathbf{D}^{-1}$-orthogonality constraint, the quadratic form $\|\mathbf{X}^T\mathbf{f}\|_Q^2$.
- The vectors $\mathbf{g}^1, \mathbf{g}^2, \ldots, \mathbf{g}^r$ successively maximize, under the $\mathbf{Q}^{-1}$-orthogonality constraint, the quadratic form $\|\mathbf{Xg}\|_D^2$.
- If we search for a pair of vectors **b** (a **Q**-normalized vector of $\mathbb{R}^p$) and **a** (a **D**-normalized vector of $\mathbb{R}^n$) that maximize the inner product $\langle \mathbf{XQb}|\mathbf{a}\rangle_D = \langle \mathbf{X}^T\mathbf{Da}|\mathbf{b}\rangle_Q$, the solution is unique. It is obtained for $\mathbf{b} = \mathbf{b}^1$ and $\mathbf{a} = \mathbf{a}^1$, and the maximum is equal to $\sqrt{\lambda_1}$. Under the orthogonality constraint, these results can be extended for the other pairs.

If **D** is diagonal, we can compute the total inertia for the cloud of row vectors (in $\mathbb{R}^p$) as follows:

$$\text{inertia}(\mathbf{X},\mathbf{Q},\mathbf{D}) = \sum_{i=1}^{n} d_{ii}\|\mathbf{x}_i\|_Q^2 = \text{trace}(\mathbf{XQX}^T\mathbf{D}) = \sum_{i=1}^{r} \lambda_i$$

where d_{ij} is the element in the ith row and jth column of **D**, and $\mathbf{x}_i$ is the ith row of matrix **X**. The rows of **X** can be projected onto a **Q**-normalized vector **b**, and the projected inertia is then equal to

$$\text{inertia}(\mathbf{b}) = \mathbf{b}^T\mathbf{QX}^T\mathbf{DXQb} = \|\mathbf{XQb}\|_D^2$$

Hence, it appears that the diagonalization of the diagram consists of identifying a set of **Q**-normalized vectors (the principal axes) that maximize the projected inertia. The inertia projected onto the principal axis $\mathbf{b}^k$ is equal to λ_k.

18.2 Playing with Correspondence Analysis

The duality diagram is very general, which enables each analysis to be defined as a particular choice for matrices **X**, **Q**, and **D**. To illustrate its use, I consider the case of the correspondence analysis of an $n \times p$ contingency table $\mathbf{N} = [n_{ij}]$, where n_{ij} is the count for the ith row and jth column. From the correspondence matrix $\mathbf{P} = \mathbf{N}/n_{++}$ (where n_{++} is the grand total of the contingency table), two vectors $\mathbf{r} = \mathbf{P1}_p$ ($n \times 1$) and $\mathbf{c} = \mathbf{P}^T\mathbf{1}_n$ ($p \times 1$) of row and column masses are derived. The diagonal matrices of the row and column weights are

$$\mathbf{D}_r = diag(\mathbf{r}) \text{ and } \mathbf{D}_c = diag(\mathbf{c})$$

Lastly, the matrix **P** is doubly centred, such that $\mathbf{P}_0 = \mathbf{P} - \mathbf{D}_r\mathbf{1}_n\mathbf{1}_p^T\mathbf{D}_c$. Correspondence analysis is the analysis of the triplet $(\mathbf{D}_r^{-1}\mathbf{P}_0\mathbf{D}_c^{-1}, \mathbf{D}_c, \mathbf{D}_r)$, and the associated diagram is

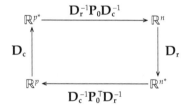

This diagram is equivalent to

$$\mathbb{R}^{p^*} \xrightarrow{\mathbf{D}_c^{-1}} \mathbb{R}^{p^*} \xrightarrow{\mathbf{P}_0} \mathbb{R}^n \xrightarrow{\mathbf{D}_r^{-1}} \mathbb{R}^n$$

with $\mathbf{D}_c$ on the left and $\mathbf{D}_r$ on the right, and

$$\mathbb{R}^p \xleftarrow{\mathbf{D}_c^{-1}} \mathbb{R}^p \xleftarrow{\mathbf{P}_0^T} \mathbb{R}^{n^*} \xleftarrow{\mathbf{D}_r^{-1}} \mathbb{R}^{n^*}$$

Applying the general formulas of the duality diagram to the CA triplet allows the definition of several properties. This analysis searches for a principal axis **b** maximizing

$$\left\| \mathbf{D}_r^{-1}\mathbf{P}_0\mathbf{D}_c^{-1}\mathbf{D}_c\mathbf{b} \right\|_{\mathbf{D}_r}^2 = \left\| \mathbf{D}_r^{-1}\mathbf{P}_0\mathbf{b} \right\|_{\mathbf{D}_r}^2$$

The matrix $\mathbf{D}_r^{-1}\mathbf{P}_0$ contains the centred row profiles such that the product $\mathbf{D}_r^{-1}\mathbf{P}_0\mathbf{b}$ places rows at the barycentres (weighted averages) of the column points, and thus the quantity maximized is simply a variance between rows. Hence, in $\mathbb{R}^p$, columns have a unit variance score **b** that maximizes the variance between the row barycentres. In $\mathbb{R}^n$, CA searches for a principal axis **a** maximizing

$$\left\| \mathbf{D}_c^{-1} \mathbf{P}_0^T \mathbf{D}_r^{-1} \mathbf{D}_r \mathbf{a} \right\|_{\mathbf{D}_c}^2 = \left\| \mathbf{D}_c^{-1} \mathbf{P}_0^T \mathbf{a} \right\|_{\mathbf{D}_c}^2$$

By symmetry, the matrix $\mathbf{D}_c^{-1} \mathbf{P}_0^T$ contains the centred column profiles such that the product $\mathbf{D}_c^{-1} \mathbf{P}_0^T \mathbf{a}$ places columns at the barycentres (weighted averages) of the row points (**a**). Hence, the rows are placed by a unit variance score **a** that maximizes the variance between column barycentres ($\left\| \mathbf{D}_c^{-1} \mathbf{P}_0^T \mathbf{a} \right\|_{\mathbf{D}_c}^2$). It can be demonstrated (see, e.g., Greenacre, 1984, p. 92) that replacing $\mathbf{P}_0$ with $\mathbf{P}$ produces the same results, except that one trivial dimension with an eigenvalue equal to one is produced.

These two viewpoints show that CA treats the rows and columns of the contingency table simultaneously and in a symmetric manner. Hence, analysing $\mathbf{N}$ or $\mathbf{N}^T$ produces the same results. Manipulating the original duality diagram allows the highlighting of two different geometrical interpretations for rows and columns. We can rewrite the CA diagram and thus obtain a new statistical triplet (represented by the dashed rectangle):

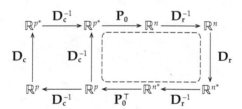

Hence, CA also corresponds to the triplet $(\mathbf{D}_r^{-1} \mathbf{P}_0, \mathbf{D}_c^{-1}, \mathbf{D}_r)$. In this case, the analysis considers the centred row profiles $(\mathbf{D}_r^{-1} \mathbf{P}_0)$, with weights $(\mathbf{D}_r)$ and χ^2 metrics $(\mathbf{D}_c^{-1})$. Note that in $\mathbb{R}^n$, the analysis of the row profiles returns exactly the same principal components as the analysis of the original diagram. CA can also be rewritten as follows:

In this case, CA corresponds to the analysis of the centred column profiles $(\mathbf{D}_c^{-1} \mathbf{P}_0^T)$, with weights $(\mathbf{D}_c)$ and χ^2 metrics $(\mathbf{D}_r^{-1})$. Note that, in $\mathbb{R}^p$, the analysis of the column profiles returns exactly the same principal axes as the analysis of the original diagram. Hence, manipulating the original CA diagram shows that it corresponds to two analyses of two sets of points (row or column) with two different metrics and weighting matrices. These two viewpoints correspond to two discriminant analyses that identify linear combinations of columns (or rows) that maximize the separation of the rows (or columns).

Lastly, a third viewpoint can be identified that corresponds to the triplet $(\mathbf{P}_0, \mathbf{D}_c^{-1}, \mathbf{D}_r^{-1})$:

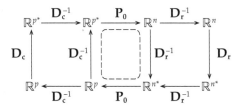

To simplify the presentation, I will consider the triplet $(\mathbf{P}, \mathbf{D}_c^{-1}, \mathbf{D}_r^{-1})$, which is equivalent but produces an additional trivial dimension associated with the eigenvalue $\lambda_1 = 1$ and eigenvectors $\mathbf{a}_1 = \mathbf{1}_n$, $\mathbf{b}_1 = \mathbf{1}_p$. The contingency table $\mathbf{N}$ is the result of the crossing of two qualitative variables. The information is encoded as dummy variables indicating to which categories of the two variables each individual belongs:

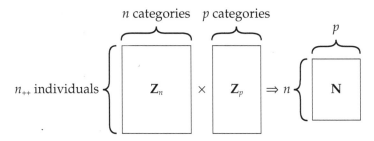

Let $\mathbf{D}_z = diag(1/n_{++}, \cdots, 1/n_{++})$ be a weighting matrix for the n_{++} individuals. We have the following relationships:

$$\mathbf{P} = \mathbf{Z}_n^T \mathbf{D}_z \mathbf{Z}_p, \quad \mathbf{D}_r = \mathbf{Z}_n^T \mathbf{D}_z \mathbf{Z}_n \text{ and } \mathbf{D}_c = \mathbf{Z}_p^T \mathbf{D}_z \mathbf{Z}_p$$

Using these relationships, we can then rewrite the analysis of $(\mathbf{P}, \mathbf{D}_c^{-1}, \mathbf{D}_r^{-1})$ as the analysis of $\left(\mathbf{Z}_n^T \mathbf{D}_z \mathbf{Z}_p, (\mathbf{Z}_p^T \mathbf{D}_z \mathbf{Z}_p)^{-1}, (\mathbf{Z}_n^T \mathbf{D}_z \mathbf{Z}_n)^{-1}\right)$.

In this analysis, the orthogonality constraint on the principal cofactor **f** leads to

$$\|\mathbf{f}\|^2_{(\mathbf{Z}_p^T \mathbf{D}_z \mathbf{Z}_p)^{-1}} = \mathbf{f}^T \mathbf{Z}_p^T \mathbf{D}_z \mathbf{Z}_p \mathbf{f} = \|\mathbf{Z}_p \mathbf{f}\|^2_{\mathbf{D}_z} = 1$$

Hence, the principal cofactors can be viewed as coefficients for the p dummy variables, allowing the computation of a linear combination of unit variance $\mathbf{Z}_p\mathbf{f}$. In contrast, we obtain a linear combination of the n dummy variables ($\mathbf{Z}_n\mathbf{g}$) due to the orthogonality constraint on the principal factor **g**:

$$\|\mathbf{g}\|^2_{(\mathbf{Z}_n^T \mathbf{D}_z \mathbf{Z}_n)^{-1}} = \mathbf{g}^T \mathbf{Z}_n^T \mathbf{D}_z \mathbf{Z}_n \mathbf{g} = \|\mathbf{Z}_n \mathbf{g}\|^2_{\mathbf{D}_z} = 1$$

Using the transition formulas, the inner product maximized by the analysis can be rewritten as

$$\langle \mathbf{Z}_p^T \mathbf{D}_z \mathbf{Z}_n (\mathbf{Z}_n^T \mathbf{D}_z \mathbf{Z}_n)^{-1} \mathbf{a} | \mathbf{b} \rangle_{(\mathbf{Z}_n^T \mathbf{D}_z \mathbf{Z}_n)^{-1}} = \mathbf{b}^T (\mathbf{Z}_n^T \mathbf{D}_z \mathbf{Z}_n)^{-1T} \mathbf{Z}_p^T \mathbf{D}_z \mathbf{Z}_n (\mathbf{Z}_n^T \mathbf{D}_z \mathbf{Z}_n)^{-1} \mathbf{a}$$

$$= \mathbf{g}^T \mathbf{Z}_p^T \mathbf{D}_z \mathbf{Z}_n \mathbf{f} = cor(\mathbf{Z}_p \mathbf{g}, \mathbf{Z}_n \mathbf{f})$$

Hence, it appears that CA is a particular case of canonical correlation analysis (Hotelling, 1936) that searches for a linear combination of rows ($\mathbf{Z}_n\mathbf{f}$) and a linear combination of columns ($\mathbf{Z}_p\mathbf{g}$) of maximal correlation.

The duality diagram appears to be a powerful and unifying tool to easily describe the various properties of an analysis. This tool provides a mathematical framework that facilitates the development and comparison of multivariate methods. In this study, we identify four diagrams associated with CA that were completely described by Cazes et al. (1988). The canonical correlation viewpoint is explicit in Williams (1952) and is used by Thioulouse and Chessel (1992) and Gimaret-Carpentier et al. (2003) in an ecological context. The discriminant analysis viewpoint is used in ecology by Hill (1973, 1974) and extended by ter Braak (1987) to introduce a table of explanatory variables in canonical correspondence analysis (see Chapter 5 by ter Braak in this book).

18.3 Relating Two Diagrams

In many situations, two sets of variables are measured on the same set of n individuals. This information is stored in two matrices, $\mathbf{X}\,(n \times p)$ and $\mathbf{Y}\,(n \times m)$. Each set of variables can be treated by a multivariate analysis defining two statistical triplets (**X**, **Q**, **D**) and (**Y**, **M**, **D**):

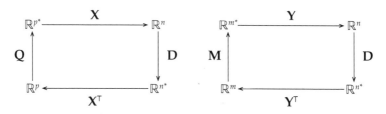

Note that **D** is common to the two triplets because we consider the same individuals in the two analyses. The individuals can be represented as a cloud of n points in $\mathbb{R}^p$ (rows of **X**) or as n points in $\mathbb{R}^n$ (rows of **Y**). Although separate analysis allows the independent study of the structures in each table, a relevant question is the evaluation of the concordance between these two configurations of individuals. To achieve this goal, the two duality diagrams must be combined into a single analysis to identify which structures are common to both data sets (i.e., co-structures):

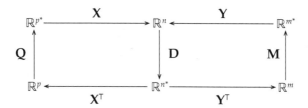

The above diagram can be rewritten as follows:

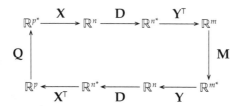

Coinertia analysis (Dolédec and Chessel, 1994) is the analysis of this diagram and thus is defined by the triplet $(\mathbf{Y}^T\mathbf{DX}, \mathbf{Q}, \mathbf{M})$. The total inertia associated with this triplet is equal to

$$\text{inertia}(\mathbf{Y}^T\mathbf{DX}, \mathbf{Q}, \mathbf{M}) = \text{trace}(\mathbf{Y}^T\mathbf{DXQX}^T\mathbf{DYM})$$

This quantity is a measure of the concordance between the two data sets and is equal to the numerator of the RV coefficient (Escoufier, 1973; see Chapter 14 by Pagès in this book), a multivariate generalization of the squared correlation coefficient. Coinertia analysis decomposes this vectorial covariance onto orthogonal axes, and the general properties of the diagram lead to the maximization of the following inner product:

$$\langle Y^TDXQb|a\rangle_M = a^TMY^TDXQb = \langle XQb|YMa\rangle_D = \sqrt{\lambda}$$

If **X** and **Y** contain centred variables, the total inertia is simply a sum of squared covariances between all combinations of variables of the two data sets

$$\left(\sum_{i=1}^{p}\sum_{j=1}^{m} cov^2(x^i, y^j)\right).$$

In this case, coinertia analysis finds two vectors of coefficients **b** and **a** to obtain linear combinations of the variable of **X** and **Y** of maximal covariance ($\langle XQb|YMa\rangle_D = cov(XQb, YMa)$). This covariance can be decomposed as a product of three factors:

$$cov(XQb, YMa) = cor(XQb, YMa) \cdot \|XQb\|_D \cdot \|YMa\|_D$$

The first term ($cor(XQb, YMa)$) is optimized by canonical correlation analysis. The second ($\|XQb\|_D$) is maximized by the analysis of **X** that aims to identify the main structures in this data set. The last term ($\|YMa\|_D$) corresponds to the simple analysis of **Y**. Hence, coinertia analysis can be viewed as a compromise between the three analyses aiming to find linear combinations of the two data sets with maximal co-structure. Unlike canonical correlation analysis, which requires many more individuals than variables, coinertia analysis is based on covariances and thus allows us to deal with tables in which the number of individuals is less than the number of variables. In this context, it shares certain similarities with the partial least-squares methods (Burnham et al., 1996; Krishnan et al., 2010).

The duality diagram of coinertia analysis is very general and encompasses several existing methods as particular cases (Chessel and Mercier, 1993). If **X** and **Y** are analysed by a normed principal component analysis, coinertia analysis corresponds to Tucker's (1958) interbattery analysis. It is also similar to Procrustes rotation (Dray et al., 2003b; Gower, 1971) and two-block partial least squares (Rohlf and Corti, 2000). If $M = (Y^TDY)^{-1}$ and $Q = (X^TDX)^{-1}$, coinertia analysis is equivalent to canonical correlation analysis. If only $Q = (X^TDX)^{-1}$, it corresponds to principal component analysis with instrumental variables (Rao, 1964), also known as redundancy analysis (van den Wollenberg, 1977), which aims to study the variation in **Y** explained by **X**. When **Y** is a contingency table analysed by correspondence analysis and $Q = (X^TDX)^{-1}$, it is similar to canonical correspondence analysis (ter Braak, 1987). If both **X** and **Y** contain qualitative variables and are analysed by multiple correspondence analysis, coinertia analysis corresponds to the correspondence analysis of the Burt matrix. The only difference is that coinertia analysis preserved the original structure of the data, whereas the correspondence analysis viewpoint does not consider the individuals (i.e., the rows of **X** and **Y**). It is thus linked to the works of Leclerc (1975) and Benzécri (1982a) if a table contains only one categorical variable, whereas Cazes (1980) provides insights into positioning the rows as supplementary points. Lastly, we consider a situation that is often encountered in community ecology: the abundance of q species (the columns of **Y**) is sampled in n sites (rows) for which qualitative environmental variables (columns of **X**) are measured. A common practice is to construct a table of ecological profiles summarizing the distribution of species in the different environmental classes (e.g., Sabatier et al., 1997). Correspondence analysis of the ecological profiles table (Bonin and Roux, 1978; Romane, 1972) is equivalent to coinertia analysis in which **Y** is analysed by a correspondence analysis and **X** is treated by multiple correspondence analysis (Mercier et al., 1992).

Coinertia analysis is based on a very general principle that has been extended to several situations (Dray et al., 2003a) to analyse a series of tables (Chessel and Hanafi, 1996) or to link external information on both rows and columns of a contingency table (Dolédec et al., 1996). The presentation based on the duality diagram allows us to summarize the various properties of a method and thus to simplify the comparison among methods. These abilities would be helpful in identifying concordances between methods that have been developed in different fields with few connections but similar methodological questions. For instance, I recently discovered the singular value decomposition (SVD) method described by Bretherton et al. (1992) in a review of two-table methods in climatology. This method is similar to a coinertia analysis between two centred principal component analyses. The use of a common mathematical language, as provided by the duality diagram theory, would probably improve exchanges between statisticians working in psychometrics, chemometrics, ecology, climatology, and other fields.

References

Abdessemed, L., and B. Escofier. 1996. Analyse factorielle multiple de tableaux de fréquences; Comparaison avec l'analyse canonique des correspondances. *Journal de la Société Statistique de Paris*, 137(2), 3–18.

Açar, E., and B. Yener. 2009. Unsupervised multiway data analysis: A literature survey. *IEEE Transactions on Knowledge and Data Engineering*, 21, 6–20.

Ackoff, R. L. 1958. Towards a behavioral theory of communication. *Management Science*, 4(3), 218–234.

Ackoff, R. L. 1974. *Redesigning the future: Systems approach to societal problems*. New York: Wiley.

Afmult. SAS macro written by B. Gelein et O. Sautory. Information on website de B. Gelein. http://www.ensai.fr/brigitte-gelein-rub,42.html (accessed March 19, 2013).

Afriat, S. N. 1957. Orthogonal and oblique projectors and the characteristics of pairs of vector spaces. *Proceedings of the Cambridge Philosophical Society*, 53, 800–816.

Agresti, A. 2007. *An introduction to categorical data analysis*. New York: Wiley.

Albers, C., F. Critchley, and J. C. Gower. 2011. Quadratic minimisation products in statistics. *Journal of Multivariate Analysis*, 102, 698–713.

Allaby, M. 1998. Shelford's law of tolerance. *A dictionary of ecology*. http://www.encyclopedia.com.

Anderberg, M. R. 1973. *Cluster analysis for applications*. New York: Academic Press.

Anderson, C. 2008. The end of theory: The data deluge makes the scientific method obsolete. *Wired Magazine*. http://www.wired.com/science/discoveries/magazine/16-07/pb.

Anderson, J. A. 1984. Regression and ordered categorical variables. *Journal of the Royal Statistical Society B*, 46, 1–30.

Anderson T. W. 1958 (1984). *An introduction to multivariate statistical analysis*. New York: Wiley.

Andersson, C. A., and R. Bro. 2000. The N-way toolbox for MATLAB. *Chemometrics and Intelligent Laboratory Systems*, 52, 1–4.

Appellof, C. J., and E. R. Davidson. 1981. Strategies for analysing data from video fluorometric monitoring of liquid chromatographic effluents. *Analytical Chemistry*, 53, 2053–2056.

Appellof, C. J., and E. R. Davidson. 1983. Three-dimensional rank annihilation for multi-component determinations. *Analytica Chimica Acta*, 146, 9–14.

Arcelay, A. R., R. T. Ross, and B. M. Ezzeddine. 1988. Photosystem I generates a free-energy change of 0.7 electron volts or less. *Biochimica et Biophysica Acta*, 936, 199–207.

Armatte, M. 2008. Histoire et préhistoire de l'analyse des données par J. P. Benzécri: Un cas de généalogie retrospective. *Electronic Journal for History of Probability and Statistics*, 4, 2.

Arroyo, J., and C. Maté. 2009. Forecasting histogram time series with k-nearest neighbours methods. *International Journal of Forecasting*, 25, 192–207.

Aşan, Z., and M. Greenacre. 2010. Biplots of fuzzy coded data. *Fuzzy Sets and Systems*, 183, 57–71.

Autonne, L. 1915. Sur les matrices hypohermitiennes et sur les matrices unitaires. *Annales de L'Université de Lyon*, 1(38), 1–77.

Bader, B. W., T. G. Kolda et al. 2012. *MATLAB tensor toolbox*. Version 2.5. Livermore, CA: Sandia National Laboratory. www.sandia.gov/~tgkolda/TensorToolbox.

Badiou, A. 2007. *Being and event*. London: Continuum.

Balbi, A., and A.-M. Guerry. 1829. *Statistique comparee de l'etat de l'instruction et du nombre des crimes dans les divers arrondissements des academies et des cours royales de France*. BL: Tab.597.b.(38); BNF: Ge C 9014. Paris: Jules Renouard.

Batagelj, V. 1988. Generalized ward and related clustering problems. In *Classification and related methods of data analysis*, ed. H. H. Bock, 67–74. Amsterdam: North-Holland.

Batchelder, W. H., A. Strashny, and A. K. Romney. 2010. Cultural consensus theory: Aggregating continuous responses in a finite interval. In *Advances in social computing*, eds. S.-K. Chai, J. J. Salerno, and P. L. Mabry, 98–107. Berlin: Springer.

Bécue, M., and J. Pagès. 2008. Analysis of a mixture of quantitative, categorical and frequency data through an extension of multiple factor analysis. Application to survey data. *Computational Statistics and Data Analysis*, 52(6), 3255–3268.

Bécue-Bertaut, M., B. Kostov, A. Morin, and G. Naro. 2013. Rhetorical strategy in forensic speeches. Multidimensional statistics-based methodology. *Journal of Classification*, in press.

Bécue-Bertaut, M., and J. Pagès. 2004. A principal axes method for comparing multiple contingency tables: MFACT. *Computational Statistics and Data Analysis*, 45, 481–503.

Beh, E. J., and R. Lombardo. 2012. A genealogy of correspondence analysis. *Australian and New Zealand Journal of Statistics*, 54, 137–168.

Beh, E. J., R. Lombardo, and B. Simonetti. 2008. Model-based methods for performing multi-way non-symmetric correspondence analysis. In *Proceedings of MTIDS2008: Conference on Methods, Models and Information Technologies for Decision Support Systems*, 329–332. Lecce, Italy: Università de Salento.

Bekker, P., and J. de Leeuw. 1988. Relation between variants of nonlinear principal component analysis. In *Component and correspondence analysis*, eds. J. L. A. van Rijckevorsel and J. de Leeuw, 1–31. Chichester, England: Wiley.

Beltrani, E. 1873. Sulle funzioni bilineari. *Giornali I Matematiche*, 11, 98–106.

Bener, A. 1981. *Étude par l'analyse des correspondances des interactions dans un tableau ternair. Applications à des données linguistiques* [A study of the interactions in a three-way contingency table by means of correspondence analysis applied to linguistic data]. Unpublished doctoral thesis, Université Pierre et Marie Curie, Paris.

Bener, A. 1982. Décomposition des interaction dans un correspondance multiple [The decomposition of interactions in multiple correspondence analysis]. *Les Cahiers de l'Analyse des Données*, 7, 25–32.

Bennett, T., M. Savage, E. Silva, A. Warde, M. Gayo-Cal, and D. Wright. 2009. *Culture, class, distinction*. London: Routledge.

Bentley, J. L., B. W. Weide, and A. C. Yao. 1980. Optimal expected time algorithms for closest point problems. *ACM Transactions on Mathematical Software*, 6, 563–580.

Benzécri, F. 1985. Introduction à la classification ascendante hiérarchique d'après un exemple de données économiques. *Les Cahiers de l'Analyse des Données*, 10(3), 279–302.

Benzécri, J.-P. 1964. *Cours de linguistique mathématique*. Faculté des Sciences de Rennes.

Benzécri, J. -P. 1969. Statistical analysis as a tool to make patterns emerge from data. In *Methodologies of pattern recognition*, ed. S. Watanabe. New York: Academic Press.

Benzécri, J.-P. 1972. *Sur l'analyse des tableaux binaires associés à une correspondance multiple*. Publication of the Laboratoire de Statistique Mathématique, Université Pierre et Marie Curie.

Benzècri, J. -P. 1976. *L'Analysis des données, tome 1, La Taxinomie*, 2nd ed. Dunod.

Benzécri, J.-P. 1977a. Choix des unités et des poids dans un tableau en vue d'une analyse de correspondance. *Cahiers de l'Analyse des Données*, 2, 333–352.

Benzécri, J.-P. 1977b. Histoire et préhistoire de l'analyse des données. Partie V: L'analyse des correspondances. *Les Cahiers de l'Analyse des Données*, 2(1), 9–40.

Benzécri, J.-P. 1977c. Sur l'analyse des tableaux binaires associés à une correspondance multiple. *Les Cahiers de l'Analyse des Données*, 2, 55–71.

Benzécri, J.-P. 1979. Sur le calcul des taux d'inertia dans l'analyse d'un questionnaire. Addendum et erratum [BIN.MULT]. *Les Cahiers de l'Analyse des Données*, 4, 377–378.

Benzécri, J.-P. 1980. The soul at the razor's edge. *Les Cahiers de l'Analyse des Donnes*, 2, 229–242.

Benzécri, J.-P. 1982a. Sur la généralisation du tableau de burt et son analyse par bandes. *Les Cahiers de l'Analyse des Données*, 7, 33–43.

Benzécri, J.-P. 1982b. *Histoire et préhistoire de l'analyse des données*. Paris: Dunod.

Benzécri, J. P. 1983. L'avenir de l'analyse des données. *Behaviormetrika*, 10, 1–11.

Benzécri, J.-P. 1992. *Correspondence analysis handbook*. New York: Dekker.

Benzécri, J. P. 2006. L'analyse des données: Histoire, bilan, projects, ..., perspective. In memoriam: Pierre Bourdieu. *Revue MODULAD*, 35, 1–5.

Benzécri, J. P. 2007. *Si j'avais un laboratoire*. www.correspondances.info.

Benzécri, J.-P. et al. 1973. *L'analyse des données. Tome 1: La taxinomie. Tome 2: L'analyse des correspondances* (2nd ed., 1976). Paris: Dunod.

Benzécri, J.-P. et al. 1981. *Pratique de l'analyse des données: Linguistique and lexicologie*. Vol. 3. Paris: Dunod.

Bertin, J. 1967. *Semiologie graphique: Les diagrammes, les reseaux, les cartes*. Paris: Gauthier-Villars.

Bertin, J. 1983 (2010). *Semiology of graphics*. Madison: University of Wisconsin Press.

Bertrand, P., and F. Goupil. 2000. Descriptive statistics for symbolic data. In *Analysis of symbolic data, exploratory methods for extracting statistical information from complex data*, eds. H.-H. Bock and E. Diday, 106–124. Berlin: Springer-Verlag.

Bienaise, S. 2013. Méthodes d'inférence combinatoire sur un nuage euclidien/ Etude statistique de la cohorte EPIEG. PhD thesis, Université Paris, Dauphine, CEREMADE.

Billard, L. 2004. Dependencies in bivariate interval-valued symbolic data. In *Classification, clustering and data mining application*, eds. D. Banks, L. House, F. R. McMorris, P. Arabie, and W. Gaul, 319–324. Berlin: Springer-Verlag.

Billard, L. 2011. Brief overview of symbolic data and analytic issues. *Statistical Analysis and Data Mining*, 4(2), 149–156.

Billard, L., and E. Diday. 2003. From the statistics of data to the statistics of knowledge: Symbolic data analysis. *Journal of the American Statistical Association*, 98(462), 470–487.

Billard, L., and E. Diday. 2006a. Descriptive statistics for interval-valued observation in presence of rules. *Computational Statistics*, 21, 187–210.

Billard, L., and E. Diday. 2006b. *Symbolic data analysis: Conceptual statistics and data mining*. Chichester, UK: Wiley.

Blashfield, R. K. 1980. The growth of cluster analysis: Tryon, Ward and Johnson. *Multivariate Behavioral Research*, 15, 429–458.

Blasius, J. 1994. Correspondence analysis in social science research. In *Correspondence analysis in the social sciences. Recent developments and applications*, eds. M. Greenacre and J. Blasius, 23–52. London: Academic Press.

Blasius, J., and J. Friedrichs. 2008. Lifestyles in distressed neighborhoods. A test of Bourdieu's 'taste of necessity' hypothesis. *Poetics*, 36, 24–44.

Blasius, J., and A. Mühlichen. 2010. Identifying audience segments applying the "social space" approach. *Poetics* 38, 69–89.

Blasius, J., and V. Thiessen. 2012. *Assessing the quality of survey data*. London: Sage.

Blossfeld, H.-P., and A. Schmitz. 2011. Online dating: Social innovation and a tool for research on partnership formation. *Journal of Family Research*, 3, 263–267.

Blyshak, L. A., G. Patonay, and I. M. Warner. 1989. Multidimensional fluorescence analysis of cyclodextrin solvent—extraction systems. In *Luminescence applications*, ed. M. Goldberg, Chapter 10. *ACS Symposium Series*, 383, 167–179. Washington, DC: The American Chemical Society.

Bock, R. D. 1960. *Methods and applications of optimal scaling*. Psychometric Laboratory Report 25. L. L. Thurstone Psychometric Laboratory, University of North Carolina, Chapel Hill.

Bock, H.-H. 2007. Clustering methods: A history of k-means algorithms. In *Selected contributions in data analysis and classification*, eds. P. Brito, G. Cucumel, P. Bertrand, and F. de Carvalho, 161–172. Berlin: Springer.

Bock, H.-H., and E. Diday. 2000. *Analysis of symbolic data*. Berlin: Springer.

Böckenholt, U., and I. Böckenholt. 1990. Canonical analysis of contingency tables with linear constraints. *Psychometrika*, 55, 633–639.

Bonin, G., and M. Roux. 1978. Utilisation de l'analyse factorielle des correspondances dans l'étude phyto-écologique de quelques pelouses de l'Apennin lucano-calabrais. *Oecologia Plantaruin*, 13, 121–138.

Borcard, D., P. Legendre, and P. Drapeau. 1992. Partialling out the spatial component of ecological variation. *Ecology*, 73, 1045–1055.

Bordet, C. 1973. *Etudes de données géophysiques*. Doctoral thesis (3ème cycle), Université de Paris VI, France.

Borg, I., and P. J. F. Groenen. 1997. Multitrait-multimethod by multidimensional scaling. In *SoftStat '97*, eds. F. Faulbaum and W. Bandilla, 59–66. Stuttgart: Lucius.

Borg, I., and P. J. F. Groenen. 1998. Regional interpretations in multidimensional scaling. In *Visualization of categorical data*, eds. J. Blasius and M. Greenacre, 347–364. New York: Academic Press.

Borg, I., and P. J. F. Groenen. 2005. *Modern multidimensional scaling*. 2nd ed. New York: Springer.

Borg, I., P. J. F. Groenen, K. A. Jehn, W. Bilsky, and S. H. Schwartz. 2011. Embedding the organizational culture profile into Schwartz's theory of universals in values. *Journal of Personnel Psychology*, 10, 1–12.

Borg, I., and S. Shye. 1995. *Facet theory: Form and content*. Newbury Park, CA: Sage.
Boudon, R. 1974. *The logic of sociological explanations*. London: Penguin.
Bourdieu, P. 1979. *La distinction*. Paris: Les éditions de Minuit. (1984. *Distinction*. Cambridge, MA: Harvard University Press.)
Bourdieu, P. 1983. The forms of capital. In *Handbook of theory and research for the sociology of education*, ed. J. Richardson, 241–258. New York: Greenwood Press.
Bourdieu, P. 1984. *Homo academicus*. Paris: Les éditions de Minuit. (1988. *Homo academicus*. Stanford, CA: Stanford University Press.)
Bourdieu, P. 1990. *The logic of practice*. Stanford, CT: Stanford University Press.
Bourdieu, P. 1998. *Practical reason: On the theory of action*. Stanford, CA: Stanford University Press.
Bourdieu, P. 1999. Une revolution conservatrice dans l'édition. *Actes de la Recherche en Sciences Sociales*, 126–127, 3–28.
Bourdieu, P. 2001. *Masculine domination*. Oxford: Polity Press.
Bourdieu, P., and J.-C. Passeron. 1977. Reproduction. In *Education, society, culture*. Beverly Hills, CA: Sage.
Bourdieu, P., and Md. de Saint Martin. 1976. Anatomie du goût. *Actes de la Recherche en Sciences Sociales*, 2, 5, 2–81.
Bourdieu, P., and Md. de Saint Martin. 1978. Le patronat. *Actes de la Recherche en Sciences Sociales*, 20–21, 3–82.
Bourland Jr., D. D., and P. D. Johnston. 1991. *To be or not: An E-prime anthology*. San Francisco: International Society for General Semantics.
Box, J. F. 1978. *R. A. Fisher, the life of a scientist*. New York: Wiley.
Breiman, L., and J. H. Friedman. 1985. Estimating optimal transformations for multiple regression and correlation. *Journal of the American Statistical Association*, 80, 580–619.
Bretherton, C., C. Smith, and J. Wallace. 1992. An intercomparison of methods for finding coupled patterns in climate data. *Journal of Climate*, 5(6), 541–560.
Brier, A., and B. Hopp. 2011. Computer assisted text analysis in the social sciences. *Quality and Quantity*, 45, 103–128.
Brito, P. 1994. Use of pyramids in symbolic data analysis. In *New approaches in classification and data analysis*, eds. E. Diday et al., 378–386. Berlin: Springer-Verlag.
Brito, P. 1995. Symbolic objects: Order structure and pyramidal clustering. *Annals of Operations Research*, 55, 277–297.
Brito, P., and F. De Carvalho. 2002. Symbolic clustering of constrained probabilistic data. In *Exploratory data analysis in empirical research, series studies in classification, data analysis and knowledge organization*, eds. O. Opitz and M. Schvaiger, 12–21. Heidelberg: Springer Verlag.
Brito, P., and A. P. Duarte Silva. 2012. Modelling interval data with normal and skew-normal distributions. *Journal of Applied Statistics*, 39(1), 3–20.
Bro, R. 1997. PARAFAC. Tutorial and applications. *Chemometrics and Intelligent Laboratory Systems*, 38, 149–171.
Bro, R. 1998. *Multi-way analysis in the food industry. Models, algorithms, and applications*. Unpublished doctoral dissertation, University of Amsterdam, Amsterdam. http://www.models.kvl.dk/sites/default/files/brothesis_0.pdf.
Bro, R. 2006. Review on multiway analysis in chemistry—2000–2005. *Critical Reviews in Analytical Chemistry*, 36, 279–293.
Bro, R., and H. A. L. Kiers. 2003. A new efficient method for determining the number of components in PARAFAC models. *Journal of Chemometrics*, 17, 274–286.

Bro, R., J. J. Workman Jr., P. R. Mobley, and B. R. Kowalski. 1997. Review of chemometrics applied to spectroscopy: 1985–1995. Part III: Multi-way analysis. *Applied Spectroscopy Reviews*, 32, 237–261.

Brusco, M. J. 2001. A simulation annealing heuristic for unidimensional and multidimensional (city-block) scaling of symmetric proximity matrices. *Journal of Classification*, 18, 3–33.

Bruynooghe, M. 1977. Méthodes nouvelles en classification automatique des données taxinomiques nombreuses. *Statistique et Analyse des Données*, 3, 24–42.

Buache, P. 1752. *Essai de geographie physique. Memoires de l'academie royale des sciences.* BNF: Ge.FF-8816-8822.

Buache, P. 1770. *Profils representants la crue et la diminution des eaux de la seine et des rivieres qu'elle recoit dans le paris-haut au dessus de paris.* G. de L'Isle et P. Buache.

Buja, A., B. F. Logan, J. R. Reeds, and L. A. Shepp. 1994. Inequalities and positive definite functions arising from a problem in multidimensional scaling. *Annals of Statistics*, 22, 406–438.

Buja, A., and D. F. Swayne. 2002. Visualization methodology for multidimensional scaling. *Journal of Classification*, 19, 7–44.

Buja, A., D. F. Swayne, M. L. Littman, N. Dean, H. Hofmann, and L. Chen. 2008. Data visualization with multidimensional scaling. *Journal of Computational and Graphical Statistics*, 17, 444–472.

Burnham, A. J., R. Viveros, and J. F. MacGregor. 1996. Frameworks for latent variable multivariate regression. *Journal of Chemometrics*, 10(1), 31–45.

Burt, C. 1949. Alternative methods of factor analysis and their relations to Pearson's method of "principle axes." *British Journal of Statistical Psychology*, 2, 98–121.

Burt, C. 1950. The factorial analysis of qualitative data. *British Journal of Statistical Psychology*, 3, 166–185.

Burt, C. 1953. Scale analysis and factor analysis. Comments on Dr. Guttman paper. *British Journal of Statistical Psychology*, 6, 5–20.

Cailliez, F., and J.-P. Pagès. 1976. *Introduction à l'analyse des données*. Paris: Societé de Mathématiques Appliquées et de Sciences Humaines.

Carlier, A., and P. M. Kroonenberg. 1993. Biplots and decompositions in three-way correspondence analysis. In *49th Session of the International Statistical Institute, Firenze, Italy, 25 August–2 September 1993*, 209–210. Voorburg, Netherlands: ISI.

Carlier, A., and P. M. Kroonenberg. 1996. Decompositions and biplots in three-way correspondence analysis. *Psychometrika*, 61, 355–373.

Carlier, A., and P. M. Kroonenberg. 1998. Three-way correspondence analysis: The case of the French cantons. In *Visualization of categorical data*, eds. J. Blasius and M. Greenacre, 253–275. New York: Academic Press.

Carroll, J. D. 1968. A generalization of canonical correlation analysis to three or more sets of variables. In *Proceedings of the 76th Annual Convention of the American Psychological Association*, 227–228. Washington, DC: American Psychological Association.

Carroll, J. D., and J. J. Chang. 1970. Analysis of individual differences in multidimensional scaling via an n-way generalization of "Eckart-Young" decomposition. *Psychometrika*, 35, 283–320.

Carroll, J. D., and J. J. Chang. 1972. IDIOSCAL. A generalization of INDSCAL allowing idiosyncratic reference systems as well as an analytic approximation to INDSCAL. Presented at the spring meeting of the Classification Society of North America, Princeton, NJ.

Casanovas, P., E. Ardévol, M. Cachón, and C. Riba. 1995. *Videos as tribunals de justícia*. Barcelona: Publicacions ICE-UCattell, R. B. 1944. Parallel proportional profiles and other principles for determining the choice of factors by rotation. *Psychometrika*, 9, 267–283.

Cattell, R. B. 1944. Paralell proportional profiles and other principles for determining the choice of factors by rotation. *Psychometrika*, 9, 267–283.

Cattell, R. B. 1946. *Description and measurement of personality*. Oxford: World Book Company.

Cattell, R. B. 1952. The three basic factor-analytic research designs—Their interrelations and derivatives. *Psychological Bulletin*, 49, 499–520.

Cattell, R. B. 1966. The data box: Its ordering of total resources in terms of possible relational systems. In *Handbook of multivariate experimental psychology*, ed. R. B. Cattell, 67–128. Chicago: Rand-McNally.

Cattell, R. B., and A. K. S. Cattell. 1955. Factor rotation for proportional profiles: Analytical solution and an example. *The British Journal of Statistical Psychology*, 8, 83–92.

Cauchy, A. 1829–1830. Ancient exercises. In *Collected Works II*, 174–195. Vol. 9.

Cauchy, A. L. 1830. Mémoire sur l'équation qui a pour racines les moments d'inertie principaux d'un corps solide et sur diverses équations du même genre. *Mémoires de l'Académie des Sciences*, t. IX, 111 (presented in 1826).

Cayley, G. 1858. A memoir on the theory of matrices. *Philosophical Transactions of the Royal Society of London*, 148, 24.

Cazes, P. 1970. *Application de l'analyse des données au traitement de problèmes géologiques*. Thèse de 3ème cycle, Faculté des Sciences de Paris.

Cazes, P. 1977a. Note sur les éléments supplémentaires en analyse des corresponddances. *Les Cahiers de l'Analyse des Données*, 1, 9–23; 2, 133–154. (Collection of notes published in 1982.)

Cazes, P. 1977b. Etude des propriétés extrêmales des sous-facteurs issus d'un soustableau d'un tableau de Burt. *Les Cahiers de l'Analyse des Données*, 2, 143–160.

Cazes, P. 1980. Analyse de certains tableaux rectangulaires décomposés en blocs: Généralisation des propriétés rencontrées dans l'étude des correspondances multiples. I. Définitions et applications à l'analyse canonique des variables qualitatives. *Les Cahiers de l'Analyse des Données*, 5, 145–161, 387–403.

Cazes, P., D. Chessel, and S. Dolédec. 1988. L'analyse des correspondances internes d'un tableau partitionné: Son usage en hydrobiologie. *Revue de Statistique Appliquée*, 36, 39–54.

Cazes P., A. Chouakria, E. Diday, and Y. Schektman. 1997. Extension de l'analyse en composantes principales a des données de type intervalle. *Revue Statistique Appliquée*, 45(3), 5–24.

Ceulemans, E., and I. van Mechelen. 2004. Tucker2 hierarchical classes analysis. *Psychometrika*, 69, 375–399.

Ceulemans, E., and I. van Mechelen. 2005. Hierarchical class models for three-way three-mode binary data: Interrelations and model selection. *Psychometrika*, 70, 461–480.

Ceulemans, E., I. van Mechelen, and I. Leenen. 2003. Tucker3 hierarchical classes analysis. *Psychometrika*, 68, 413–433.

Chakraborty, P. 2005. *Looking through newly to the amazing irrationals*. Technical report. http://arxiv.org/pdf/math.HD/0502049.pdf.

Chavent, M., F. A. T. De Carvalho, Y. Lechevallier, and R. Verde. 2006. New clustering methods for interval data. *Computational Statistics*, 21(2), 211–230.

Chavent, M., and Y. Lechevallier. 2002. Dynamical clustering algorithm of interval data: Optimization of an adequacy criterion based on Hausdorff distance. In *Classification, clustering and data analysis (IFCS2002)*, eds. H. H. Sokolowsky, K. Bock, and A. Jaguja, 53–59. Berlin: Springer.

Chessel, D., A.-B. Dufour, and J. Thioulouse. 2004. The ade4 package. I. One-table methods. *R News*, 4, 5–10.

Chessel, D., and M. Hanafi. 1996. Analyse de la co-inertie de {K} nuages de points. *Revue de Statistique Appliquée*, 44(2), 35–60.

Chessel, D., J.-D. Lebreton. and R. Prodon. 1982. Mesures symétriques d'amplitude d'habitat et de diversité, intra-échantillon dans un tableau espèces-relevés: Cas d'un gradient simple. *Comptes Rendus de l'Academie Sciences Paris*, III, 295, 83–88.

Chessel, D., J. B. Lebreton, and N. Yoccoz. 1987. Propriétés de l'analyse canonique des correspondances; une illustration en hydrobiologie. *Revue de Statistique Appliquée*, 35, 55–72.

Chessel, D., and P. Mercier. 1993. Couplage de triplets statistiques et liaisons es pèces-environnement. In *Biométrie et Environnement*, eds. J. D. Lebreton and B. Asselain, 15–43. Paris: Masson.

Choulakian, V. 1988a. Analyse factorielle des correspondances de tableaux multiples [Correspondence analysis of multiway tables]. *Revue de Statistiques Appliquées*, 36(4), 33–41.

Choulakian, V. 1988b. Exploratory analysis of contingency tables by loglinear formulations and generalizations of correspondence analysis. *Psychometrika*, 53, 235–250.

Clark, H. H., and T. B. Carlson. 1982. Hearers and speech act. *Language*, 58(2), 332–373.

Classification Society. 1964. Record of the inaugural meeting, held at the offices of Aslib, 3 Belgrave Square, London S.W. 1 at 2:30 p.m. on Friday, 17 April, 1964. http://brclasssoc.org.uk/membership/BCS-history/overview.html.

Clinton, J., S. Jackman, and D. Rivers. 2004. The statistical analysis of roll call data. *American Political Science Review*, 98, 355–370.

Clobert, J., and J. D. Lebreton. 1985. Dépendance de facteurs de milieu dans les estimations de taux de survie par capture-recapture. *Biometrics*, 41, 1031–1037.

Clogg, C. C. 1982. Some models for the analysis of association in multiway cross-classifications having ordered categories. *Journal of the American Statistical Association*, 77, 803–815.

Comon, P. 2006. Tensor decompositions: State of the art and applications. In *Mathematics in signal processing V*, eds. J. G. McWhirter and I. K. Proudler, 1–24. Oxford: Oxford University Press.

Comon, P., X. Luciani, and A. L. F. De Almeida. 2009. Tensor decompositions, alternating least squares and other tales. *Journal of Chemometrics*, 23, 393–405.

Coombs, C. H., and G. S. Avrunin. 1977. Single-peaked functions and the theory of preference. *Psychological Review*, 84, 216–230.

Coombs, C. H., and R. C. Kao. 1955. *Nonmetric factor analysis*. Engineering Research Bulletin 38. Ann Arbor: Engineering Research Institute, University of Michigan.

Coombs, C. H., and R. C. Kao, eds. 1960. *On a connection between factor analysis and multidimensional unfolding*. New York: Wiley.

Coppi, R., and S. Bolasco, eds. 1989. *Multiway data analysis*. Amsterdam: North Holland.

Cordier, B. 1965. *L'analyse factorielle des correspondances*. Thèse de Troisieme Cycle, Université de Rennes.

Coulangeon, P., and Y. Lemel. 2007. Is 'distinction' really outdated? Questioning the meaning of the omnivorization of musical taste in contemporary France. *Poetics*, 35, 93–111.

Cuadras, C. M., J. Fortiana, and M. J. Greenacre. 1999. Continuous extensions of matrix formulations in correspondence analysis, with applications to the FGM family of distributions. In *Innovations in multivariate statistical analysis*, eds. R. D. H. Heijmans, D. S. G. Pollock, and A. Satorra, 101–116. Dordrecht: Kluwer.

D'Ambra, L., and N. C. Lauro. 1989. Non symmetrical analysis of three-way contingency tables. In *Multiway data analysis*, eds. R. Coppi and S. Bolasco, 301–315. Amsterdam: North Holland.

D'Alfonso, S. 2011. On quantifying semantic information. *Information*, 2, 1, 61–101.

Day, W. H. E., and H. Edelsbrunner. 1984. Efficient algorithms for agglomerative hierarchical clustering methods. *Journal of Classification*, 1, 7–24.

Dazy, F., J.-F. Le Barzic, G. Saporta, and F. Lavallard. 1996. *L'analyse des données évolutives: Méthodes et applications*. Paris: Technip.

De Boor, C. 1978. *A practical guide to splines*. New York: Springer.

De Carvalho, F. A. T., R. Verde, and Y. Lechevallier. 1999. A dynamical clustering of symbolic objects based on a context dependent proximity measure. In *Proceedings of the IX International Symposium—ASMDA '99*, eds. H. Bacelar-Nicolau, F. C. Nicolau, and J. Janssen, 237–242. LEAD, University of Lisboa.

De Carvalho, F. A. T., M. Csernel, and Y. Lechevallier. 2009a. Clustering constrained symbolic data. *Pattern Recognition Letters*, 30(11), 1037–1045.

De Carvalho, F. A. T., and Y. Lechevallier. 2009b. Partitional clustering algorithms for symbolic interval data based on single adaptive distances. *Pattern Recognition*, 42(7), 1223–1236.

De Carvalho, F. A. T., Y. Lechevallier, and F. M. de Melo. 2012. Partitioning hard clustering algorithms based on multiple dissimilarity matrices. *Pattern Recognition*, 45(1), 447–464.

Defays, D. 1977. An efficient algorithm for a complete link method. *Computer Journal*, 20, 364–366.

Defays, D. 1978. A short note on a method of seriation. *British Journal of Mathematical and Statistical Psychology*, 31, 49–53.

De la Cruz, O., and S. Holmes. 2011. The duality diagram in data analysis: Examples of modern applications. *Annals of Applied Statistics*, 5(4), 2266–2277.

De Leeuw, J. 1968. *Canonical discriminant analysis of relational data. Research Note 007-68*. Leiden, Netherlands: Department of Data Theory. http://www.stat.ucla.edu/~deleeuw/janspubs/1968/reports/deleeuw_R_68e.pdf.

De Leeuw, J. 1973. *Canonical analysis of contingency tables*. Leiden, Netherlands: Leiden University. (Reprinted in 1984. Leiden, Netherlands: DSWO Press.)

De Leeuw, J. 1977. Applications of convex analysis to multidimensional scaling. In *Recent developments in statistics*, eds. J. R. Barra, F. Brodeau, G. Romier, and B. van Cutsem, 133–145. Amsterdam: North-Holland.

De Leeuw, J. 1982. Nonlinear principal component analysis. In *COMPSTAT 1982*, eds. H. Caussinus, P. Ettinger, and R. Tomassone, 77–86. Vienna: Physika.

De Leeuw, J. 1983. On the prehistory of correspondence analysis. *Statistica Neerlandica*, 37, 161–164.

De Leeuw, J. 1988a. Convergence of the majorization method for multidimensional scaling. *Journal of Classification*, 5, 163–180.
De Leeuw, J. 1988b. Multivariate analysis with optimal scaling. In *Proceedings of the international conference on advances in multivariate statistical analysis*, eds. S. Das Gupta and J. K. Ghosh, 127–160. Calcutta: Indian Statistical Institute.
De Leeuw, J. 1988c. Multivariate analysis with optimal scaling. In *Proceedings of the international conference on advances in multivariate statistical analysis*, eds. S. Das Gupta and J. K. Ghosh, 127–160. Calcutta: Indian Statistical Institute.
De Leeuw, J. 1993. *Fitting distances by least squares*. Technical Report 130. Los Angeles: Interdivisional Program in Statistics, UCLA.
De Leeuw, J. 1994. Block relaxation algorithms in statistics. In *Information systems and data analysis*, eds. H.-H. Bock, W. Lenski, and M. M. Richter, 308–317, 324. Berlin: Springer.
De Leeuw, J. 2003. Homogeneity analysis using Euclidean minimum spanning trees. Unpublished. http://www.stat.ucla.edu/~deleeuw/janspubs/2003/notes/deleeuw_U_03d.pdf.
De Leeuw, J. 2005. *Gifi goes logistic*. Preprint Series 449. Department of Statistics, UCLA. http://www.stat.ucla.edu/~deleeuw/janspubs/2005/reports/deleeuw_R_05a.pdf.
De Leeuw, J. 2006a. Principal component analysis of binary data by iterated singular value decomposition. *Computational Statistics and Data Analysis*, 50(1), 21–39.
De Leeuw, J. 2006b. Nonlinear principal component analysis and related techniques. In *Multiple correspondence analysis and related methods*, eds. M. Greenacre and J. Blasius, 107–133. Boca Raton, FL: Chapman & Hall.
De Leeuw, J., and W. J. Heiser. 1977. Convergence of correction-matrix algorithms for multidimensional scaling. In *Geometric representations of relational data*, eds. J. C. Lingoes, E. E. Roskam, and I. Borg, 735–752. Ann Arbor, MI: Mathesis Press.
De Leeuw, J., and W. J. Heiser. 1980. Multidimensional scaling with restrictions on the confguration. In *Multivariate analysis*, vol. V, ed. P. R. Krishnaiah, 501–522. Amsterdam: North-Holland.
De Leeuw, J., and P. Mair. 2009a. Homogeneity analysis in R: The package homals. *Journal of Statistical Software*, 31(4), 1–21.
De Leeuw, J., and P. Mair. 2009b. Simple and canonical correspondence analysis using the R package anacor. *Journal of Statistical Software*, 31, 1–18.
De Leeuw, J., and P. Mair. 2009c. Multidimensional scaling using majorization: SMACOF. *Journal of Statistical Software*, 31(3), 1–30.
De Leeuw, J., G. Michailidis, and D. Y. Wang. 1999. Correspondence analysis techniques. In *Multivariate analysis, design of experiments, and survey sampling*, ed. S. Ghosh, 523–547. New York: Marcel Dekker.
De Leeuw, J., and J. L. A. van Rijckevorsel. 1980. *Homals and princals: Some generalizations of principal components analysis*. Amsterdam: North Holland.
De Ligny, C. L., M. C. Spanjer, J. C. Van Houwelingen, and H. M. Weesie. 1984. Three-mode factor analysis of data on retention in normal-phase high-performance liquid chromatography. *Journal of Chromatography*, 301, 311–324.
Delsaut, Y., and M.-C. Rivière. 2002. *Bibliographie des travaux de Pierre Bourdieu*. Paris: Le Temps des Cerises.
De Nautonier, G. 1602–1604. *Mecometrie de l'eymant, c'est a dire la maniere de mesurer les longitudes par le moyen de l'eymant*. BL: 533.k.9; BNF: RES-V-432; LC: QB225.N3. Paris: R. Colomies.

Denoeux, T., and M. Masson. 2000. Multidimensinal scaling of interval-valued dissimilarity data. *Pattern Recognition Letters*, 21, 83–92.
Dequier, A. 1973. *Contributions à l'étude des tables de contingence entre trois caractères* [A contribution to the study of three-way contingency tables]. Unpublished doctoral thesis, Université de Paris VI, Paris. http://three-mode.leidenuniv.nl.
De Rham, C. 1980. La classification hiérarchique ascendante selon la méthode des voisins réciproques. *Les Cahiers de l'Analyse des Données*, 5(2), 135–144.
De Rooij, M. 2007. The distance perspective of generalized biadditive models: Scalings and transformations. *Journal of Computational and Graphical Statistics*, 16, 210–227.
De Rooij, M., and W. J. Heiser. 2005. Graphical representations and odds ratios in a distance-association model for the analysis of cross-classified data. *Psychometrika*, 70, 99–122.
Derrida, J. 1981. *Plato's pharmacy dissemination*. London: Athlone Press.
De Tienne, A. 2006. Peirce's logic of information. http://www.unav.es/gep/SeminariodeTienne.html.
Deutsch, S. B., and J. J. Martin. 1971. An ordering algorithm for analysis of data arrays. *Operations Research*, 19, 1350–1362.
Diaconis, P., and B. Efron. 1983. Computer intensive methods in statistics. *Scientific American*, 248, 116–130.
Diday, E. 1988. The symbolic approach in clustering and related methods of data analysis: The basic choices. In *Classification and related methods of data analysis*, Proceedings of IFCS'87, ed. H.-H. Bock, 673–684. Amsterdam: North Holland.
Diday, E. 1989. Introduction à l'approche symbolique en analyse des données, RAIRO. *Recherche Opérationnelle*, 23(2), 193–236.
Diday, E. 1992. Des objets de l'analyse des données à ceux de l'analyse des connaissances. In *Inférences Inductives à Partir de Données Numériques et Symboliques*, eds. Y. Kodratoff and E. Diday, 9–75. Toulouse: CEPADUES Publisher.
Diday, E., and M. Noirhomme-Fraiture, eds. 2008. *Symbolic data analysis and the SODAS software*. Chichester, UK: Wiley.
D'Ocagne, M. 1885. *Coordonnées paralleles et axiales: Méthode de transformation geometrique et procede nouveau de calcul graphique deduits de la consideration des coordonnees parallelles*. Paris: Gauthier-Villars.
D'Ocagne, M. 1899. *Traite de nomographie: Theorie des abaques, applications pratiques*. Paris: Gauthier-Villars.
Dolédec, S., and D. Chessel. 1994. Co-inertia analysis: An alternative method for studying species-environment relationships. *Freshwater Biology*, 31, 277–294.
Dolédec, S., D. Chessel, C. J. F. ter Braak, and S. Champely. 1996. Matching species traits to environmental variables: A new three-table ordination method. *Environmental and Ecological Statistics*, 3, 143–166.
Dorans, N. 2004. *A conversation with Ledyard R Tucker*. Princeton: Educational Testing Service. Retrieved February 14, 2012, from www.ets.org/research/dload/tucker.pdf. (Parts are also available as: Dorans, N. 2004. In memoriam: Ledyard R Tucker (1910–2004). *Journal of Educational and Behavioral Statistics*, 29, 377–378.)
Dray, S., D. Chessel, and J. Thioulouse. 2003a. Co-inertia analysis and the linking of ecological data tables. *Ecology*, 84, 3078–3089.
Dray, S., D. Chessel, and J. Thioulouse. 2003b. Procrustean co-inertia analysis for the linking of multivariate data sets. *Ecoscience*, 10(1), 110–119.

Dray, S., and A. B. Dufour. 2007. The ade4 package: Implementing the duality diagram for ecologists. *Journal of Statistical Software*, 22, 1–20.

Dray, S., A. B. Dufour, and D. Chessel. 2007. The ade4 package. II. Two-table and K-table methods. *R News*, 7(2), 4752.

Dray, S., and T. Jombart. 2011. Revisiting Guerry's data: Introducing spatial constraints in multivariate analysis. *Annals of Applied Statistics*, 5(4), 2278–2299.

Duarte Silva, A. P., and P. Brito. 2006. Linear discriminant analysis for interval data. *Computational Statistics*, 21(2), 289–308.

Dupin, C. 1826. *Carte figurative de l'instruction populaire de la France*. Jobard. BNF: Ge C 6588.

Eckart, C., and G. Young. 1936. The approximation of one matrix by another of lower rank. *Psychometrika*, 1, 211–218.

Edgerton, H. A., and L. E. Kolbe. 1936. The method of minimum variation for the combination of criteria. *Psychometrika*, 1, 183–187.

Efron, B. 1979. Bootstrap methods: Another look at the jackknife. *Annals of Statistitics*, 7, 1–26.

Eisenstein, E. L. 1979. *The printing press as an agent of change: Communications and cultural transformations in early modern Europe* (2 vol. ed.). Cambridge, UK: Cambridge University Press.

Ellenberg, H. 1948. Unkrautgesellschaften als Maß für den Säuregrad, die Verdichtung und andere Eigenschaften des Ackerbodens. *Berichte der Landtechnik*, 4, 130–146.

Ellenberg, H., H. E. Weber, R. Dull, V. Wirth, W. Werner, and D. Paulissen. 1991. *Indicator values of plants in Central Europe*. Scripta Geobotanica 18:0-0. Transition formulae.

Escofier, B. 1978. Analyse factorielle et distances répondant au principe d'équivalence distributionnelle. *Revue de Statistique Appliquée*, 26(4), 29–37.

Escofier, B. 1979a. Traitement simultané de variables qualitatives et quantitatives. *Les Cahiers de l'Analyse des Données*, 4(2), 137–146.

Escofier, B. 1979b. Une représentation des variables dans l'analyse des correspondances multiples. *Revue de Statistique Appliquée*, 27(4), 37–47.

Escofier, B. 2003. *Analyse des correspondances. Recherches au coeur de l'analyse des données*. Rennes, France: Presses Universitaires de Rennes.

Escofier, B., and B. Le Roux. 1976. Influence d'un élément sur les facteurs en analyse des correspondances. *Les Cahiers de l'Analyse des Données*, 3, 297–318.

Escofier, B., and J. Pagès. 1982. *Comparaison de groupes de variables définies sur le même ensemble d'individus*. Rapport de Recherche INRIA, no. 149.

Escofier, B., and J. Pagès. 1984a. L'analyse factorielle multiple. *Cahiers du BURO* (Bureau Universitaire de Recherche Opérationnelle), 42, 3–68.

Escofier, B., and J. Pagès. 1984b. L'analyse factorielle multiple: Une méthode de comparaison de groupes de variables [Multiple factor analysis: A method to compare groups of variables]. In *Data analysis and informatics III*, ed. E. L. Diday, 41–55. Amsterdam: Elsevier.

Escofier, B., and J. Pagès. 1988 (1990, 1998, 2008). *Analyses factorielles simples et multiples; objectifs, méthodes et interprétation*. Paris: Dunod.

Escofier, B., and J. Pagès. 1989. Multiple factor analysis: Results of a three-year utilization. In *Multiway data analysis*, eds. R. Coppi and S. Bolasco, 277–285. Amsterdam: North-Holland.

Escofier, B., and J. Pages. 1994. Multiple factor analysis (Afmult package). *Computational Statistics and Data Analysis*, 18, 121–140.

Escofier-Cordier, B. 1969. L'analyse factorielle des correspondances. *Cahiers du Bureau Universitaire de Recherche Opérationnelle Paris*, 13.

Escoufier, Y. 1973. Le traitement des variables vectorielles. *Biometrics*, 29, 750–760.

Escoufier, Y. 1982. L'analyse des tableaux de contingence simple at multiples. *Metron*, 40, 53–77.

Escoufier, Y. 1987. The duality diagram: A means of better practical applications. In *Developments in numerical ecology*, eds. P. Legendre and L. Legendre, 139–156. Vol. 14. Berlin: Springer.

Faber, N. M., R. Bro, and P. K. Hopke. 2003. Recent developments in CANDECOMP/PARAFAC algorithms: A critical review. *Chemometrics and Intelligent Laboratory Systems*, 65, 119–137.

Farris, J. S. 1969. On the cophenetic correlation coefficient. *Systematic Zoology*, 18, 279–285.

Few, S. 2009. *Now you see it: Simple visualization techniques for quantitative analysis*. Oakland, CA: Analytics Press.

Fisher, L., and van Ness J. W. 1971. Admissible clustering procedures. *Biometrika*, 58(1), 91–104.

Fisher, R. A. 1924. The distribution of the partial correlation. *Metron*, iii, 329–332.

Fisher, R. A. 1936. The use of multiple measurements in taxonomic problems. *Annals of Eugenics*, 6, 179–188.

Fisher, R. A. 1938. *Statistical methods for research workers*. London: Oliver and Boyd.

Fisher, R. A. 1940. The precision of discriminant functions. *Annals of Eugenics*, 10, 422–429.

Fisher, R. A., and W. A. Mackenzie. 1923. Studies in crop variation: The manural response of different potato varieties. *Journal of Agricultural Science*, 13, 311–320.

Fletcher, R. 1991. *Science, ideology and the media: The Cyril Burt scandal*. New Brunswick, NJ: Transaction Publishers.

Florida, R. 2005. *The flight of the creative class: The new global competition for talent*. New York: HarperCollins.

Frère de Montizon, A. J. 1830. *Carte philosophique figurant la population de la France*. BNF.

Friendly, M. 1994. Mosaic displays for multi-way contingency tables. *Journal of the American Statistical Association*, 89, 190–200.

Friendly, M. 2005. Milestones in the history of data visualization: A case study in statistical historiography. In *Classification: The ubiquitous challenge*, eds. C. Weihs and W. Gaul, 34–52. New York: Springer.

Friendly, M. 2007a. A.-M. Guerry's *moral statistics of France*: Challenges for multivariable spatial analysis. *Statistical Science*, 22(3), 368–399.

Friendly, M. 2007b. HE plots for multivariate general linear models. *Journal of Computational and Graphical Statistics*, 16(2), 421–444.

Friendly, M. 2008. The golden age of statistical graphics. *Statistical Science*, 23(4), 502–535.

Friendly, M., and D. Denis. 2001. Milestones in the history of thematic cartography, statistical graphics, and data visualization. http://datavis.ca/gallery/milestone/.

Friendly, M., and E. Kwan. 2011. Comment (graph people versus table people). *Journal of Computational and Graphical Statistics*, 20(1), 18–27.

Friendly, M., G. Monette, and J. Fox. 2013. Elliptical insights: Understanding statistical methods through elliptical geometry. *Statistical Science*, 28(1), 1–39.

Friendly, M., M. Sigal, and D. Harnanansingh. 2013. The milestones project: A database for the history of data visualization. In *Visible numbers*, eds. M. Kimball and C. Kostelnick, chap. 1. London: Ashgate Press. In press.

Froh, J. J., J. Fan, R. A. Emmons, G. Bono, E. S. Huebner, and P. Watkins. 2011. Measuring gratitude in youth: Assessing the psychometric properties of adult gratitude scales in children and adolescents. *Psychological Assessment*, 23, 311–324.

Gabriel, K. R. 1971. The biplot graphic display of matrices with application to principal component analysis. *Biometrika*, 58(3), 453–467.

Galton, F. 1863. *Meteorographica, or methods of mapping the weather*. BL: Maps.53.b.32. London: Macmillan.

Galton, F. 1886. Regression towards mediocrity in hereditary stature. *Journal of the Anthropological Institute*, 15, 246–263.

Galton, F. 1889. *Natural inheritance*. London: MacMillan and Co.

Gantmacher, F. 1959. *Theory of matrices*. Providence, RI: AMS Chelsea Publishing. (Translation of the original 1953 Russian edition.)

Gatrell, A. C., J. Popay, and C. Thomas. 2004. Mapping the determinants of health inequalities in social space: Can Bourdieu help us? *Health and Place*, 10, 245–257.

Gauch, H. G. 1982. *Multivariate analysis in community ecology*. Cambridge: Cambridge University Press.

Gause, G. F. 1930. Studies on the ecology of the orthoptera. *Ecology*, 11, 307–325.

Gebelein, H. 1941. Das Statistische Problem der Korrelation als Variations und Eigenwertproblem und sein Zusammenhang mit der Ausgleichsrechnung. *Zeitschrift für Angewandte Mathematik und Mechanik*, 21, 364–379.

Geladi, P. 1989. Analysis of multi-way (multi-mode) data. *Chemometrics and Intelligent Laboratory Systems*, 7, 11–30.

Gifi, A. 1981. *Nonlinear multivariate analysis*. Leiden, Netherlands: DWSO Press.

Gifi, A. 1990. *Nonlinear multivariate analysis*. Chichester, UK: Wiley.

Gimaret-Carpentier, C., S. Dray, and J.-P. Pascal. 2003. Broad-scale biodiversity pattern of the endemic tree flora of the Western Ghats (India) using canonical correlation analysis of herbarium records. *Ecography*, 26, 429 444.

Glaçon, F. 1981. *Analyse conjointe de plusieurs matrices de données: Comparaison de différentes méthodes*. Unpublished doctoral thesis, Scientific and Medical University of Grenoble, Grenoble, France.

Goodall, D. W., and R. W. Johnson. 1982. Non-linear ordination in several dimensions. A maximum likelihood approach. *Vegetatio*, 48, 197–208.

Goodman, L. A. 1979. Simple models for the analysis of association in cross-classifications having ordered categories. *Journal of the American Statistical Association*, 74, 537–552.

Goodman, L. A. 1981. Association models and canonical correlation in the analysis of cross-classifications having ordered categories. *Journal of the American Statistical Association*, 76, 320–334.

Goodman, L. A., and W. H. Kruskal. 1954. Measures of association for cross classifications. *Journal of the American Statistical Association*, 49, 732–764.

Gordon, A. D. 1996. Hierarchical classification. In *Clustering and classification*, eds. P. Arabie, L. J. Hubert, and G. De Soete, 65–121. River Edge, NJ: World Scientific Publishing Company.

Goscinny, R., and A. Uderzo. 1989. *How Obelix fell into the magic potion when he was a little boy*. London: Hodder and Stoughton.

References

Gould, S. J. 1983. The real error of Cyril Burt. In *The mismeasure of man*. New York: W. W. Norton and Company.

Gourvénec, S., G. Tomasi, C. Durvillec, E. Di Crescenzo, C. A. Saby, D. L. Massarta, R. Bro, and G. Oppenheim. 2005. CuBatch, a MATLAB® interface for n-mode data analysis. *Chemometrics and Intelligent Laboratory Systems*, 77, 122–130. http://www.models.life.ku.dk/CuBatch.

Gowda, K. C., and T. V. Ravi. 1995. Divisive clustering of symbolic objects using the concepts of both similarity and dissimilarity. *Pattern Recognition*, 28(8), 1277–1282.

Gower, J. C. 1958. A note on an iterative method for root extraction. *Computer Journal*, 1(3), 142–143.

Gower, J. C. 1962. The handling of multiway tables on computers. *Computer Journal*, 4(4), 280–286.

Gower, J. C. 1966. Some distance properties of latent and vector methods used in multivariate analysis. *Biometrika*, 53, 325–328.

Gower, J. C. 1968. Adding a point to vector diagram in multivariate analysis. *Biometrika*, 55, 582–585.

Gower, J. C. 1971. Statistical methods of comparing different multivariate analyses of the same data. In *Mathematics in the archaeological and historical sciences*, eds. F. R. Hodson, D. G. Kendall, and P. Tautu, 138–149. Edinburgh: Edinburgh University Press.

Gower, J. C. 1977. The analysis of asymmetry and orthogonality. In *Recent developments in statistics*, eds. J. R. Barra et al., 109–123. Amsterdam: North Holland Press.

Gower, J. C. 1990. Fisher's optimal scores and multiple correspondence analysis. *Biometrics*, 46, 947–961.

Gower, J. C. 2006a. Divided by a common language: Analysing and visualising two-way arrays. In *Multiple correspondence analysis and related methods*, eds. M. Greenacre and J. Blasius, 77–105. Boca Raton, FL: Chapman & Hall.

Gower, J. C. 2006b. Statistica data analytica est et aliter. *Statistica Neerlandica*, 62, 124–134.

Gower, J. C. 2008. The biological stimulus to multidimensional data analysis. *Electronic Journal for History of Probability and Statistics: JEHPS*, 4(2). www.jehps.net.

Gower, J. C., and D. J. Hand. 1996. *Biplots*. London: Chapman & Hall.

Gower, J. C., S. Lubbe, and N. le Roux. 2011. *Understanding biplots*. Chichester, UK: Wiley.

Graffelman, J. 2000. Use of the Moore-Penrose inverse in canonical correspondence analysis—Solution. *Econometric Theory*, 16, 792–793.

Graffelman, J. 2001. Quality statistics in canonical correspondence analysis. *Environmetrics*, 12, 485–497.

Graffelman, J., and R. Tuft. 2004. Site scores and conditional biplots in canonical correspondence analysis. *Environmetrics*, 15, 67–80.

Graham, R. L., and P. Hell. 1985. On the history of the minimum spanning tree problem. *Annals of History of Computation*, 7, 43–57.

Gray, J., L. Chambers, and L. Bounegru. 2012. *The data journalism handbook*. O'Reilly Media.

Green, P. E., and F. J. Carmone. 1970. *Multidimensional scaling and related techniques in marketing research*. Boston, MA: Allyn & Bacon.

Green, P. E., and V. Rao. 1972. *Applied multidimensional scaling*. Hinsdale, IL: Dryden.

Green, R. H. 1971. A multivariate statistical approach to the Hutchinsonian niche: Bivalve molluscs of central Canada. *Ecology*, 52, 543–556.

Green, R. H. 1974. Multivariate niche analyis with temporally varying environmental factors. *Ecology*, 55, 73–83.

Greenacre, M. 1984. *Theory and applications of correspondence analysis.* London: Academic Press.

Greenacre, M. 1988a. Correspondence analysis of multivariate categorical data by weighted least squares. *Biometrika*, 75, 457–467.

Greenacre, M. 1988b. Clustering the rows and columns of a contingency table. *Journal of Classification*, 5, 39–51.

Greenacre, M. 1995. Multivariate generalisations of correspondence analysis. In *Multivariate analysis: Future directions 2*, eds. C. M. Cuadras and C. R. Rao, 327–340. Amsterdam: North Holland.

Greenacre, M. 2006. From simple to multiple correspondence analysis. In *Multiple correspondence analysis and related methods,* eds. M. Greenacre and J. B. Blasius, 41–76. Boca Raton, FL: Chapman & Hall.

Greenacre, M. 2007. *Correspondence analysis in practice.* 2nd ed. London: Chapman & Hall. http://www.multivariatestatistics.org.

Greenacre, M. 2009. Power transformations in correspondence analysis. *Computational Statistics and Data Analysis*, 53, 3107–3116.

Greenacre, M. 2010. *Biplots in practice.* Madrid: BBVA Foundation. http://www.multivariatestatistics.org.

Greenacre, M. J. 2012. Biplots: The joy of singular value decomposition. *WIREs Computational Statistics*, 4, 399–406.

Greenacre, M. 2013. Contribution biplots. *Journal of Computational and Graphical Statistics*, 22(1), 107–122.

Greenacre, M., and J. Blasius. 1994. Preface. In *Correspondence analysis in the social sciences. Recent developments and applications*, eds. M. Greenacre and J. Blasius, vii–xv. London: Academic Press.

Greenacre, M., and J. Blasius, eds. 2006. *Multiple correspondence analysis and related methods.* London: Chapman & Hall.

Greenacre, M., and P. Lewi. 2009. Distributional equivalence and subcompositional coherence in the analysis of compositional data, contingency tables and ratio-scale measurements. *Journal of Classification*, 26, 29–54.

Greenacre, M., and R. Pardo. 2006a. Subset correspondence analysis. Visualization of selected response categories in a questionnaire survey. *Sociological Methods and Research*, 35, 193–218.

Greenacre, M., and R. Pardo. 2006b. Multiple correspondence analysis of subsets of response categories. In *Multiple correspondence analysis and related methods*, eds. M. Greenacre and J. Blasius, 197–217. Boca Raton, FL: Chapman & Hall.

Griffiths, A., L. A. Robinson, and P. Willett. 1984. Hierarchic agglomerative clustering methods for automatic document classification. *Journal of Documentation*, 40, 175–205.

Groenen, P. J. F., and W. J. Heiser. 1996. The tunneling method for global optimization in multidimensional scaling. *Psychometrika*, 61, 529–550.

Groenen, P. J. F., W. J. Heiser, and J. J. Meulman. 1999. Global optimization in least-squares multidimensional scaling by distance smoothing. *Journal of Classication*, 16, 225–254.

Groenen, P. J. F., R. Mathar, and W. J. Heiser. 1995. The majorization approach to multidimensional scaling for Minkowski distances. *Journal of Classication*, 12, 3–19.

Groenen, P. J. F., and I. van der Lans. 2004. Multidimensional scaling with regional restrictions for facet theory: An application to Levy's political protest data. In *Beyond the horizon of measurement*, eds. M. Braun and P. Mohler, 41–64. Mannheim: ZUMA.

Groenen, P. J. F., and S. Winsberg. 2006. Multidimensional scaling of histogram dissimilarities. In *Data science and classification*, eds. V. Batagelj, H. Bock, A. Ferligoj, and A. Ziberna, 161–170. Berlin: Springer.

Guerry, A.-M. 1829. Tableau des variations meteorologique comparees aux phenomenes physiologiques, d'apres les observations faites a l'obervatoire royal, et les recherches statistique les plus recentes. *Annales d'Hygiene Publique et de Medecine Legale*, 1, 228–237.

Guerry, A.-M. 1832. *Statistique comparée de l'etat de l'instruction et du nombre des crimes*. Paris: Everat.

Guerry, A.-M. 1833. *Essai sur la statistique morale de la France*. Paris: Crochard. (English translation: Whitt, H. P., and V. W. Reinking 2002. Lewiston, NY: Edwin Mellen Press.)

Guerry, A.-M. 1864. *Statistique morale de l'Angleterre comparee avec la statistique morale de la France, d'apres les comptes de l'administration de la justice criminelle en Angleterre et en France, etc.* BNF: GR FOL-N-319; SG D/4330; BL: Maps 32.e.34; SBB: Fe 8586; LC: 11005911. Paris: J.-B. Bailliere et ls.

Guittonneau, A., and M. Roux. 1977. Sur la taxinomie de genre Erodium. *Les Cahiers de l'Analyse des Données*, 2, 97–113.

Guttman, L. 1941. The quantification of a class of attributes: A theory and method of scale construction. In *The prediction of personal adjustment*, ed. P. Horst, 318–348. New York: Social Science Research.

Guttman, L. 1946. An approach for quantifying paired comparisons and rank order. *Annals of Mathematical Statistics*, 17, 144–163.

Guttman, L. 1950. The principal components of scale analysis. In *Measurement and prediction*, eds. S. A. Stouffer, L. Guttman, E. A. Suchman, P. F. Lazarsfeld, S. A. Star, and J. A. Clausen. Princeton, NJ: Princeton University Press.

Guttman, L. 1953. A note on Sir Cyril Burt's factorial analysis of qualitative data. *British Journal of Statistical Psychology*, 6, 1–4.

Guttman, L. 1959. Metricizing rank-ordered or unordered data for a linear factor analysis. *Sankhya*, A21, 257–268.

Guttman, L. 1964. The structure of interrelations among intelligence tests. In *Proceedings of the 1964 Invitational Conference on Testing Problems*. Princeton, NJ: Educational Testing Service.

Guttman, L. 1968. A general nonmetric technique for finding the smallest coordinate space for a configuration of points. *Psychometrika*, 33, 469–506.

Guttman, L. 1985. Multidimensional structuple analysis (MSA-I) for the classification of cetacea: Whales, porpoises and dolphins. In *Data analysis in real life environment: Ins and outs of solving problems*, eds. J.-F. Marcotorchino, J.-M. Proth, and J. Janssen, 45–54. Advanced Series in Management 8. Amsterdam: North Holland.

Hald, A. 1990. *A history of probability and statistics and their applications before 1750*, i–xiii, 1–586. New York: Wiley.

Halley, E. 1701. *The description and uses of a new and correct sea-chart of the whole world, shewing variations of the compass*. London.

Hamrouni, A. 1977. Programme de calcul dont les contributions relatives pour plusieurs groupes homogènes de variables. *Les Cahiers de l'Analyse des Données*, 2, 353–359.

Hankins, T. L. 1999. Blood, dirt, and nomograms: A particular history of graphs. *Isis*, 90, 50–80.

Harris, Z. S. 1954. Distributional structure. *Word*, 10(23), 146–162.

Harshman, R. A. 1970. Foundation of the PARAFAC procedure: Models and conditions for an "explanatory" multi-modal factor analysis. *UCLA Working Papers in Phonetics*, 16, 1–84. http://www.psychology.uwo.ca/faculty/harshman/wpppfac0.pdf.

Harshman, R. A. 1972a. PARAFAC2: Mathematical and technical notes. *UCLA Working Papers in Phonetics*, 22, 30–44. http://www.psychology.uwo.ca/faculty/harshman/wpppfac2.pdf.

Harshman, R. A. 1972b. Determination and proofs of minimum uniqueness conditions for PARAFAC1. *UCLA Working Papers in Phonetics*, 22, 111–117. http://www.psychology.uwo.ca/faculty/harshman/wpppfac1.pdf.

Harshman, R. A. 1984. "How do I know if it's 'real'?" A catalog of diagnostics for use with three-mode factor analysis and multidimensional scaling. In *Research methods for multimode data analysis*, eds. H. G. Law, C. W. Snyder Jr., J. A. Hattie, and R. P. McDonald, 566–591. New York: Praeger.

Harshman, R. A. 2001. An index to formalism that generalizes the capabilities of matrix notation and algebra to n-way arrays. *Journal of Chemometrics*, 15, 689–714.

Harshman, R. A., and S. A. Berenbaum. 1981. Basic concepts underlying the PARAFAC-CANDECOMP three-way factor analysis model and its application to longitudinal data. In *Present and past in middle life*, eds. D. H. Eichorn, J. A. Clausen, N. Haan, M. P. Honzik, and P. H. Mussen, 435–459. New York: Academic Press.

Harshman, R. A., P. Ladefoged, and L. Goldstein. 1977. Factor analysis of tongue shapes. *Journal of the Acoustical Society of America*, 62, 693–707.

Harshman, R. A., and M. E. Lundy. 1984a. The PARAFAC model for three-way factor analysis and multidimensional scaling. In *Research methods for multimode data analysis*, eds. H. G. Law, C. W. Snyder Jr., J. A. Hattie, and R. P. McDonald, 122–215. New York: Praeger.

Harshman, R. A., and M. E. Lundy. 1984b. Data preprocessing and the extended PARAFAC model. In *Research methods for multimode data analysis*, eds. H. G. Law, C. W. Snyder Jr., J. A. Hattie, and R. P. McDonald, 216–284. New York: Praeger.

Harshman, R. A., and M. E. Lundy. 1994. PARAFAC—Parallel factor-analysis. *Computational Statistics and Data Analysis*, 18, 39–72.

Hatheway, W. H. 1971. Contingency-table analysis of rain forest vegetation. In *Statistical ecology*, eds. G. P. Patil, E. C. Pielou, and W. E. Waters. University Park: Pennsylvania State University Press.

Hathout, A., and M. Reinert. 1981. Recherches sur les profils sonores des textes poétiques. In *Pratique de l'analyse des données: Linguistique and lexicologie*, ed. J.-P. Benzécri, 203–226. Vol. 3. Paris: Dunod.

Hayashi, C. 1952. On the prediction of phenomena from qualitative data and the quantification of qualitative data from the mathematico-statistical point of view. *Annals of the Institute of Statistical Mathematics*.

Hayashi, C. 1954. Multidimensional quantification I. *Proceedings of the Japan Academy*, 30, 61–65. Available at http://www.ism.ac.jp/editsec/aism/pdf/005_2_0121.pdf.

Hayashi, C. 1956. Theory and examples of quantification (II). *Proceedings of the Institute of Statistical Mathematics*, 4(2), 19–30.

Healy, M. J. R. 1978. Is statistics a science? *Journal of the Royal Statistical Society, Series A*, 141(3), 385–393.

Hearnshaw, L. S. 1979. *Cyril Burt, psychologist*. London: Hodder and Stoughton.

Heiser, W. J. 1981. *Unfolding analysis of proximity data*. Thesis, University of Leiden, Leiden.

Heiser, W. J. 1987. *Joint ordination of species and sites: The unfolding technique*. Berlin: Springer.

Heiser, W. J. 1988. Multidimensional scaling with least absolute residuals. In *Classification and related methods*, ed. H. H. Bock, 455–462. Amsterdam: North-Holland.

Henrion, R. 1994. N-way principal component analysis. Theory, algorithms and applications. *Chemometrics and Intelligent Laboratory Systems*, 25, 1–23.

Herrmann, W. M., J. Röhmel, B. Streitberg, and J. Willman. 1983. Example for applying the comstat multimodal factor analysis algorithm to eeg data to describe variance sources. *Neuropsychobiology*, 10, 164–172.

Herschel, J. F. W. 1833. On the investigation of the orbits of revolving double stars. *Memoirs of the Royal Astronomical Society*, 5, 171–222.

Hill, M. O. 1973. Reciprocal averaging: An eigenvector method of ordination. *Journal of Ecology*, 61, 237–249.

Hill, M. O. 1974. Correspondence analysis: A neglected multivariate method. *Applied Statistics*, 3, 340–354.

Hill, M. O. 1979a. *Decorana—A FORTRAN program for detrended correspondence analysis and reciprocal averaging. Ecology and Systematics*. Ithaca, NY: Cornell University.

Hill, M. O. 1979b. *Twinspan—A FORTRAN program for detrended correspondence analysis and reciprocal averaging*. Ithaca, NY: Cornell University.

Hill, M. O., and H. G. Gauch. 1980. Detrended correspondence analysis: An improved ordination technique. *Vegetatio*, 42, 47–58.

Hirschfeld, H. O. 1935. A connection between correlation and contingency. *Proceedings of the Cambridge Philosophical Society*, 31, 520–524.

Hirschfeld, T. 1985. Instrumentation in the next decade. *Science*, 230(4723), 286–291.

Hitchcock, F. L. 1927a. Multiple invariants and generalized rank of a p-way matrix or tensor. *Journal of Mathematics and Physics*, 7, 39–79.

Hitchcock, F. L. 1927b. The expression of a tensor or a polyadic as a sum of products. *Journal of Mathematics and Physics*, 6, 164–189.

Hjellbrekke, J., B. Le Roux, O. Korsnes, F. Lebaron, R. Lennart, and H. Rouanet. 2007. The Norwegian field of power anno 2000. *European Societies*, 9, 245–273.

Ho, C. N., G. D. Christian, and E. R. Davidson. 1978. Application of the method of rank annihilation to quantative analysis of multicomponent fluorescence data from the video fluorometer. *Analytical Chemistry*, 50, 1108–1113.

Hohn, M. E. 1979. Principal components analysis of three-way tables. *Mathematical Geology*, 11, 611–626.

Holmes, S. 2006. Multivariate analysis: The French way. In *Festschrift for David Freedman*, eds. D. Nolan and T. Speed, 219–233. Beachwood, OH: Institute of Mathematical Statistics.

Horan, C. B. 1969. Multidimensional scaling: Combining observations when individuals have different perceptual structures. *Psychometrika*, 34, 139–165.
Horst, P. 1935. Measuring complex attitudes. *Journal of Social Psychology*, 6(3), 369–374.
Horst, P. 1936. Obtaining a composite measure from a number of different measures of the same attribute. *Psychometrika*, 1, 53–60.
Horst, P. 1961. Relation among m sets of measures. *Psychometrika*, 26, 129–149.
Horst, P. 1965. *Factor analysis of data matrices*. New York: Holt, Rinehart, Winston.
Hotelling, H. 1933. Analysis of a complex of statistical variables into principal components. *Journal of Educational Psychology*, 24, 417–441, 498–520.
Hotelling, H. 1936. Relations between two sets of variates. *Biometrika*, 28, 321–377.
Hubert, L. J., and P. Arabie. 1986. Unidimensional scaling and combinatorial optimization. In *Multidimensional data analysis*, eds. J. de Leeuw, W. J. Heiser, J. J. Meulman, and F. Critchley, 181–196. Leiden, Netherlands: DSWO Press.
Hubert, L. J., P. Arabie, and M. Hesson-McInnis. 1992. Multidimensional scaling in the city-block metric: A combinatorial approach. *Journal of Classification*, 9, 211–236.
Hubert, L. J., and R. G. Golledge. 1981. Matrix reorganisation and dynamic programming: Applications to paired comparison and unidimensional seriation. *Psychometrika*, 46, 429–441.
Husson, F., J. Josse, and J. Mazet. 2013. FactoMineR: Multivariate exploratory data analysis and data mining with R. R package version 1.24. http://factominer.free.fr.
Husson, F., S. Lê, and J. Pagès. 2011. *Exploratory multivariate analysis by example using R*. London: Chapman & Hall/CRC Press.
Huynh-Armanet, V. 1976. *Recherches sur la structuration syntaxique de l'espagnol contemporain*. Paris: Honoré-Champion.
IBM Corporation. 2012. *GPL reference guide for IBM SPSS visualization designer*. IBM SPSS Statistics for Windows, Version 21.0. Armonk, NY: IBM Corp.
Ichino, M. 2008. Symbolic PCA for histogram-valued data. In *Proceedings of IASC 2008, Joint Meeting of 4th World Conference of the IASC and 6th Conference of the Asian Regional Section of the IASC on Computational Statistics and Data Analysis*, Yokohama, Japan. http://www.dspace.cam.ac.uk/bitstream/1810/236581/1/DSpace%20Dunning%20Cairns%20Russell%20Lynch%202008.pdf.
Ihm, P., and H. van Groenewoud. 1984. Correspondence analysis and Gaussian ordination. *Compstat Lectures*, 3, 5–60.
Irpino A., and R. Verde. 2006. A new Wasserstein based distance for the hierarchical clustering of histogram symbolic data. In *Data science and classification, IFCS 2006*, eds. V. Batanjeli, H.-H. Bock, A. Ferligoj, A. Ziberna, 185–192. Berlin: Spinger.
Irpino A., and R. Verde. 2010. Ordinary least squares for histogram data based on Wasserstein distance. In *Proceedings of COMPSTAT'2010*, eds. Y. Lechevallier and G. Saporta, 581–588. Heidelberg: PhysicaVerlag.
ISSP. 2002. ISSP Research Group, International Social Survey Programme (ISSP): Family and Changing Gender Roles III. Distributor: GESIS, Cologne, Germany ZA 3880, Data version 1.1.0 (2013-03-04).
Iwatsubo, S. 1974. Two classification techniques of 3-way discrete data: Quantification by means of correlation ratio and three-dimensional correlation coefficient. *Kodokeiryogaku (Japanese Journal of Behaviormetrics)*, 2, 54–65; English abstract, 79.
Jain, A. K. 2010. Data clustering: 50 years beyond k-means. *Pattern Recognition Letters*, 31, 651–666.

Jambu, M. 1978. *Classification automatique pour l'analyse des données, Tôme 1*. Paris: Dunod. (1983. *Cluster analysis and data analysis*. Amsterdam: North-Holland.)
Jambu, M. 1989. *Exploration informatique et statistique des données*. Paris: Dunod.
Jaynes, J. 1978. *The origin of consciousness in the breakdown of the bicameral mind*. London: Houghton Mifflin.
Johnson, K. W., and N. S. Altman. 1999. Canonical correspondence analysis as an approximation to Gaussian ordination. *Environmetrics*, 10, 39–52.
Johnson, S. C. 1967. Hierarchical clustering schemes. *Psychometrika*, 32, 241–254.
Jongman, R. H. G., C. J. F. ter Braak, and O. F. R. van Tongeren. 1987. *Data analysis in community and landscape ecology*. Wageningen, Netherlands: Pudoc.
Jongman, R. H. G., C. J. F. ter Braak, and O. F. R. van Tongeren. 1995. *Data analysis in community and landscape ecology*. Cambridge: Cambridge University Press.
Jordan, C. 1870. Traite des substitutions et des equations algebraiques. *Livre*, 2, 88–249.
Jordan, C. 1874. Mémoire sur les formes bilineares. *Journal de Mathématiques Pures et Appliquées*, 19, 35–54.
Josse, J., M. Chavent, B. Liquet, and F. Husson. 2012. Handling missing values with regularized iterative multiple correspondence analysis. *Journal of Classification*, 29(1), 91–116.
Joynson, R. B. 1989. *The Burt affair*. New York: Routledge.
Juan, J. 1982a. Le programme HIVOR de classification ascendante hiérarchique selon les voisins réciproques et le critère de la variance. *Les Cahiers de l'Analyse des Données*, VII, 173–118.
Juan, J. 1982b. Programme de classification hiérarchique par l'algorithme de la recherche en chaîne des voisins réciproques. *Les Cahiers de l'Analyse des Données*, VII, 219–225.
Kapteyn, A., H. Neudecker, and T. Wansbeek. 1986. An approach to n-mode components analysis. *Psychometrika*, 51, 269–275.
Katz, J. 2012. *Designing information*. Hoboken, NJ: John Wiley & Sons.
Kendall, M. G., and A. Stuart. 1961. *The advanced theory of statistics, 2*. London: Griffin.
Kettenring, R. J. 1971. Canonical analysis of several sets of variables. *Biometrika*, 58(3), 433–450.
Kiers, H. A. L. 1988. Comparison of "Anglo-Saxon" and "French" three-mode methods. *Statistique et Analyse des Données*, 13, 14–32.
Kiers, H. A. L. 1991. Hierarchical relations among three-way methods. *Psychometrika*, 56, 449–470.
Kiers, H. A. L. 1998a. Recent developments in three-mode factor analysis: Constrained three-mode factor analysis and core rotations. In *Data science, classification, and related methods*, eds. C. Hayashi, N. Ohsumi, K. Yajima, Y. Tanaka, H.-H. Bock, and Y. Baba, 563–574. Tokyo: Springer.
Kiers, H. A. L. 1998b. Joint orthomax rotation of the core and component matrices resulting from three-mode principal components analysis. *Journal of Classification*, 15, 245–263.
Kiers, H. A. L. 2000a. Towards a standardized notation and terminology in multiway analysis. *Journal of Chemometrics*, 14, 105–122.
Kiers, H. A. L. 2000b. Some procedures for displaying results from three-way methods. *Journal of Chemometrics*, 14, 151–170.
Kiers, H. A. L., J. M. F. ten Berge, and R. Bro. 1999. PARAFAC2. Part I. A direct fitting algorithm for the PARAFAC2 model. *Journal of Chemometrics*, 13, 275–294.

Kiers, H. A. L., and I. van Mechelen. 2001. Three-way component analysis: Principles and illustrative application. *Psychological Methods*, 6, 84–110.

Klauser, H. A. 1987. *Writing on both sides of the brain: Breakthrough techniques for people who write*. London: HarperOne.

Kloek, T., and H. Theil. 1965. International comparisons of prices and quantities consumed. *Econometrica*, 33, 535–556.

Kolda, T. G., and B. W. Bader. 2009. Tensor decompositions and applications. *SIAM Review*, 51, 455–500.

Kooijman, S. A. L. M. 1977. Species abundance with optimum relations to environmental factors. *Annals of System Research*, 6, 123–138.

Kooijman, S. A. L. M., and R. Hengeveld. 1979. The description of a non-linear relationship between some carabid beetles and environmental factors. In *Contemporary quantitative ecology and related econometrics*, eds. G. P. Patil and M. L. Rosenzweig, 635–647. Fairland, MD: International Co-operative Publishing House.

Koyak, R. 1987. On measuring internal dependence in a set of random variables. *Annals of Statistics*, 15, 1215–1228.

Krantz, D. H., R. D. Luce, P. Suppes, and A. Tversky. 1971. *Foundations of measurement*. Vol. 1. New York: Academic.

Krasner, M. 1944. Nombres semi-réels et espaces ultramétriques. *Comptes-Rendus de l'Académie des Sciences*, II, 219–433.

Krishnan, A., L. J. Williams, A. R. McIntosh, and H. Abdi. 2010. Partial least squares (PLS) methods for neuroimaging: A tutorial and review. *Neuroimage*, 56(2), 455–475.

Kroonenberg, P. M. 1983a. *Three-mode principal component analysis*. Leiden, Netherlands: DSWO Press.

Kroonenberg, P. M. 1983b. Annotated bibliography of three-mode factor analysis. *British Journal of Mathematical and Statistical Psychology*, 36, 81–113.

Kroonenberg, P. M. 1983c. Three-mode correspondence analysis. Presented at the 3rd European Meeting of the Psychometric Society, Jouy-en-Josas, France.

Kroonenberg, P. M. 1989. Singular value decompositions of interactions in three-way contigency tables. In *Multiway data analysis*, eds. R. Coppi and S. Bolasco, 169–184. Amsterdam: North Holland.

Kroonenberg, P. M. 2008a. *Applied multiway data analysis*. Hoboken, NJ: Wiley.

Kroonenberg, P. M. 2008b. *3WayPack: Three-mode correspondence analysis*. Leiden, Netherlands: Three-Mode Company, Leiden University. http://three-mode.leidenuniv.nl/.

Kroonenberg, P. M. 2012. Multiway data in the social and behavioral sciences: Practice, challenges and prospects. Presented at Tricap 2012, Brugge, Belgium.

Kroonenberg, P. M., and C. J., Anderson. 2008. Additive and multiplicative models for three-way contingency tables: Darroch (1974) revisited. In *Multiple correspondence analysis and related methods*, eds. M. Greenacre and J. Blasius, 455–486. Boca Raton, FL: Chapman & Hall.

Kroonenberg, P. M., and J. de Leeuw. 1977. *Tuckals2: A principal component analysis of three mode data*. Research Bulletin RB 001-'77. Department of Data Theory, Leiden University, Leiden, Netherlands.

Kroonenberg, P. M., and J. de Leeuw. 1980. Principal component analysis of three-mode data by means of alternating least squares algorithms. *Psychometrika*, 45, 69–97.

Kroonenberg, P. M., and Y. de Roo. 2010. *3WayPack. A program suite for three-way analysis*. Leiden, Netherlands: Three-Mode Company, Leiden University. http://three-mode.leidenuniv.nl.

Kruskal, J. B. 1964a. Multidimensional scaling by optimizing goodness of fit to a nonmetric hypothesis. *Psychometrika*, 29, 1–27.

Kruskal, J. B. 1964b. Nonmetric multidimensional scaling: A numerical method. *Psychometrika*, 29, 115–129.

Kruskal, J. B. 1977. Three-way arrays: Rank and uniqueness of trilinear compositions, with application to arithmetic complexity and statistics. *Linear Algebra and Its Applications*, 18, 95–138.

Kruskal, J. B., and R. N. Shepard. 1974. A nonmetric variety of linear factor analysis. *Psychometrika*, 39, 123–157.

Kuyumcuyan, A. 1999. Hétérogénéité textuelle: L'exemple de la fable. *Cahiers de Linguistique Française*, 21, 151–179.

Lalanne, L. 1844. *Abaque, ou Compteur univsersel, donnant a vue a moins de 1/200 pres les resultats de tous les calculs d'arithmetique, de geometrie et de mecanique practique*. Paris: Carilan-Goery et Dalmont.

Lalanne, L. 1845. Appendice sur la representation graphique des tableaux meteorologiques et des lois naturelles en general. In *Cours Complet de Meteorologie*, ed. L. F. Kaemtz, trans. and annot. C. Martins, 1–35. Paris: Paulin.

Lallemand, C. 1885. *Les abaques hexagonaux: Nouvelle methode generale de calcul graphique, avec de nombreux exemples d'application*. Paris: Ministere des travaux publics, Comité du nivellement general de la France.

Lancaster, H. O. 1951. Complex contingency tables treated by the partition of χ^2. *Journal of the Royal Statistical Society B*, 13, 242–249.

Lancaster, H. O. 1957. The structure of bivariate distributions, *Annals of Mathematical Statistics*, 29, 719–736.

Lancaster, H. O. 1963. Canonical correlation and partition of χ^2. *Quarterly Journal of Mathematics*, 14, 220–224.

Lancaster, H. O. 1969. *The chi-squared distribution*. New York: Wiley.

Lance, G. N., and W. T. Williams. 1967. A general theory of classificatory sorting strategies. 1. Hierarchical systems. *Computer Journal*, 9(4), 373–380.

Lastovicka, J. L. 1981. The extension of component analysis to four-mode matrices. *Psychometrika*, 46, 47–57.

Lauro, N. C., R. Verde, and A. Irpino, 2008a. Factorial discriminant analysis. In *Symbolic data analysis and the sodas software*, eds. E. Diday and M. Noirhomme-Fraiture, 341–358. Chichester, UK: Wiley.

Lauro, N. C., R. Verde, and A. Irpino. 2008b. Principal component analysis of symbolic data described by intervals. In *Symbolic data analysis and the Sodas software*, eds. E. Diday and M. Noirhomme-Fraiture, 279–312. Chichester, UK: Wiley.

Lauro, N. C., R. Verde, and A. Irpino. 2008c. Generalized canonical analysis. In *Symbolic data analysis and the Sodas software*, eds. E. Diday and M. Noirhomme-Fraiture, 313–330. Chichester, UK: Wiley.

Lauro, N. C., R. Verde, and F. Palumbo. 2000. Factorial discriminant analysis on symbolic objects. In *Analysis of symbolic data, exploratory methods for extracting statistical information from complex data*, eds. H.-H. Bock and E. Diday, 212–233. Heidelberg: Springer.

Lavit C., Y. Escoufier, R. Sabatier, and P. Traissac. 1994. The ACT (STATIS method). *Computational Statistics and Data Analysis*, 18, 97–119.

Lavit, C. 1988. *Analyse conjointe de tableaux quantitatifs*. Paris: Masson.
Lavorel, S., C. Rochette, and J. D. Lebreton. 1999. Functional groups for response to disturbance in Mediterranean old fields. *Oikos*, 84, 480–498.
Law Jr., H. G., C. W. Snyder, Jr., J. A. Hattie, and R. P. McDonald, eds. 1984. *Research methods for multimode data analysis*. New York: Praeger.
Lê, S., J. Josse, and F. Husson. 2008. FactoMineR: An R package for multivariate analysis. *Journal of Statistical Software*, 25(1), 1–18.
Lebaron, F. 2000. Economists and the economic order. The field of economists and the field of power in France. *European Societies*, 3, 91–110.
Lebaron, F. 2009. How Bourdieu "quantified" Bourdieu: The geometric modeling of data. In *Quantifying theory: Pierre Bourdieu*, eds. K. Robson and C. Sanders, 11–29. New York: Springer.
Lebart, L. 1974. On the Benzécri's method for finding eigenvectors by stochastic approximation. In *Proceedings in Computational Statistics (COMPSTAT)*, 202–211. Vienna: Physica.
Lebart, L. 1975. L'orientation du dépouillement de certaines enquêtes par l'analyse des correspondances multiples. *Consommation*, 2, 73–96.
Lebart, L. 2003. Analyse des données textuelles. In *Analyse des Données*, ed. G. Govaert, 151–168. Paris: Lavoisier.
Lebart, L., and N. Tabard. 1973. *Recherches sur la description automatique des données socio-economiques*. Rapport CORDES-CREDOC. Convention de Recherche 13/1971.
Lebart, L., A. Morineau, and N. Tabard. 1977. *Techniques de la description statistique*. Paris: Dunod.
Lebart L., A. Morineau, and K. Warwick. 1984. *Multivariate descriptive statistical analysis*. New York: Wiley.
Lebart, L., A. Morineau, and M. Piron. 1995 (1997, 2006). *Statistique exploratoire multidimensionnelle*. Paris: Dunod.
Lebart, L., A. Salem, and L. Berry. 1998. *Exploring textual data*. Dordrecht: Kluwer.
Lebreton, J.-D., K. P. Burnham, J. Clobert, and D. R. Anderson. 1992. Modelling survival and testing biological hypotheses using marked animals: A unified approach with case studies. *Ecological Monographs*, 62, 67–118.
Lebreton, J. D., D. Chessel, R. Prodon, and N. Yoccoz. 1988a. L'analyse des relations espèces-milieu par l'analyse canonique des correspondances. I. Variables de milieu quantitatives. *Acta Oecologia Generalis*, 9, 53–67.
Lebreton, J. D., D. Chessel, M. Richardot-Coulet, and N. Yoccoz. 1988b. L'analyse des relations espèces-milieu par l'analyse canonique des correspondances. II. Variables de milieu qualitatives. *Acta Oecologia Generalis*, 9, 137–151.
Lebreton, J. D., R. Sabatier, G. Banco, and A. M. Bacou. 1991. Principal component and correspondence analysis with respect to instrumental variables: An overview of their role in studies of structure-activity and species-environment relationships. In *Applied multivariate analysis in SAR and environmental studies*, eds. J. Devillers and W. Karcher, 85–114. Dordrecht: Kluwer.
Leclerc, A. 1975. L'analyse des correspondances sur juxtaposition de tableaux de contingence. *Revue de Statistique Appliquée*, 23(3), 516.
Le Dien, S., and J. Pagès. 2003a. Analyse factorielle multiple hiérarchique. *Revue Statistique Appliquée*, LI(2), 47–73.
Le Dien, S., and J. Pagès. 2003b. Hierarchical multiple factor analysis: Application to the comparison of sensory profiles. *Food Quality and Preference*, 14(5–6), 397–403.

Lee, J. A., and M. Verleysen. 2007. *Nonlinear dimensionality reduction*. New York: Springer.

Legendre, P., and L. Legendre. 1998. *Numerical ecology*. 2nd ed. Amsterdam: Elsevier.

Leibovici, D. G. 2010. Spatio-temporal multiway data decomposition using principal tensor analysis on k-modes: The R package PTAk. *Journal of Statistical Software*, 34(10), 1–34.

Leinster, T., and C. A. Cobbold. 2012. Measuring diversity: The importance of species similarity. *Ecology*, 93(3), 477–489.

Lengen, C., and J. Blasius. 2007. Constructing a Swiss health space model of self-perceived health. *Social Science and Medicine*, 65, 80–94.

Lengen, C., J. Blasius, and T. Kistemann. 2008. Self-perceived health space and geographic areas in Switzerland. *International Journal of Hygiene and Environmental Health*, 211, 420–431.

Lerman, I. C. 1981. *Classification et l'analyse ordinale des données*. Paris: Dunod.

Le Roux, B. 1991. Sur la construction d'un protocole additif de référence. *Mathématiques Informatique et Sciences Humaines*, 114, 57–62.

Le Roux, B. 2014. *Analyse géométrique des données multidimensionnelles*. Paris: Dunod.

Le Roux, B., and H. Rouanet. 1983. L'analyse statistique des protocoles multidimensionnels: Analyse des comparaisons (nuage pondéré sur le croisement de deux facteurs). *Publications de l'Institut de Statistique de Paris*, 28, 7–70.

Le Roux, B., and H. Rouanet. 1984. L'analyse multidimensionnelle des données structurées. *Mathématiques Informatique et Sciences humaines*, 85–18.

Le Roux, B., and H. Rouanet. 1998. Interpreting axes in multiple correspondence analysis: Method of the contributions of points and deviations. In *Visualization of categorical data*, eds. J. Blasius and M. Greenacre, 197–220. San Diego: Academic Press.

Le Roux, B., and H. Rouanet. 2004. *Geometric data analysis. From correspondence analysis to structured data analysis*. Dordrecht: Kluwer.

Le Roux, B., and H. Rouanet. 2010. *Multiple correspondence analysis*. Thousand Oaks, CA: Sage.

Le Roux, B., H. Rouanet, M. Savage, and A. Warde. 2008. Class and cultural division in the UK. *Sociology*, 42(6), 1049–1071.

Levin, J. 1965. Three-mode factor analysis. *Psychological Bulletin*, 64, 442–452.

L'Hermier des Plantes, H. 1976. *Structuration des tableaux à trois indices de la statistique*. Thèse de troisième cycle, Université de Montpellier.

L'Hermier des Plantes, H., and B. Thiébaut. 1977. Étude de la pluviosité au moyen de la méthode S. T. A. T. I. S. *Revue de Statistique Appliquée*, 25(2), 57–81.

Lingoes, J. C. 1968. The multivariate analysis of qualitative data. *Multivariate Behavioral Research*, 3, 61–94.

Lingoes, J. C. 1973. *The Guttman-Lingoes nonmetric program series*. Pittsburgh, PA: Mathesis Press.

Lingoes, J. C., and I. Borg. 1978. A direct approach to individual differences scaling using increasingly complex transformations. *Psychometrika*, 43, 491–519.

Lingoes, J. C., and L. Guttman. 1967. Nonmetric factor analysis: A rank seducing alternative to linear factor analysis. *Multivariate Behavioral Research*, 2, 485–505.

Lloyd, D. B. 1964. Data retrieval. *Computer Journal*, 7(2), 110–113.

Lombardo, R. 1994. *Modelli di decomposizione per l'analisi della dipendenza nelle tabelle di contingenza a tre vie* [Decomposition models for the analysis of three-way contingency tables]. Tesi dottorato in Statisica Computazionale et Applicazioni VI cicio, Università di Napoli, Italy.

Lombardo, R., E. J. Beh, and L. D'Ambra. 2007. Non-symmetric correspondence analysis with ordinal veriables using orthogonal polynomials. *Computational Statistics & Data Analysis*, 52, 566–578.

Lombardo, R., A. Carlier, and L. D'Ambra. 1996. Non-symmetric correspondence analysis for three-way contingency tables. *Methodologica*, 4, 59–80.

López Sintas, J., and E. García Álvarez. 2002. The consumption of cultural products: An analysis of the Spanish social space. *Journal of Cultural Economics*, 26, 115–138.

Lorber, A. 1984. Quantifying chemical composition from two-dimensional data arrays. *Analytica Chimica Acta*, 164, 293–297.

Luhmann, N. 2000. *The reality of the mass media*. Stanford, CA: Stanford University Press.

MacDonell, W. R. 1902. On criminal anthropometry and the identification of criminals. *Biometrika*, 1, 177–227.

Macduffee, C. C. 1946. *The theory of matrices*. New York: Chelsea Publishing Company.

Madeira, S. C., and A. L. Oliveira. 2004. Biclustering algorithms for biological data analysis: A survey. *IEEE/ACM Transactions on Computational Biology and Bioinformatics*, 1, 24–45.

Mair, P., and J. de Leeuw. 2010. A general framework for multivariate analysis with optimal scaling: The R package aspect. *Journal of Statistical Software*, 32(9), 1–23.

March, S. T. 1983. Techniques for structuring database records. *Computing Surveys*, 15, 45–79.

Marey, E.-J. 1885. *La méthode graphique dans les sciences expérimentales*. Paris: Masson.

Maung, K. 1941a. Discriminant analysis of Tocher's eye colour data for Scottish school children. *Annals of Eugenics*, 11, 64–76.

Maung, K. 1941b. Measurement of association in a contingency table with special reference to the pigmentation of hair and eye colours of Scottish school children. *Annals of Eugenics*, 11, 189–223.

McCormick, W. T., P. J. Schweitzer, and T. J. White. 1972. Problem decomposition and data reorganization by a clustering technique. *Operations Research*, 20, 993–1009.

Mercier, P., D. Chessel, and S. Dolédec. 1992. Complete correspondence analysis of an ecological profile data table: A central ordination method. *Acta Oecologica—International Journal of Ecology*, 13(1), 25–44.

Meulman, J. 1982. *Homogeneity analysis of incomplete data*. Leiden, Netherlands: DSWO Press.

Meulman, J. 1986. *A distance approach to nonlinear multivariate analysis*. Leiden, Netherlands: DSWO Press.

Meulman, J. J. 1992. The integration of multidimensional scaling and multivariate analysis with optimal transformations. *Psychometrika*, 57, 539–565.

Meulman, J. J., W. J. Heiser, and L. Hubert. 1999. Three-way correspondence analysis through individual differences models. Presented at the 11th European Meeting of the Psychometric Society, Lünenburg, Germany.

Meulman, J. J., W. J. Heiser, and SPSS. 1999. *SPSS categories 10.0*. Chicago: SPSS.

Michailidis, G., and J. de Leeuw. 2005. Homogeneity analysis using absolute deviations. *Computational Statistics and Data Analysis*, 48, 587–603.

Minard, C. J. 1845. *Tableau figuratif du mouvement commercial du canal du Centre en 1844 dresse d'apre les renseignements de M. Comoy.* Lith. (n.s.). ENPC: 5299/C307 P:35.

Minard, C. J. 1862. *Carte figurative et approximative des quantités de coton en laine importées en Europe en 1858 et en 1861.* Lith. (868 × 535). ENPC: Fol 10975.

Minard, C. J. 1863. *Carte figurative et approximative des grands ports du globe, 2 ed. Corrigee et augmentee de 26 ports.* Lith. (765 × 540). ENPC: Fol 10975.

Minard, C. J. 1869. *Carte figurative des pertes successives en hommes de l'armee qu'Annibal conduisit d'Espagne en Italie en traversant les Gaules (selon Polybe). Carte figurative des pertes successives en hommes de l'armee francaise dans la campagne de Russie, 1812–1813.* Lith. (624 × 207, 624 × 245). ENPC: Fol 10975, 10974/C612.

Mochmann, I. C., and Y. El-Menouar. 2005. Lifestyle groups, social milieus and party preference in Eastern and Western Germany: Theoretical considerations and empirical results. *German Politics*, 14, 417–437.

Möcks, J. 1988a. Decomposing event-related potentials: A new topographic components model. *Biological Psychology*, 26, 199–215.

Möcks, J. 1988b. Topographic components model for event-related potentials and some biophysical considerations. *IEEE Transactions on Biomedical Engineering*, 35, 482–484.

Morand, E., and J. Pagès. 2006. Procrustes multiple factor analysis to analyse the overall perception of food products. *Food Quality and Preference*, 17(1–2), 36–42.

Morand, E., and J. Pagès. 2007. L'analyse factorielle multiple procustéenne. *Journal de la Société Française de Statistique*, 148(2), 65–97.

Muller, Ch. 1977. *Principes et méthodes de la statistique lexicale.* Paris: Hachette.

Murillo, A., J. F. Vera, and W. J. Heiser. 2005. A permutation-translation simulated annealing algorithm for l1 and l2 unidimensional scaling. *Journal of Classification*, 22, 119–138.

Murtagh, F. 1983a. A survey of recent advances in hierarchical clustering algorithms. *Computer Journal*, 26, 354–359.

Murtagh, F. 1983b. Expected-time complexity results for hierarchic clustering algorithms which use cluster centres. *Information Processing Letters*, 16, 237–241.

Murtagh, F. 1984. Complexities of hierarchic clustering algorithms: State of the art. *Computational Statistics Quarterly*, 1, 101–113.

Murtagh, F. 1985. *Multidimensional clustering algorithms.* Würzburg, Germany: Physica.

Murtagh, F. 2000. Multivariate data analysis software and resources. http://astro.u-strasbg.fr/~fmurtagh/mda-sw.

Murtagh, F. 2005. *Correspondence analysis and data coding with Java and R.* London: Chapman & Hall.

Murtagh, F. 2008. Origins of modern data analysis linked to the beginnings and early development of computer science and information engineering. *Electronic Journal for History of Probability and Statistics*, 4(2).

Murtagh, F., and M. J. Kurtz. 2013. A history of cluster analysis using the Classification Society's bibliography over four decades. Submitted.

Murtagh, F., and P. Legendre. 2014. Ward's hierarchical clustering method: Clustering criterion and agglomerative algorithm. *Journal of Classification* (in press).

Nakache, J. P. 1973. Influence du codage des données en analyse factorielle des correspondances. Etude d'un exemple pratique médical. *Revue de Statistique Appliquée*, 21(2), 55–70.

Naouri, J. C. 1971. *Analyse factorielle des correspondances continues.* Thèse d'etat, Université Pierre et Marie Curie, Paris.

Nauta, D. 1970. *The meaning of information*. Hague: Mouton.
Needham, R. 1965. Applications of the theory of clumps. *Mechanical Translation*, 8, 113–127.
Needham, R. 1967. Automatic classification in linguistics. *The Statistician*, 17, 45–54.
Nenadić, O., and M. Greenacre. 2007. Correspondence analysis in R, with two- and three-dimensional graphics: The ca package. *Journal of Statistical Software*, 20(3), 1–13.
Nightingale, F. 1858. *Notes on matters affecting the health, efficiency, and hospital administration of the British army*. London: Harrison and Sons.
Nishisato, S. 1980. *Analysis of categorical data. Dual scaling and its application*. Toronto: University of Toronto Press.
Nishisato, S. 1994. *Elements of dual scaling: An introduction to practical data analysis*. Hillsdale, NJ: Lawrence Erlbaum.
Nishisato, S. 2007. *Multidimensional nonlinear descriptive analysis*. Boca Raton, FL: Chapman & Hall.
Noirhomme-Fraiture, M., and M. Rouard. 2000. Visualizing and editing symbolic objects. In *Analysis of symbolic data, exploratory methods for extracting statistical information from complex data*, eds. H.-H. Bock and E. Diday, 125–138. Berlin: Springer.
Nomikos, P., and J. F. McGregor. 1994. Monitoring batch processes using multiway principal component analysis. *Aiche Journal*, 40, 1361–1375.
Okland, R. H., and O. Eilertsen. 1994. Canonical correspondence analysis with variation partitioning: Some comments and an application. *Journal of Vegetation Science*, 5, 117–126.
Oksanen, J., F. G. Blanchet, R. Kindt, P. Legendre, R. B. O'Hara, G. L. Simpson, P. Solymos, M. H. H. Stevens, and H. Wagner. 2011. Vegan: Community ecology package. R package version 1.17-12. http://CRAN.R-project.org/package=vegan.
Oldenburger, R. 1934. Composition and rank of n-way matrices and multilinear forms. *The Annals of Mathematics, 2nd Series*, 35, 622–653.
Orlóci, L. 1967. An agglomerative method for classification of plant communities. *Journal of Ecology*, 55, 193–206.
Paatero, P. 1999. The multilinear engine—A table-driven, least squares program for solving multilinear problems, including the n-way parallel factor analysis model. *Journal of Computational and Graphical Statistics*, 8, 854–888.
Pagès, J. 1996. Eléments de comparaison entre l'analyse factorielle multiple et la méthode STATIS. *Revue Statistique Appliqué*, XLIV(4), 81–95.
Pagès, J. 2002. Analyse factorielle multiple appliquée aux variables qualitatives et aux données mixtes. *Revue Statistique Appliquée*, L(4), 5–37.
Pagès, J. 2004. Analyse factorielle de données mixtes. *Revue Statistique Appliquée*, LII(4), 93–111.
Pagès, J. 2005a. Collection and analysis of perceived product inter-distances using multiple factor analysis; application to the study of ten white wines from the Loire valley. *Food Quality and Preference*, 16(7), 642–649.
Pagès, J. 2005b. Analyse factorielle multiple et analyse procustéenne. *Revue Statistique Appliquée*, LIII(4), 61–86.
Pagès, J. 2013. *Analyse factorielle multiple avec R*. Paris: EDP Sciences.
Pagès, J., M. Cadoret, and S. Lê. 2010. The sorted napping: A new holistic approach in sensory evaluation. *Journal of Sensory Studies*, 25(5), 637–658.

Pagès, J.-P., Y. Escoufier, and P. Cazes. 1976. Opérateurs et analyse de tableaux à plus de deux dimensions. *Cahiers du BURO*, 61–89.
Pagès, J., and F. Husson. 2005. Multiple factor analysis with confidence ellipses: A methodology to study the relationships between sensory and instrumental data. *Journal of Chemometrics*, 19(3), 138–144.
Palmer, M. W. 1993. Putting things in even better order: The advantages of canonical correspondence analysis. *Ecology*, 74, 2215–2230.
Palsky, G. 1999. The debate on the standardization of statistical maps and diagrams (1857–1901). *Cybergeo*, 65. http://cybergeo.revues.org/148.
Palumbo, F., and C. L. Lauro. 2003. A PCA for interval-valued data based on midpoints and radii. In *New developments on psychometrics*, eds. H. Yanai et al., 641–648. Tokyo: Springer.
Patil, G. P. 1995. Editorial: Statistical ecology and related ecological statistics—25 years. *Environmental and Ecological Statistics*, 2, 81–89.
Pearson, K. 1894. Contribution to the mathematical theory of evolution. *Philosophical Transactions of the Royal Society of London A*, 185, 71–110. (Reprinted in Karl Pearson's early statistical papers by E. S. Pearson, ed. 1948. Cambridge: Cambridge University Press.)
Pearson, K. 1901. On lines and planes of closest fit to systems of points in space. *Philosophical Magazine*, 2(6), 559–572.
Pearson, K. 1906. On certain points connected with scale order in the case of a correlation of two characters which for some arrangement give a linear regression line. *Biometrika*, 5, 176–178.
Pearson, K. 1920. Notes on the history of correlation. *Biometrika*, 13(1), 25–45.
Perriere, G., J. R. Lobry, and J. Thioulouse. 1996. Correspondence discriminant analysis: A multivariate method for comparing classes of protein and nucleic acid sequences. *Computer Applications in the Biosciences*, 12, 519–524.
Pham, T. D., and J. Möcks. 1992. Beyond principal component analysis: A trilinear decomposition model and least squares estimation. *Psychometrika*, 57, 203–215.
Playfair, W. 1801. *Statistical breviary; shewing, on a principle entirely new, the resources of every state and kingdom in Europe*. London: Wallis. (Republished 2005 in *The commercial and political atlas and statistical breviary*, eds. H. Wainer and I. Spence. Cambridge: Cambridge University Press.)
Pliner, V. 1996. Metric, unidimensional scaling and global optimization. *Journal of Classification*, 13, 3–18.
Poole, K. T., and H. Rosenthal. 1985. A spatial model for legislative roll call analysis. *American Journal of Political Science*, 29(2), 357–384.
Porter, T. M. 1995. *Trust in numbers*. Princeton, NJ: Princeton University Press.
Pratt, A. D. 1977. *Information of the image*. Greenwich, CT: Ablex.
Prentice, I. C. 1977. Non-metric ordination methods in ecology. *Journal of Ecology*, 65, 85–94.
Prentice, I. C. 1980a. Multidimensional scaling as a research tool in quaternary palynology: A review of theory and methods. *Review of Palaeobotanay and Palynology*, 31, 71–104.
Prentice, I. C. 1980b. Vegetation analysis and order invariant gradient models. *Vegetatio*, 42, 27–34.
Prodon, R., and J.-D. Lebreton. 1981. Breeding avifauna of a Mediterranean succession: The holm oak and cork oak series in the eastern Pyrenees, 1. Analysis and modelling of the structure gradient. *Oikos*, 37, 21–38.

Prodon, R., and J.-D. Lebreton. 1994. Analyses multivariees des relations especes-milieu: Structure et interpretation ecologique. *Vie Milieu*, 44, 69–91.

Purdom, E. 2011. Analysis of a data matrix and a graph: Metagenomic data and the phylogenetic tree. *Annals of Applied Statistics*, 5(4), 2326–2358.

R Development Core Team. 2012. *R: A language and environment for statistical computing*. Vienna: R Foundation for Statistical Computing.

Ramsay, J., and J. de Leeuw. 1983. Review. *Psychometrika*, 48(1), 147–151.

Ramsay, J. O. 1977. Maximum likelihood estimation in multidimensional scaling. *Psychometrika*, 42, 241–266.

Rao, C. R. 1964. The use and interpretation of principal component analysis in applied research. *Sankhya Series A*, 26, 329–357.

Renyi, A. 1959. On measures of dependence. *Acta Mathematica Academy Sciences Hungary*, 10, 441–451.

Richardson, M., and G. F. Kuder. 1933. Making a rating scale that measures. *Personnel Journal*, 12, 36–40.

Robert, P., and Y. Escoufier. 1976. A unifying tool for linear multivariate methods: The RV coefficient. *Applied Statistics*, 25(3), 257–265.

Robson, K., and C. Sanders, eds. 2009. *Quantifying theory: Pierre Bourdieu*. New York: Springer.

Rodriguez, O., E. Diday, and S. Winsberg. 2001. Generalization of the principal component analysis to histogram data. In *Workshop on symbolic data analysis of the 4th European conference PKDD: Principles and practice of knowledge discovery in data bases*, eds. D. A. Zighed, H. J. Komorowski, and Jan M. Zytkow. Lecture Notes in Computer Science. Berlin: Springer. http://link.springer.com/article/10.1007/s11634-012-0108-0#page-1.

Rodriguez, O., and A. Pacheco. 2004. Applications of histogram principal components analysis. Presented at the 15th European conference on machine learning (ECML) and the 8th European conference on principles and practice of knowledge discovery in databases (PKDD), Pisa. http://www.predisoft.com/la_es/Articulos_Matematicos_files/PCA-H_Applications_of_Histogram_Principal_Components_Analysis_Rodriguez-Pacheco.pdf.

Rohlf, F. J. 1973. Algorithm 76: Hierarchical clustering using the minimum spanning tree. *Computer Journal*, 16, 93–95.

Rohlf, F. J., and M. Corti. 2000. Use of two-block partial least-squares to study covariation in shape. *Systematic Biology*, 49(4), 740–753.

Röhmel, J., B. Streitberg, and W. Herrmann. 1983. The COMSTAT algorithm for multimodal factor analysis: An improvement of Tucker's three-mode factor analysis method. *Neuropsychobiology*, 10, 157–163.

Romane, F. 1972. Utilisation de l'analyse multivariable en phytoécologie. *Investigacion Pesquera*, 36, 131–139.

Romney, A. K., C. C. Moore, and T. J. Brazill. 1998. Correspondence analysis as a multidimensional scaling technique for non-frequency similarity matrices. In *Visualization of categorical data*, eds. J. Blasius and M. Greenacre, 529–546. New York: Academic Press.

Rose, M. J. 1964. Classification of a set of elements. *Computer Journal*, 7(3), 208–211.

Rosenlund, L. 1996. Cultural changes in a Norwegian urban community: Applying Pierre Bourdieu's approach and analytical framework. *International Journal of Contemporary Sociology*, 33, 211–236.

Roskam, E. E. 1968. *Metric analysis of ordinal data in psychology.* PhD thesis, University of Leiden, Netherlands.

Ross, G. 2007. Earlier days of computer classification. Vote of thanks. Inaugural lecture, F. Murtagh, "Thinking ultrametrically: Understanding massive data sets and navigating information spaces," Royal Holloway, University of London, February 22, 2007. http://www.multiresolutions.com/strule/inaugural.

Rouanet, H. 2006. The geometric analysis of structured individuals × variables tables. In *Multiple correspondence analysis and related methods,* eds. M. Greenacre and J. Blasius, 137–159. Boca Raton, FL: Chapman & Hall.

Rouanet, H., W. Ackermann, and B. Le Roux. 2000. The geometric analysis of questionnaires: The lesson of Bourdieu's "la distinction." *Bulletin de Méthodologie Sociologique,* 65, 5–18.

Rouanet, H., and D. Lépine. 1976. Structures linéaires et analyse des comparaisons. *Mathématiques et Sciences Humaines,* 56, 5–46.

Roux, G., and M. Roux. 1967. A propos de quelques méthodes de classification en phytosociologie. *Revue Statistique Appliqué,* 15, 59–72.

Sabatier, D., M. Grimaldi, M. F. Prévost, J. Guillaume, M. Godron, M. Dosso, and P. Curmi. 1997. The influence of soil cover organization on the floristic and structural heterogeneity of a Guianan rain forest. *Plant Ecology,* 13.

Sabatier, R., J.-D. Lebreton, and D. Chessel. 1989. Multivariate analysis of composition data accompanied by qualitative variables describing a structure. In *Multiway data tables,* eds. R. Coppi and S. Bolasco, 341–352. Amsterdam: North-Holland.

Sainte-Marie, P. 1973. *Le vocabulaire du théâtre classique français.* Thèse de 3ème cycle, Paris.

Sallaz, J. J., and J. Zavisca. 2007. Bourdieu in American sociology, 1980–2004. *Annual Review of Sociology,* 33, 21–41.

Sanchez, E., and B. R. Kowalski. 1986. Generalized rank annihilation factor analysis. *Analytical Chemistry,* 58, 496–499.

Sanchez, E., and B. R. Kowalski. 1990. Tensorial resolution: A direct trilinear decomposition. *Journal of Chemometrics,* 4, 29–45.

Sands, R., and F. W. Young. 1980. Component models for three-way data: Alscomp3, an alternating least squares algorithm with optimal scaling features. *Psychometrika,* 45, 39–67.

Saporta, G. 1977. Une méthode et un programme d'analyse discriminante sur variables qualitatives. In *Premières Journées Internationales Analyse des Données et informatiques.* Paris: INRIA, Rocquencourt.

Saporta, G. 1990. Simultaneous analysis of qualitative and quantitative data. *Atti Della XXXV Riunione Scientifica; Società Italiana di Statistica,* 63–72.

SAS. 1992. *SAS/STAT software: Changes and enhancements.* Technical Report P-229. Cary, NC: SAS Institute, Inc.

Savage, M., F. Devine, N. Cunningham, M. Taylor, Y. Li, J. Hjellbrekke, B. Le Roux, S. Friedman, and A. Miles. 2013. A new model of social class? Findings from the BBC's great British class survey experiment. *Sociology,* 47, 219–250.

Schaffers, A. P., I. P. Raemakers, K. V. Sỳkora, and C. J. F. ter Braak. 2008. Arthropod assemblages are best predicted by plant species composition. *Ecology,* 89, 782–794.

Schmidt, E. 1907. Zur Theorie der linearen und nichtlinearen Integralgleichungen. 1. Teil, Entwicklung willkürlicher Funktionen nach System vorgeschrieben. *Mathematische Annalen,* 63, 433–476.

Schmitz, A. 2012. Elective affinities 2.0? A Bourdieusian approach to couple formation and the methodology of E-dating. *Recherche en Sciences Sociales sur Internet (RESET)*, 1, 175–202.

Schönemann, P. H., and A. Lazarte. 1987. Psychophysical maps for subadditive dissimilarity ratings. *Perception & Psychophysics*, 42, 342–354.

Schwartz, S. H. 1992. Universals in the content and structure of values: Theoretical advances and empirical tests in 20 countries. In *Advances in experimental social psychology*, ed. M. P. Zanna, Vol. 25, pp. 1–65. New York: Academic Press.

Shepard, R. N. 1958. Stimulus and response generalization: Tests of a model relating generalization to distance in psychological space. *Journal of Experimental Psychology*, 55, 509–523.

Shepard, R. N. 1962a. The analysis of proximities: Multidimensional scaling with an unknown distance function, I. *Psychometrika*, 27, 125–140.

Shepard, R. N. 1962b. The analysis of proximities: Multidimensional scaling with an unknown distance function, II. *Psychometrika*, 27, 219–246.

Sibson, R. 1973. SLINK: An optimally efficient algorithm for the single link cluster method. *Computer Journal*, 16, 30–34.

Sidiropoulos, N. D., G. B. Giannakis, and R. Bro. 2000. Blind PARAFAC receivers for DS-CDMA systems. *IEEE Transactions on Signal Processing*, 48, 810–823.

Silva, E., and B. Le Roux. 2011. Cultural capital of couples: Tensions of elective affinities. *Poetics*, 30, 547–565.

Smilde, A. K. 1990. *Multivariate calibration of reversed-phase chromatographic systems*. Unpublished doctoral thesis, University of Groningen, Groningen, Netherlands. http://dissertations.ub.rug.nl/faculties/science/1990/a.k.smilde/.

Smilde, A. K. 1992. Three-way analyses. Problems and prospects. *Chemometrics and Intelligent Laboratory Systems*, 15, 143–157.

Smilde, A. K., R. Bro, and P. Geladi. 2004. *Multi-way analysis: Applications in the chemical sciences*. Chichester, UK: Wiley.

Sneath, P. H. A., and R. R. Sokal. 1973. *Numerical taxonomy*. San Francisco: Freeman.

Snelders, H. M., J. J. Dirk, and M. J. W. Stokmans. 1994. Product perception and preferences in consumer decision-making. In *Correspondence analysis in the social sciences*, eds. M. Greenacre and J. Blasius, 324–349. London: Academic Press.

Sokal, R. R., and C. D. Michener. 1958. A statistical method for evaluating systematic relationships. *University of Kansas Scientific Bulletin*, 38, 1409–1438.

Sokal, R. R., and P. H. A. Sneath. 1963. *Principles of numerical taxonomy*. San Francisco: Freeman and Co.

Sparck Jones, K. 1965. Experiments in semantic classification. *Mechanical Translation*, 8, 97–112.

Sparck Jones, K. 1970. Some thoughts on classification for retrieval. *Journal of Documentation*, 26, 89–101 (reprinted in *Journal of Documentation*, 2005).

Sparck Jones, K. 1971. *Automatic keyword classification for information retrieval*. Butterworths.

Sparck Jones, K., and R. M. Needham. 1964. Keywords and clumps. *Journal of Documentation*, 20, 5–15.

Sparck Jones, K., and R. M. Needham. 1968. Automatic term classifications and retrieval. *Information Storage and Retrieval*, 4, 91–100.

Sparck Jones, K., and D.M. Jackson. 1970. The use of automatically-obtained keyword classifications for information retrieval. *Information Storage and Retrieval*, 5, 175–201.

SPAD. *Système portable pour l'analyse des données.* Software distributed by Cohéris, Paris.
SPSS. 1989. *SPSS Categories.* Chicago, IL: SPSS.
Steinley, D. 2006. K-means clustering: A half-century synthesis. *British Journal of Mathematical and Statistical Psychology,* 59, 1–34.
Stéphane, V., G. Hebrail, and Y. Lechevallier. 2000. Generation of symbolic objects from relational databases. In *Analysis of symbolic data, exploratory methods for extracting statistical information from complex data,* eds. H.-H. Bock and E. Diday, 78–105. Berlin: Springer.
Stewart, G. W. 1993. On the early history of the singular value decomposition. *SIAM Review,* 35, 551–566.
Sylvester, J. J. 1852. A demonstration of the theorem that every homogeneous quadratic polynomial is reducible by real othogonal substitutions to the form of a sum of positive and negative squares. *Philosophical Magazine (ser. 4),* 4(23), 138–142.
Sylvester, J. J. 1884. *Academy of Sciences, Paris I,* 472–474.
Sylvester, J. J. 1889. *Messenger of Mathematics,* 19, 42–46. (Quoted by Eckart and Young, 1939.)
Székely, G. J., and M. L. Rizzo. 2005. Hierarchical clustering via joint between-within distances: Extending Ward's minimum variance method. *Journal of Classification,* 22(2), 151–183.
Takane, Y., H. Bozdogan, and T. Shibayama. 1987. Ideal point discriminant analysis. *Psychometrika,* 52, 371–392.
Takane, Y., and M. A. Hunter. 2001. Constrained principal component analysis: A comprehensive theory. *Applicable Algebra in Engineering, Communications and Computing,* 12, 391–419.
Takane, Y., F. W. Young, and J. de Leeuw. 1977. Nonmetric individual differences multidimensional scaling: An alternating least-squares method with optimal scaling features. *Psychometrika,* 42, 7–67.
Tauler, R. 1995. Multivariate curve resolution applied to second order data. *Chemometrics and Intelligent Laboratory Systems,* 30, 133–146.
Ten Berge, J. M. F. 2000. The typical rank of tall three-way arrays. *Psychometrika,* 65, 525–532.
Ten Berge, J. M. F., J. de Leeuw, and P. M. Kroonenberg. 1987. Some additional results on principal components analysis of three-mode data by means of alternating least squares algorithms. *Psychometrika,* 52, 183–191.
Ten Berge, J. M. F., and H. A. L. Kiers. 1999. Simplicity of core arrays in three-way principal component analysis and the typical rank of px qx2 arrays. *Linear Algebra and Its Applications,* 294, 169–179.
Tenenbaum, J. B., V. De Silva, and J. C. Langford. 2000. A global geometric frame work for nonlinear dimension reduction. *Science,* 290, 2319–2323.
Tenenhaus, A., and M. Tenenhaus. 2011. Regularized generalized canonical correlation analysis. *Psychometrika,*76, 1–28.
Tenenhaus, M., and F. W. Young. 1985. An analysis and synthesis of multiple correspondence analysis, optimal scaling, dual scaling, homogeneity analysis and other methods for quantifying categorical multivariate data. *Psychometrika,* 50(1), 91–119.
Ter Braak, C. J. F. 1983. Principal components biplots and alpha and beta diversity. *Ecology,* 64, 454–462.

Ter Braak, C. J. F. 1985. Correspondence analysis of incidence and abundance data: Properties in terms of a unimodal response model. *Biometrics*, 41, 859–873.
Ter Braak, C. J. F. 1986. Canonical correspondence analysis: A new eigenvector technique for multivariate direct gradient analysis. *Ecology*, 67, 1167–1179.
Ter Braak, C. J. F. 1987. The analysis of vegetation-environment relationships by canonical correspondence analysis. *Vegetatio*, 69, 69–77.
Ter Braak, C. J. F. 1988a. *CANOCO—A FORTRAN program for canonical community ordination by [partial] [detrended] [canonical] correspondence analysis, principal components analysis and redundancy analysis*. Version 2.1, Report LWA-88-02. Wageningen, Netherlands: Agricultural Mathematics Group.
Ter Braak, C. J. F. 1988b. CANOCO—An extension of DECORANA to analyze species-environment relationships. *Vegetatio*, 75, 159–160.
Ter Braak, C. J. F. 1988c. Partial canonical correspondence analysis. In *Classification and related methods of data analysis*, ed. H. H. Bock, 551–558. Amsterdam: North-Holland.
Ter Braak, C. J. F. 1990. *Update notes: CANOCO version 3.1*. Wageningen, Netherlands: Agricultural Mathematics Group.
Ter Braak, C. J. F. 1995. Canonical correspondence analysis and related multivariate methods in aquatic ecology. *Aquatic Sciences*, 57(3), 1015–1621.
Ter Braak, C. J. F., and L. G. Barendregt. 1986. Weighted averaging of species indicator values: Its efficiency in environmental calibration. *Mathematical Biosciences*, 78, 57–72.
Ter Braak, C. J. F., and S. de Jong. 1998. The objective function of partial least squares regression. *Journal of Chemometrics*, 12, 41–54.
Ter Braak, C. J. F., and N. J. M. Gremmen. 1987. Ecological amplitudes of plant species and the internal consistency of Ellenberg's indicator values for moisture. *Vegetatio*, 69, 79–87.
Ter Braak, C. J. F., and C. W. N. Looman. 1986. Weighted averaging, logistic regression and the Gaussian response model. *Vegetatio*, 65, 3–11.
Ter Braak, C. J. F., and C. W. N. Looman. 1994. Biplots in reduced-rank regression. *Biometrical Journal*, 36, 983–1003.
Ter Braak, C. J. F., and I. C. Prentice. 1988. A theory of gradient analysis. *Advances in Ecological Research*, 18, 271–317.
Ter Braak, C. J. F., and A. P. Schaffers. 2004. Co-correspondence analysis: A new ordination method to relate two community compositions. *Ecology*, 85, 834–846.
Ter Braak, C. J. F., and P. Šmilauer. 2012. *Canoco reference manual and user's guide: Software for ordination*. Version 5.0. Ithaca, NY: Microcomputer Power.
Ter Braak, C. J. F., and H. van Dam. 1989. Inferring pH from diatoms: A comparison of old and new calibration methods. *Hydrobiologia*, 178, 209–223.
Ter Braak, C. J. F., and P. F. M. Verdonschot. 1995. Canonical correspondence analysis and related multivariate methods in aquatic ecology. *Aquatic Sciences*, 57, 255–289.
Thioulouse, J. 2011. Simultaneous analysis of a sequence of paired ecological tables: A comparison of several methods. *Annals of Applied Statistics*, 5(4), 2300–2325.
Thioulouse, J., and D. Chessel. 1992. A method for reciprocal scaling of species tolerance and sample diversity. *Ecology*, 73, 670–680.
Thioulouse, J., D. Chessel, S. Dolédec, and J. M. Olivier. 1997. Ade-4: A multivariate analysis and graphical display software. *Statistics and Computing*, 7, 75–83.
Thurstone, L. L. 1947. *Multiple factor analysis*. Chicago: University of Chicago Press.

Torgerson, W. S. 1958. *Theory and methods of scaling*. New York: Wiley.
Torres, A., and M. Greenacre. 2002. Dual scaling and correspondence analysis of preferences, paired comparisons and ratings. *International Journal of Research in Marketing*, 19, 401–405.
Tucker, L. R 1958. An inter-battery method of factor analysis. *Psychometrika*, 23(2), 111–136.
Tucker, L. R 1963. Implications of factor analysis of three-way matrices for measurement of change. In *Problems in measuring change*, ed. C. W. Harris, 122–137. Madison: University of Wisconsin Press.
Tucker, L. R. 1964. The extension of factor analysis to three-dimensional matrices. In *Contributions to mathematical psychology*, eds. H. Gulliksen and N. Frederiksen, 110–127. New York: Holt, Rinehart and Winston.
Tucker L. R. 1966. Some mathematical notes on three-mode factor analysis. *Psychometrika*, 31, 279–311.
Tucker, L. R. 1972. Relations between multidimensional scaling and three-mode factor-analysis. *Psychometrika*, 37, 3–27.
Tufte, E. R. 1997. *Visual explanations: Images and quantities evidence and narrative*. Cheshire, CT: Graphics Press.
Tukey, J. W. 1977. *Exploratory data analysis*. Reading, MA: Addison-Wesley.
UNDP Human Development Report. 2010. *The real wealth of nations: Pathways to human development.* United Nations. http://hdr.undp.org/en/reports/global/ndr2010/chapters.
Van Dam, H., G. Suurmond, and C. J. F. ter Braak. 1981. Impact of acidification on diatoms and chemistry of Dutch moorland pools. *Hydrobiologia*, 83, 425–459.
Van den Wollenberg, A. L. 1977. Redundancy analysis, an alternative for canonical analysis. *Psychometrika*, 42(2), 207–219.
Van der Burg, E., and J. de Leeuw. 1983. Non-linear canonical correlation. *British Journal of Mathematical and Statistical Psychology*, 36, 54–80.
Van der Burg, E., and J. de Leeuw. 1990. Non-linear redundancy analysis. *British Journal of Mathematical and Statistical Psychology*, 43, 217–230.
Van der Heijden, P. G. M., A. de Falguerolles, and J. de Leeuw 1989. A combined approach to contingency table analysis using correspondence-analysis and log-linear analysis (with discussion). *Applied Statistics*, 38, 249–292.
Van der Heijden, P. G. M., and B. Escofier. 2003. *Multiple correspondence analysis with missing data*. Rennes, France: Presse Universitaire de Rennes.
Van der Heijden, P. G. M., A. Mooijaart, and Y. Takane. 1994. Correspondence analysis and contingency models. In *Correspondence analysis in the social sciences*, eds. M. J. Greenacre and J. Blasius, 79–111. New York: Academic Press.
Van Eck, N. J., and L. Waltman. 2010. Software survey: Vosviewer, a computer program for bibliometric mapping. *Scientometrics*, 84, 523–538.
Van Herk, H., and M. van de Velden. 2007. Insight into the relative merits of rating and ranking in a cross-national context using three-way correspondence analysis. *Food Quality and Preference*, 18, 1096–1105.
Van Mechelen, I., H.-H. Bock, and P. De Boeck. 2004. Two-mode clustering methods: A structured overview. *Statistical Methods in Medical Research*, 13, 363–394.
Van Mechelen I., L. Lombardi, and E. Ceulemans. 2007. Hierarchical classes modeling of rating data. *Psychometrika*, 72, 475–488.
Van Rijsbergen, C.J. 1970. A clustering algorithm. *Computer Journal*, 13, 113–115.

Van Rijsbergen, C.J. 1974. Further experiments with hierarchic clustering in document retrieval. *Information Storage and Retrieval*, 10, 1–14.
Van Rijsbergen, C.J. 1977. A theoretical basis for the use of co-occurrence data in information retrieval. *Journal of Documentation*, 33, 106–119.
Van Rijsbergen, C.J., and N. Jardine. 1971. The use of hierarchic clustering in information retrieval. *Information Storage and Retrieval*, 7, 217–240.
Van Rooij, A. C. M. 1978. *Non-Archimedean functional analysis*. New York: Marcel Dekker.
Vera, J. F., W. J. Heiser, and A. Murillo. 2007. Global optimization in any Minkowski metric: A permutation-translation simulated annealing algorithm for multidimensional scaling. *Journal of Classification*, 24, 277–301.
Vercellone, C., A. Fumagalli, and V. Cvijanovic. 2010. *Cognitive capitalism and its reflections in South-Eastern Europe*. Frankfurt: Peter Lang.
Verde, R. 1997. *Symbolic object decomposition by factorial techniques*. Indo-French Meeting, LISE-CEREMADE, Université Paris IX Dauphine.
Verde, R. 1999. Generalised canonical analysis on symbolic objects. In *Classification and data analysis, theory and application*, eds. M. Vichi and O. Opitz, 195–202. Heidelberg: Springer.
Verde, R., and A. Irpino. 2007. Dynamic clustering of histogram data: Using the right metric. In *Selected contributions in data analysis and classification*, eds. P. Brito, P. Bertrand, G. Cucumel, and F. De Carvalho, 123–134. Berlin: Springer.
Vitali, G. 1929. Bollettino delle unione. *Matematica Italiana*, 7, 1–7.
Volle, M. 1985. *Analyse des données. Economie et statistiques avancées*. Paris: Economica.
Von Humboldt, A. 1817. Sur les lignes isothermes. *Annales de Chimie et de Physique*, 5, 102–112.
Von Mises, R., and H. Pollackzek-Geiringer. 1929. Praktische Verfahren der Gleichungsauflösung. *Zeitschrift für Angewandte Mathematik und Mechanik*, 9, 58–79, 152–164.
Wansbeek, T., and J. Verhees. 1990. The algebra of multimode factor analysis. *Linear Algebra and Its Applications*, 127, 631–639.
Ward Jr., J. H. 1963. Hierarchical grouping to optimize an objective function. *Journal of the American Statistical Association*, 58, 236–244.
Wartenberg, D., S. Ferson, and F. J. Rohlf. 1987. Putting things in order: A critique of detrended correspondence analysis. *American Naturalist*, 129, 434–448.
Weesie, J., and H. Van Houwelingen. 1983. *Gepcam users' manual. Generalized principal component analysis with missing values*. Technical report. Utrecht, Netherlands: Institute of Mathematical Statistics, University of Utrecht.
Weierstrass, K. 1868. Zur Theorie der bilinearen und quadratischen Formen. *Monatsberichte der Akademie der Wissenschaften, Berlin*, 310–338.
Whittaker, R. H. 1956. Vegetation of the great Smoky Mountains. *Ecological Monographs*, 26, 1–80.
Wickham, H. 2009. *ggplot2: Elegant graphics for data analysis*. New York: Springer.
Wilkinson, L. 2005. *The grammar of graphics*. 2nd ed. New York: Springer.
Wilks, S. S. 1938. Weighting systems for linear functions of correlated variables when there is no dependent variable. *Psychometrika*, 3, 23–40.
Wilkes, M., P. Hammersley, C. J. van Rijsbergen, and F. Murtagh. 2007. Editorial on the occasion of volume 50 of the *Computer Journal*. *Computer Journal*, 50, 4–6.
Williams, E. J. 1952. Use of scores for the analysis of association in contingency tables. *Biometrika*, 39, 274–289.
Winer, B. J. 2004. *Statistical principles in experimental design*. New York: McGraw-Hill.

Winsberg, S., and J. O. Ramsay. 1983. Monotone spline transformations for dimension reduction. *Psychometrika*, 48, 575–595.
Wishart, D. 1969. An algorithm for hierachical classifications. *Biometrics*, 25, 165–170.
Wishart, J. 1928. The generalized product moment distribution in samples from a normal multivariate distribution. *Biometrika*, 20A, 32–52.
World Bank. 2010. *World development indicators.* Washington, DC: World Bank.
Yau, N. 2013. *Data points.* Indianapolis: John Wiley & Sons.
Yoccoz, N. G., and D. Chessel. 1988. Ordination sous contraintes de relevés d'avifaune: Elimination d'effets dans un plan d'observation à deux facteurs. *Compte-Rendus de l'Académie des Sciences, Paris, Série III*, 307, 189–194.
Yoshizawa, T. 1975. Models for quantification techniques in multiple contingency tables: The theoretical approach (in Japanese). *Kodokeiryugaku (Japanese Journal of Behaviormetrics)*, 3, 1–11.
Yoshizawa, T. 1976. A generalized definition of interactions and singular value decomposition of multiway arrays (in Japanese). *Kodokeiryogaku (Japanese Journal of Behaviormetrics)*, 4, 32–42. English abstract, 87.
Yoshizawa, T. 1988. Singular value decomposition of multiarray data and its applications. In *Recent developments in clustering and data analysis*, eds. C. Hayashi, E. Diday, M. Jambu, and N. Ohsumi, 241–257. New York: Academic Press.
Young, F. W., J. de Leeuw, and Y. Takane. 1976. Regression with qualitative and quantitative variables: An alternating least squares method with optimal scaling features. *Psychometrika*, 41, 505–529.
Young, F. W., Y. Takane, and J. de Leeuw. 1978. The principal components of mixed measurement level multivariate data: An alternating least squares method with optimal scaling features. *Psychometrika*, 45, 279–281.
Yule, G. U. 1912. On the methods of measuring association between two attributes. *Journal of the Royal Statistical Society*, 75, 107–170.
Zeileis, A., C. Strobl, and F. Wickelmaier. 2012. *psychotools: Infrastructure for psychometric modeling.* R package version 0.1-4. http://CRAN.R-project.org/package=psychotools.
Zelinka, M., and P. Marvan. 1961. Zur Präzisierung der biologischen Klassifikation der Reinheit fliessender Gewässer. *Archiv für Hydrobi.*
Zhang, Y., and O. Thas. 2012. Constrained ordination analysis in the presence of zero inflation. *Statistical Modelling*, 12, 463–485.
Zhu, M., T. J. Hastie, and G. Walther. 2005. Constrained ordination analysis with flexible response functions. *Ecological Modelling*, 187, 524–536.

Index

3WayPack, 87

A

ACE (alternating conditional expectation), 49
ACT, 89
Additive decomposition, 91–92
Additive effects
    clouds and, 197–200
    example, 202–203
ade4, 237
AFM. *See* MFA
AFMULT, 89
Agglomerations, 126–127
Agglomerative clustering criterion, 124–125
Agglomerative procedures, 271
Aggregated lexical table, application of to CA (CA-ALT), 158
Album de Statistique Graphique, 12
Algebra, history of, 17–18
Algebraic decompositions, 18–19
ALSOS (alternating least squares with optimal scaling), 49
anacor, 69
Analyse conjointe de tableaux (ACT), 89
Analyse des correspondances, 35
Analyse des données, 29, 32, 121
    development of in France, 206
Analyse factorielle multiple (AFMULT), 89
Analysis of comparisons, 188
Analysis of variance. *See* ANOVA
ANOVA
    algebraic aspects of, 188
    cloud of points, 189
    contract and comparison of with GDA, 190
    double breakdown of inertias, 191
    notations, 189–190
Arch effect, 66. *See also* Guttman effect
Arches, 48

Argumentation, application of CA to, 159–162
Arrays, three-way, 79
ASLIB, 119
aspect, 56
Aspects, use of NLPCA with, 56–59
Association of Special Libraries and Information Bureaux. *See* ASLIB
Asymmetric maps, 146
Average linkage, 271
Axis calibration, 11

B

Balbi, Adriano, 8
Bar-chart variable, 259
Barycentre, 139, 146. *See also* Centre of gravity
Barycentric representation, 172
Benzécri, Jean-Paul, 35, 91, 119, 137–138, 205–207
    data analysis project of, 130–131
    vision of science, 121–122
Bertin, Jacques, 7
Between cloud, 192
Between-cluster inertia, 271
Biadditive models, 24–25
Bilinear forms, 18–19
Binary data, logit and probit PCA of, 59
Biplots
    CCA and, 70–71
    nonlinear calibrations, 10
Bivariate distribution, canonical analysis of, 36
Bivariate symbolic interval variables, 257
Blind source separation, 84
Bootstrapping, 43–44
Bourdieu, Pierre, 42, 186, 206–207
    La Distinction, 205
    social space approach, 207–215
Brain function, lateralization of, 4
Bro, R., 85

339

Buache, Phillipe, 8
Burt table, 47–48, 148, 172
    CA on, 177–178
Burt, Cyril, 40–41

C

CA, 3, 27–28, 47, 131, 137–138
    application of to argumentation, 159–162
    application of to social space, 212–214, 217–218
    discretization of a continuous bivariate distribution, 36
    doubled data, 242–243
    doubling of preferences and paired comparisons, 245–246
    duality diagram and, 294–297
    eigenvalues in, 147–148
    French tradition of, 206
    geometric approach to, 35–36
    history of, 31–32
    history of in ecology, 63–67
    linguistics and transition relationships, 156–157
    nonsymmetric, 92
    notation for, 143–144
    quantification of categorical variables, 32
    reciprocal averaging, 34–35
    recoding of data, 239
    selecting words and sequences, 157–159
    simple form of, 32–33
    three-way, 90–92
    use of with textual data, 149–150
Cailliez, F., 290
CANALS, 73
CANDECOMP, 78, 86
CANDECOMP/PARAFAC (CP), 80–81
CANOCO, 66, 69
Canonical analysis
    bivariate distribution, 36
    Gifi loss function in, 50
Canonical correlation, 25
Canonical correlation analysis, 34
Canonical correspondence analysis. *See* CCA
Canonical forms, 18–23
Canonical polyadic decomposition (CP decomposition), 80
Canonical variate analysis. *See* CVA
Capture-mark-recapture sampling, 69
CARME
    data and information visualization, 6–12
    history of, 3–5
Carroll, J.D., 78, 107, 109
Cartes figuratives, 12
Categorical data, multiway, 89–92
Categorical multivalued variables, coding of, 260
Categorical variables, 47
    optimal scaling of, 32
Categories
    cloud of, 173–176, 214
    interpreting cloud of individuals from, 170–173
CATPCA, 49
Cattell, R.B., 82
Cazes, P., 290
CCA, 3
    history of, 67–69
    overview of, 62–63
    relation of CVA with, 71–73
    SVD and, 74–75
    triplets and biplots in, 70–71
Centre of gravity, 139, 145. *See also* Barycentre
Centred cloud, 139
Centroid, 145. *See also* Barycentre
Chang, Jih-Jie, 107, 109
Chemistry, use of multiway analysis in, 83–84, 89
Chessel, Daniel, 67, 69
Cheysson, Émile, 7
Chi-square distance
    cloud of categories, 173–174
    cloud of individuals, 167
    use of linguistics with, 155
Chi-squared statistic, 34
Choropleth maps, 6, 8
Chronological clustering (CC), 159
Classical MDS, 104–105
Classification, 27
Classification Society, establishment of, 118–119

Index

Cloud of categories, 173–174, 214
    fitting, 174–176
Cloud of individuals, 166–168, 214, 220–221
    fitting, 168–169
    interpreting from categories, 170–173
    interpreting using variables, 169–170
Cloud of points, 189
Cloud of variables, 214
Cloud of weighted points, factorial analysis of, 138–143
Clouds
    main and within effects, 196–197
    nesting of two factors between and within, 192
Cluster analysis
    Benzécri's and Hayashi's vision of science, 121–122
    historical, 118–122
    language and, 119–120
    role of in science, 131–132
    symbolic data, 258
    UPGMA, 272–277
Clustering techniques
    three-way, 78
    ultrametric matrices, 129
    UPGMA *vs.* Ward's algorithm, 281–285
    Ward's algorithm, 277–281
Co-correspondence analysis, 73
Co-inertia analysis, 73
Co-occurrences, linguistic units, 150–152
Codage flou. *See* Fuzzy coding
Coding phase, 260
Cohesion constraint, symbolic GCA under, 263–265
Coinertia analysis, duality diagram and, 297–300
Column profiles
    factorial analysis of, 146
    PCA of weighted, 36
    spaces of, 153–154
Common graphic forms, 7
Communication tools, data as, 6
Compactness criterion, 124–125
Comparative moral maps, 8
Complete disjunctive binary table, 37.
    See also Indicator matrices

Complete disjunctive coding, 47–48, 138
Complete linkage algorithm, 159, 271
Complex data, 256
Composition of capitals, 208
Compound signal processing, use of multiway techniques for, 84
Computational diagrams, 10–11
Computer Journal, 120
Computer science, clustering in, 120
Computers, early data analysis and, 26–29
COMSTAT, 85–86
Conditioned responses, 101–102
Confidence areas, 159
Confirmatory data analysis (CDA), 29
Confirmatory MDS (CMDS), 103
Constant dissimilarities, 113
Constrained MDS, 109–110
Content information, 106–107
Content words, 152
Contingency tables, 36
    analysis of using CA, 32–33
    interactions in three-way, 91
Continuous bivariate distribution, CA as a discretization of, 36
Continuous variables, fuzzy coding of, 247–252
Contour lines, 6
Contour maps, 8
Contribution biplots, 10, 211
Coomb's unfolding model, 100
Cophenetic correlation coefficient, 282
Cordier, B., 35
Corpus
    encoding of, 150–152
    sequencing, 158
Correlation coefficients, 282
Correlation ellipsoid, 14, 47
Correlation matrix, aspects and, 58
Correspondence analysis. *See* CA
Correspondence analysis and related methods. *See* CARME
Covariation chart, 82–83
Crisp coding, 138
Crossing of two factors, 196
    example, 197–202
    main and within effects and clouds, 196–197
CuBatch, 87

Cultural capital, 208–209
CVA, relation of CCA with, 71–73

D

d'Ocagne, Maurice, 10–11
Dasymetric maps, 6
Data
 mixed, 235–237
 multiple factor analysis example, 224–231
 recoding of, 239–240
 symbolic, 115–116, 256
Data analysis, 29
 clustering and, 118–122
 impact of computers on, 26–29
 language and, 119–120
 multiway, 82–84 (See also Multiway analyses)
 use of MDS for, 102–103
Data analytics, antecedents of, 130–132
Data coding, 163. See also Encoding
Data displays
 conventions for, 6–7
 visual language and, 6–12
Data mining, 121
de Castelfranc, Guillaume de Nautonier, 8
de Leeuw, J., 110
Decomposition
 canonical polyadic, 80
 Lancaster additive, 91–92
 rectangular matrices, 22–23
 square matrices, 19–20
DECORANA, 66, 69
Dédoublement. See Doubling
Dendograms, 123
Dependence measures, 257
Dependent variables, 186
Detrended correspondence analysis, 35
Diagonalization of a symmetric matrix, 19
 two-sided eigendecomposition, 20
Diagrams, 10–12
Discriminant analysis, 121, 131
Discrimination from a set of qualitative variables (DISQUAL), 44
Dissimilarity, hierarchical clustering and, 123

Dissimilarity judgements, distance formula as a psychological model of, 99–100
Dissimilarity matrix, 115
Dissimilarity update formula, 125–126
Distance between categories, 173–176
Distance between individuals, 166–168
Distance models, 103
Distance-based multivariate analysis, 112
Distributional equivalence principle, 146–147, 150, 158–159
 use of linguistics with, 154–155
Distributional linguistics, 150–152
Distributional synonyms, 153–154
Dot symbols, 6
Doubling, 28, 240
 preferences and paired comparisons, 245–246
 ratings, 240–245
Dual scaling, 28, 35, 41
Duality, 174, 176–177
 principles of, 10–11
Duality diagrams
 CA and, 294–297
 coinertia analysis, 297–300
 comparison of, 297–300
 definition, 290–293
 properties, 293
Dummy variables, 248–249
 coding of, 138
Dupin, Charles, 8
Dynamic MDS, 114

E

Ecology, history of CA in, 63–67
Economic capital, 208–209
Eigenvalues, two-sided problem, 25–26
Encoding, 150–152
Escoufier, Yves, 69, 223
Essai sur la Stastique Morale de la France (Guerry), 8–9
Euclidian clouds, 185–187
Euclidian distance formula, 99
Euclidian distances, 111
Eugenics, 23
Expected time algorithm, 126
Experimental factors, 186

Index

Explained inertia, 252–253
Exploration and analysis, 6
Exploratory data analysis (EDA), 29
Exploratory factorial techniques, 257

F

Facet theory, 106–107
FactoMineR, 217, 237
Factor analysis for mixed data (FAMD), 235
Factorial analysis, cloud of weighted points, 138–143
Factorial axial vectors, 145
Factorial design, 186
Factorial representation of symbolic data, 265–270
Fibre profiles, 92
Fisher, R.A., 24, 34
Flattening, 83
Flow maps, 6
French plot, 209. *See also* Symmetric maps
Full-dimensional scaling, local minima, 114
Function words, 152
Functional blocks, 151
Fuzzy coding, 138, 240
  continuous variables, 247–252
  interval bounds, 261
  symbolic variables, 258

G

Galton, Francis, 23–24
  visual discoveries of, 14–15
GCA, 258
  under a cohesion constraint, 263–265
  strategy for symbolic data, 259–265
GDA, 186
  contrast and comparison of with ANOVA, 190
  double breakdown of inertias, 191
  notations, 189–190
Geladi,P., 85
Generalized canonical analysis. *See* GCA
Generalized linear models (GLIMs), 28
Geography, early MDS in, 98–99

Geometric approach to correspondence analysis, 35–36
Geometric data analysis. *See* GDA
Ggplot2, 7
Gifi loss function, 50
Gifi project, 28, 49, 59
  NLPCA in, 49–51
Goodman–Kruskal coefficient, 282
Gower, John, 118
Gradient-based minimization, 105
Grammar of Graphics (Wilkinson), 7
Graphical language, standardization of, 6
Graphics Programming Language (GPL), 7
Graphs, 10–12
Green, R.H., 102
Group average linkage algorithm, 272–277
Groups of variables. *See also* Variables
  balance between, 226–228
  overall comparison of, 231–233
  superimposed representation, 228–231
Guerry, André-Michel, 8–10
Guttman effect, 35, 48, 66
Guttman, Louis, 40–41, 47, 101

H

Harshman, R.A., 78
Hartley, H.O., 25, 32, 34
Harzing, Anne-Wil, 78
Hayashi,Chikio, 47
  vision of science, 121
Heiser, Willem, 28
Hellinger distances, 147
Hermitian matrices, 20–22
HICLAS models, 89
Hierarchical classes models. *See* HICLAS models
Hierarchical classification, 27
Hierarchical clustering, 123, 271
  algorithms, 126–127
  embedding of observations in ultrametric topology, 127–130
  Lance–Williams dissimilarity update formula, 125–126
  Ward's agglomerative method of, 124–125

Hierarchical multiple factor analysis (HMFA), 224
Higher-order tensors, 84
Hill, Mark, 67
Hill's scaling, 73
Hirschfeld, Hermann Otto, 25, 34
Histogram dissimilarities, 115–116
Histogram variables, 256, 259
Hitchcock, F.L., 82
homals, 49
   use of for NLPCA, 50–51
Homogeneity analysis, 41, 50
Horst, P., 34–35
Hotelling, H., 24

I

IDOSCAL, 109
Illustrational variables, 210. *See also* Supplementary variables
Impositions, 7
Incidence matrices, 67
Inconsistent aggregation, example of using Ward's clustering method, 286–287
Incorporated cultural capital, 208
Independent variables, 186
Indicator matrices, 37
  application of MCA to, 166
  explained inertia, 252–253
  use of with social space approach, 218–221
Individual Differences in Orientation Scaling (IDOSCAL), 109
Individual differences scaling, 78
Individual weights, comparison of, 100
Individuals
  cloud of, 214, 220–221
  fitting the cloud of, 168–169
  interpreting the cloud of from categories, 170–173
  interpreting the cloud of using variables, 169–170
  studying, 166–168
INDSCAL, 78, 81–82, 100, 107
Inertia, 189–190, 271
  double breakdown of, 191
  total, 139–140, 145, 167–168
Inertias, explained, 252–253

Information retrieval, 119–120
Institutionalized cultural capital, 208
Interaction effects
  clouds and, 197–200
  example, 202–203
Interactive computing, 26–27
Interval bounds, fuzzy coding of, 261
Interval variable, coding of, 261
Interword dependencies, 150–152
Isolines, 8
ISOMAP, 115
Isopleth lines, 14
Iwatsubo, S., 91

J

JCA, 24, 27
Joint correspondence analysis. *See* JCA
Jordan form, 19–20

K

Kiers, H.A.L., 85
Kroonenberg, Pieter M., 85, 92
Kruskal, Joseph B., 78, 101, 105

L

L'espace social. See Social space
La Distinction (Bourdieu), 205–206. *See also* Bourdieu, Pierre
Lalanne, Léon, 10–11
Lallemand, Charles, 8
Lancaster additive decomposition, 91–92
Lance–Williams dissimilarity update formula, 125–126
  generalization of, 126
Language, clustering and data analysis, 119–120
Large-scale MDS, 114–115
Le Roux, Brigitte, 206
Learning, use of ordinal MDS to study generalization gradients in, 101–102
Least-squares criterion, 282
Least-squares equations, 17–18
  use of in multiway data analysis, 86
Lebart, Ludovic, 41–42, 206, 215
Lebreton, Jean-Dominique, 67, 69

Lellemand, Charles, 11
Lemmas, 152
Levasseur, Émile, 7
Levels of variables, 7
Lexical table, use of with correspondence analysis (CA-ALT), 158
Liebig's law, 64–65
Lifestyle approach, 208. *See also* Social space
Linear discriminant analysis, 71–73. *See also* CVA
Linear principal components analysis, 46
Linear systems, estimating parameters of, 17–18
Linear transformations, 18–19
Linguistics
  correspondence analysis and transition relationships in, 156–157
  distributional, 150–152
Local minima, 113–114
Logit PCA, 59
Loss functions
  Gifi, 50
  measurement of pavings and, 54–56
  MULTISCALE, 112
  S-stress, 112
  strain, 109
  stress, 105–106

M

Machine learning, 121
MacKenzie, W.A., 24
Main effects, 196–197
Majorization, 59, 110–112
  function, 110
Maps, 8–10
Marginal frequencies, 32
Matrices
  decomposition of square, 19–20
  diagonalization of symmetric, 19
  Hermitian, 20–22
  orthogonal, 20–22
  skew-symmetric, 20–22
  unitary, 21
Matrix algebra, 18

Maximal correlation, 47
MCA, 3, 24, 27–28, 37–39, 47–48, 165–166
  application of to cloud of categories, 173–176
  application of to cloud of individuals, 166–173
  application of to Life Story Survey, 178–183
  application of to social space, 212–214, 218–221
  dissemination of, 42
  doubled data, 242–243
  factor analysis using, 235
  first implementations, 41–42
  formulas and methodology, 40–41
  history of, 31–32
  multiway analyses and, 42–43
  preliminary bases of, 39–40
  recoding of, 239
  social space approach and, 207–208
  specific, 214, 250–252
  stability, validation, and resampling in, 43–44
  use of in linguistic analysis, 158
MDC, facet theory and regional interpretation in, 106–107
mdrace, 49
MDS, 3, 47, 96
  basic ideas of, 96–97
  classical, 104–105
  constant dissimilarities, 113
  constrained, 109–110
  current utilization of, 103–104
  future of, 114–116
  general data analysis using, 102–103
  history of, 97–103
  local minima, 113–114
  ordinal, 100–101
  three-way models, 107–110
MDSCAL, 102
Mean-square contingency coefficient, 34
Measurements, psychological, 100–101
Membership functions, 247
Method of reciprocal averaging, 34–35
Meulman, Jacqueline, 28
MFA
  comparison with STATIS, 234
  example, 224–231

representation of groups using, 231–233
MFACT, 224
Michigan–Israel–Niimegen Integrated Smallest Space Analysis. *See* MINISSA
Milestones Project, 4–5
Minard, Charles Joseph, graphic vision of, 12–14
Minimum variance compactness criterion, 124–125
MINISSA, 101
Minkowski distances, 99, 111
Missing value passive analysis, 214, 250–252
Missing values, 212–214
Mixed data, qualitative variables and, 235–237
Modal variable, 259
  coding of, 260
Moral statistics, 8–10
Multidimensional scaling. *See* MDS
Multigraphic nomogram, 11
Multiple correspondence analysis. *See* MCA
Multiple factor analysis. *See* MFA
Multiple factor analysis for contingency tables. *See* AFMULT
MULTISCALE loss function, 112
Multivariate analysis, distance-based, 112
Multivariate generalizations, 26
Multivariate statistical distributions, derivations of, 26
Multiway analyses, 82
  applications, 87–88
  computational methods, 86
  current status, 88–89
  early contributors, 85–86
  MCA and, 42–43
  motivations for, 82–84
  software, 87
Multiway categorical data, 89–90
  three-way CA, 90–92
Multiway continuous data
  INDSCAL, 81–82
  PARAFAC, 80–81
  relationships between models of, 81–82

Tucker2 model of, 79–80
Tucker3 model of, 79
MULTM, 215

N

N-way Toolbox, 87
Native symbolic data, 256
Nearest-neighbour chains, 127
Needham, Roger, 119
Nesting of two factors, 191
  between and within clouds, 192
  example, 192–196
Neural network research, 131
Neuromimetic computing, 131
Neyman, Jerzy, 25
Niche of a taxon, 64–65
Nishisato, S., 28, 35
Nivellement Général de la France, 11
NLPCA, 10, 45, 48
  example, 51–54
  Gifi project, 49–51
  optimal scaling with, 48–49
  software, 49
  use of with aspects, 56–59
  using pavings with, 54–56
Nomograms, 10–11
Nonlinear principal components analysis. *See* NLPCA
Nonmetric MDS, 100–101. *See also* Ordinal MDS
Nonsymmetric correspondence analysis, 92
Nonsymmetric data, 103

O

$O(n)$ expected time algorithm, 126
Objective cultural capital, 208
Oldenburger, R., 82
Optimal scaling, 48
  categorical variables, 32
  NLPCA with, 48–49
Optimality, 140
Ordinal MDS
  psychological measurements and, 100–101
  use of to study generalization gradients in learning, 101–102

Index

Ordinal transformation, 105
Ordination, 35
Orthogonal matrices, 20–22
OVERALS, 73

P

Pagés, Jean-Paul, 290
Paired comparisons, doubling of preferences and, 245–246
PARAFAC, 78, 80–81, 85
    use of in chemistry, 84
Parallel coordinates plots, 10
Parallel factors model, 80–81
Partial bootstrap, 44
Partial individuals, 228–231
Partial least-squares (PLS), 73
Partial order scalogram analysis, 101
Partial transition relationship, 229–230
Passive point, 142
Passive variables. *See* Supplementary variables
Pattern recognition, 131
Pavings, use of NLPCA with, 54–56
PCA, 3, 24, 27, 32–33, 36
    balance between groups of variables, 226–228
    doubled data, 242–243
    factor analysis using, 235
    linear, 46
    nonlinear (*See* NLPCA)
PCA-OS, 49
Pearson, Karl, 23–24, 34
Perceptron, 131
PERMAP, 116
Phonemes, distribution of, 151
Placket, R.L., 67
Playfair, William, 6
PLS_Toolbox, 87
Preferences, doubling of, 245–246
Prentice, Colin, 67
Presence-absence data, 66–67
Presentation, 6
PRINCALS, 49
Principal axes, 14, 292
Principal axis analyses, 39
Principal cofactors, 292
Principal components, 292
Principal components analysis. *See* PCA

Principal coordinates, 141, 145, 190
Principal factors, 292
PRINCIPALS, 49
PRINQUAL, 49
Probit PCA, 59
Procrustean Individual Difference Scaling (PINDIS), 100
PROXSCAL, 103, 111
Pseudobarycentric graphical representation, 176–177
Psychology
    dissimilarity judgements, 99–100
    ordinal MDS and measurements in, 100–101
    use of ordinal MDS to study generalization gradients in learning, 101–102
Psychometrics, multiway data analysis, 82
psychotools, 51
PTAk package, 87

Q

Qualitative variables, mixed data and, 235–237
Quantification of data, 121
Questionnaire data
    application of MCA to, 178–183
    use of MCA to analyse, 165–166

R

Rank annihilation factor analysis (RAFA), 83–84
Rank image principle, 49, 101
Ratings, doubling of, 240–245
Reciprocal averaging, 34–35, 63–67. *See also* CA
Recoding data, 239–240
Reconstitution formula, 142–143, 147
Rectangular matrices, singular value decomposition (SVD) of, 22–23
Reduced rank model, 70, 110
Reducibility property, 127
Redundancy analysis, 69
Regional interpretation, 106–107
Repeated segments, 152
Resampling, MCA and, 43–44

Residual cloud, 192
Response variables, 186
Retinal variables, 7
Rhetoric, 6
Rohlf, James, 120
Rosenblatt, Frank, 131
Ross, Gavin, 118
Rouanet, Henry, 206
Row profiles, 145
   factorial analysis of, 146
   PCA of, 32
   PCA of weighted, 36
   spaces of, 153–154
RV coefficient, 223

S

S-stress loss function, 112
Sabatier, Robert, 69
Sample correlation coefficient,
   derivation of the distribution
   of, 26
Sample surveys, 42
Satellite Program in Statistical Ecology, 67
Schwartz, S.H., 102–103
Sémiologie graphique, 7
Sequencing, 157–159
SGCA, under a cohesion constraint, 263–265
Shading, 6
Shelford's law, 64
Shepard diagram, 102
Signal processing, use of multiway techniques for, 84
Similarity judgements, distance formula as a psychological model of, 99–100
Similarity structure analysis, 101. *See also* MDS
Simple correspondence analysis, 32–33. *See also* CA
Simulated annealing, 113–114
Simultaneous diagonalization of two symmetric matrices, 20
Simultaneous linearizability, 51
Simultaneous representation, 146
Single linkage, 271
Singular value decomposition. *See* SVD

Site scores, 62–63
Skew-symmetric matrices, 20–22
Slice profiles, 92
SMACOF (Scaling by MAjorizing a COmplicated Function) algorithm, 110–111
smacof, 103, 111
Small multiples, 8
Smallest space analysis (SSA), 101. *See also* MDS
Smilde, A.K., 85
Social capital, 208–209
Social space, 205–206
   Bourdieu's approach to, 207–215
   example interpretation of, 215–221
SODAS 2, 258
Software
   multiway analysis, 87
   NLPCA, 49
Space of lifestyles, 186
SPAD, 203, 237
Spärck-Jones, Karen, 119
Specific MCA, 214, 250–252
Square matrix, decomposition of, 19–20
Stability, MCA and, 43–44
Stacked tables, 209, 217
   missing values in, 212
Standard data tables, 256
Statgraphics, 237
STATIS, 89, 223–224
   comparison with MFA, 234
   representation of groups using, 231–233
Statistical diagrams, 12
Statistical graphics, 3
Statistical triplets, 70–71
Statistics, 29. *See also* Data analysis
   innovations in, 42
   technological nature of, 37–39
Strain, 107
   loss function, 109
Stress
   criterion, 282
   facet theory and regional interpretation, 106–107
   function, 110
   loss function, 105–106
Structuration des tableaux á trois indices de la statistique. *See* STATIS

Structured data, 185–187
Structured data analysis, 186
  empirical example, 187–188
Structuring factors, 185–187
Subset correspondence analysis, 210, 250–252
Sum of squares, 189–190
Superimposed representation, 228–231
Supervised classification, 121
Suppes, Patrick, 186
Supplementary elements, 44, 142–143
Supplementary variables, 173, 186, 210
SVD, 3, 39
  CCA via, 74–75
  rectangular matrices, 22–23
Symbolic data
  factorial representation of, 265–270
  generalized canonical analysis strategy for, 259–265
Symbolic data analysis (SDA), 29, 255–258
Symbolic generalized canonical analysis. *See* SGCA
Symbolic input data, 258–259
Symbolic MDS, 115–116
Symbolic variables, 256
  coding of, 260–261
Symmetric correspondence, 148
Symmetric maps, 146, 176–177, 209, 211
Symmetric matrix
  diagonalization of, 19
  simultaneous diagonalization, 20
Synonymy relationships, 156–157

T

Tableux graphiques, 12
Taxons
  niche of, 64–65
  scores, 62–63
Tensor Toolbox, 87
ter Braak, Cajo J.F., 35
Textual data, use of CA with, 149–150
Thematic maps, 12
Theory of universals in values (TUV), 102
Three-mode analysis, 28–29, 77–78
Three-way canonical decomposition model. *See* CANDECOMP

Three-way clustering, 78
Three-way component based models, 78–79, 85–86
  INDSCAL, 81–82 (*See also* INDSCAL)
  PARAFAC, 80–81 (*See also* PARAFAC)
  relationships between, 81–82
  Tucker2, 79–80
  Tucker3, 79
Three-way correspondence analysis, 90–92
Three-way MDS models, 107–110
Three-way methods, 43
Topographic component model, 80
Torgerson formula, 143
Total inertia, 139–140, 145, 167–168
  categories, 174
Transition relationships
  linguistics, 156–157
  partial, 229–230
Triplets, 70–71
Tucker, Ledyard, 77, 82–83
Tucker2 model, 79–80
Tucker3 model, 79
Two factors
  crossing of, 196–203
  nesting of, 191–196
Two-sided eigendecomposition, 20

U

Ultrametric inequality, 277
Ultrametric matrices, properties of, 128–130
Ultrametric spaces
  geometrical properties of, 128
  hierarchical representations of, 127–128
Understanding Biplots (Gower et al.), 7
Unfolding model, 100
Unidimensional scaling, local minima, 113–114
Unimodal response, 64–65
Unitary matrices, 21–22
Universal calculator, 11
Unsupervised classification, 121
Unweighted pair group method. *See* UPGMA

UPGMA, 271
    comparison of with Ward's algorithm, 281–285
    group average linkage algorithm, 272–277
UPGMC, 127

V

Validation, MCA and, 43–44
van Rijsbergen, C.J. (Keith), 120
Variables
    balance between groups of, 226–228
    bivariate symbolic interval, 257
    categorical, 34, 47
    cloud of, 214
    coding of symbolic, 260–261
    dependent, 186
    fuzzy coding of continuous, 247–252
    histogram, 256–257
    independent, 186
    levels of, 7
    overall comparison of groups of, 231–233
    qualitative, 235–237
    response, 186
    superimposed representation of groups, 228–231
    supplementary, 173, 186, 210
    symbolic, 256, 259
    transformation of, 50–51
    use of to interpret cloud of individuals, 169–170
Visual discovery, 14–15
Visual explanation, 12–14
Visual language, 6
    graphs and diagrams, 10–12
    maps, 8–10
    rise of, 6–7
Visual thinking, 12–15

W

Ward agglomerative clustering method, 124–125
    example of inconsistent aggregation with, 286–287
    generalized, 126
Ward, Joe H., 124–125
Ward's clustering algorithm, 272, 277–281
    comparison of with UPGMA, 281–285
Weighted averaging, 66
Weighted log-ratio distances, 147
Weighted points, factorial analysis of a cloud of, 138–143
Weighting considerations, 284–285
Weights, MDS with, 114
Wilkinson, Lee, 7
Wishart distribution, 26
Within cloud, 192, 196–197
    example, 200–202
Words, 152
WPGMC, 127

X

XLSTAT, 237

Y

Yocoz, Nigel, 69
Yoshizawa,T., 91